Mathematics in Mind

AF696527

Series Editor

Marcel Danesi, Department of Anthropology, University of Toronto, Toronto, ON, Canada

Editorial Board Members

Norbert Hounkonnou, International Chair in Mathematical Physics and Applications (ICMPA-UNESCO Chair), University of Abomey-Calavi, Benin Republic

Louis H. Kauffman, Dept of Math., Statistics & Comp.Science, University of Illinois at Chicago, CHICAGO, IL, USA

Dragana Martinovic, Faculty of Education, University of Windsor, Windsor, ON, Canada

Melanija Mitrović, Faculty of Mechanical Engineering, CAM-FMEN, University of Nis, Niš, Serbia

Yair Neuman, Dept. of Brain and Cognitive Sciences, Ben-Gurion University of the Negev, Be'er Sheva, Southern, Israel

Rafael Núñez, Department of Cognitive Science, University of California, San Diego, La Jolla, CA, USA

Anna Sfard, Education, University of Haifa, Haifa, Israel

David Tall, Institute of Education, University of Warwick, Coventry, UK

Kumiko Tanaka-Ishii, Research Ctr for Advanced Sci and Tech, University of Tokyo, Tokyo, Tokyo, Japan

Shlomo Vinner, Amos De-Shalit Science Teaching Center, The Hebrew University of Jerusalem, Jerusalem, Israel

The monographs and occasional textbooks published in this series tap directly into the kinds of themes, research findings, and general professional activities of the **Fields Cognitive Science Network**, which brings together mathematicians, philosophers, and cognitive scientists to explore the question of the nature of mathematics and how it is learned from various interdisciplinary angles. Themes and concepts to be explored include connections between mathematical modeling and artificial intelligence research, the historical context of any topic involving the emergence of mathematical thinking, interrelationships between mathematical discovery and cultural processes, and the connection between math cognition and symbolism, annotation, and other semiotic processes. All works are peer-reviewed to meet the highest standards of scientific literature.

Mahouton Norbert Hounkonnou •
Dragana Martinovic • Melanija Mitrović •
Philippa Pattison
Editors

Mathematics for Social Sciences and Arts

Algebraic Modeling

Springer

Editors

Mahouton Norbert Hounkonnou
International Chair in Mathematical Physics
and Applications (ICMPA-UNESCO Chair)
University of Abomey-Calavi
Benin Republic

Melanija Mitrović
Faculty of Mechanical Engineering,
CAM-FMEN
University of Niš
Niš, Serbia

Dragana Martinovic
Faculty of Education
University of Windsor
Windsor, ON, Canada

Philippa Pattison
University of Sydney
Sydney, NSW, Australia

ISSN 2522-5405 ISSN 2522-5413 (electronic)
Mathematics in Mind
ISBN 978-3-031-37791-4 ISBN 978-3-031-37792-1 (eBook)
https://doi.org/10.1007/978-3-031-37792-1

Mathematics Subject Classification: 00A05, 00A06, 03B30, 03E15, 03E70, 03F65, 03H10, 05A15, 05C21, 05c22, 05c25, 05c50, 05c38, 05c40, 05c82, 05c85, 06A06, 06A11, 06B05, 06F05, 08A02

© The Editor(s) (if applicable) and The Author(s), under exclusive license to Springer Nature Switzerland AG 2023
This work is subject to copyright. All rights are solely and exclusively licensed by the Publisher, whether the whole or part of the material is concerned, specifically the rights of translation, reprinting, reuse of illustrations, recitation, broadcasting, reproduction on microfilms or in any other physical way, and transmission or information storage and retrieval, electronic adaptation, computer software, or by similar or dissimilar methodology now known or hereafter developed.
The use of general descriptive names, registered names, trademarks, service marks, etc. in this publication does not imply, even in the absence of a specific statement, that such names are exempt from the relevant protective laws and regulations and therefore free for general use.
The publisher, the authors, and the editors are safe to assume that the advice and information in this book are believed to be true and accurate at the date of publication. Neither the publisher nor the authors or the editors give a warranty, expressed or implied, with respect to the material contained herein or for any errors or omissions that may have been made. The publisher remains neutral with regard to jurisdictional claims in published maps and institutional affiliations.

This Springer imprint is published by the registered company Springer Nature Switzerland AG
The registered company address is: Gewerbestrasse 11, 6330 Cham, Switzerland

Paper in this product is recyclable.

Foreword

It is now well understood that the traditional silos in which disciplines find themselves constitute a straitjacket that inhibits growth of a discipline and presents a barrier to its fruitful interactions with other disciplines. While challenges in the form of interesting and important problems are to be found in all of the scientific disciplines, it is the case that many deep questions, accompanied by open problems, reside at the intersections of disciplines. This is even more evident in regard to the major challenges that face society, for example, as captured in the Sustainable Development Goals. Thus, multidisciplinary thinking and activity are essential to continued scientific progress, and in addressing global challenges.

The apparent dichotomy between fundamental and applied or goal-oriented research is one which is at best counterproductive to scientific progress. These considerations are captured by Abraham Flexner (1939) in his work "The usefulness of useless knowledge." His thesis, persuasively argued, is that the investigation of deep questions of a fundamental nature, driven solely by curiosity and without any apparent applications, often leads to the greatest scientific discoveries and, in due course, to technological breakthroughs having dramatic and lasting impact.

Inspired by the ideas of cross-pollination between disciplines, an international scientific virtual conference Mathematics for Social Sciences and Arts: Algebraic Modeling (MS2A2M 2021), was organized by higher education institutions from four continents: Europe (Center of Applied Mathematics of the Faculty of Mechanical Engineering Niš, CAM-FMEN, Serbia), Africa (University of Abomey-Calavi, Benin), Australia (University of Sydney), and America (University of Windsor, Canada). Organizers of this event were academician Mahouton Norbert Hounkonnou (the President of the Network of African Science Academies, NASAC), professor Melanija Mitrović (Head of the Center of Applied Mathematics of the Faculty of Mechnical Engineering Niš, CAM-FMEN, University of Niš), professor Philippa Pattison (Deputy Vice Chancellor (Education) at the University of Sydney), and professor Dragana Martinovic (University of Windsor; Co-director of the Cognitive Science Network of the Fields Institute).

This conference was the first in the world to gather in one place the world's leading researchers in the field of social sciences and algebra, more generally mathematics, after the merger of the former International Council for Science and the International Social Science Council (ISSC) into the International Science Council (ISC) on 4 July 2018. From that point of view, MS2A2M 2021 has a symbolic meaning.

The theme of this conference, on the interaction between mathematics, and the social sciences and arts, provides substance to the effectiveness of multidisciplinary approaches for addressing major questions; and in particular, on the influence and effectiveness of abstract mathematics – in this case, algebraic modelling – in providing the rigorous structure and underpinnings to theories in the social sciences, much as it has done in the natural sciences since time immemorial. There is a benefit in the other direction as well, in that the mathematical approach to complex questions in the social sciences inspires new mathematical theories and methods.

This conference has brought together leading scholars in the social sciences and mathematics, particularly algebraic structures, coming from renowned universities from around 25 countries from all continents. It constitutes the second in the conference series on Algebra without borders.

The conference, and this book, shed light on ways in which mathematics is learned, understood, and applied, in relation to other disciplines, particularly in the social sciences. Thus, the conference presentations provide insights into the many and varied existing and potential interactions between areas of algebra and the social sciences. The products of these valuable insights are to be found in this volume, which comprises a set of papers based on the conference talks.

Various papers provide examples of the "unreasonable effectiveness of mathematics," to quote the physicist Eugene Wigner. In one, the capacity of signs and symbols in mathematics to have multiple meanings is treated, illustrating the power of its flexibility and effectiveness in modelling the world around us. In some of the papers making up this volume, the close links between semiotics – the study of signs, symbols, and their interpretation – and the semiotic roots of algebraic methods and systems are explored, while attention is given also to the importance of promoting algebra at all levels of the mathematics curriculum, and to the task of making it more readily accessible, given its fundamental linkage to modelling.

Thus, beyond the physical sciences, we are witnessing striking examples of the effectiveness of mathematics as a language and basis for modelling: in the life and health sciences, molecular sciences, and communication and digital sciences, and of course the social sciences, in which the tools of mathematics are central to the rapid contemporary developments in these areas.

Social networks and other relationships constitute a recurring theme. Thus, the volume includes the results of work on algebraic approaches for understanding the interconnected nature of social relationships. The focus here is on structural regularities in social relational forms and the use of algebraic approaches to explore similarities and differences in form.

In three other papers, social network analysis is deployed as a modelling framework to describe the structure of relations and interactions between actors, drawing on the relevant algebraic structures. Also related is the use of algebraic structures to describe social data, with an emphasis on explanatory rather than descriptive approaches; as well as the application of network approaches to historical studies, providing insights into networks of relations in antiquity.

A further contribution studies the use of mathematics to provide a rigorous basis for decisions that affect social dynamics. Here, and in other works, aspects of education and the curricula in mathematics and the social sciences are considered, including the feasibility of using examples in the classroom.

There is also to be found an exploration of the logical complexity of vetoing in social life: this is measured via analysis of the complexity of language, and the mathematical complexity that is evident from the difficulties that appear in formulating and proving basis statements on the concept of vetoing.

One of the two parts into which the volume is organized, on algebraic thinking and modelling, is thus given sustained attention through the various presentations. The other theme, on semigroups and other algebraic structures in social sciences, is represented by a contribution on the formulation of an algebraic theory of semigroups with apartness, with order theory as a basic tool.

Mathematical logic is represented as well, through an investigation of Gödel's Incompleteness Theorems in relation to the clash between naturalism and dualism.

It is a pleasure to acknowledge the fine work of the organizers, not only in bringing together a number of experts on the very timely topic of the relationship between mathematics and the social sciences, but also, importantly, in ensuring through this book that there is a record of the outstanding research reported at the conference. The papers in this volume will no doubt serve as an invaluable resource both for established researchers, and for those, whether from mathematics or the social sciences, wishing to enter this fertile and important area of research.

Cape Town, South Africa
September 2022

B. Daya Reddy

Preface

> People took a long time to arrive at the conclusion that individuals and society could be studied scientifically. As that realization spread, nothing was more natural than to employ, in ever more sophisticated ways, a tool they had always used. Mathematics has always been an integral part of the social sciences which, by their very nature, could not exist without it. (Senn 2000, p. 285)

Roger Bacon, an English Franciscan friar, philosopher, scientist, and scholar of the thirteenth century stated that "Neglect of mathematics works injury to all knowledge, since he who is ignorant of it cannot know the other sciences or the things of the world" (1928, Capitulum 1). Bacon's premonitory statement has been confirmed over the years, as mathematics has gained an increasingly important place in all the life domains including, as Senn's (2000) analysis shows, the social sciences.

This volume brings together contemporary advances in mathematics, especially algebra, with applications in the social sciences and the arts. Its contributions seek to illuminate some of the ways in which algebra is developed, learned, understood, communicated, and applied in the social sciences and the humanities. It has its origins in a unique conference held in 2021, Mathematics for Social Sciences and Arts: Algebraic Modelling (MS^2A^2M 2021), virtually hosted by the Faculty of Mechanical Engineering, University of Niš, Serbia, 24–26 May, 2021 (http://mathsocart.masfak.ni.ac.rs/). The conference brought together scholars from different disciplines and geographic regions and was, we believe, the first to focus on the juxtaposition of algebra and social science applications in this way. It was organized by higher education institutions from four continents: Europe (University of Niš, Serbia), Africa (University of Abomey-Calavi, Benin), Australia (University of Sydney, Australia), and North America (University of Windsor, Canada). Following the conference, the organizers invited some of the leading scientists in social sciences and algebra – all of whom had participated at the conference – to contribute to this volume.

This volume has two parts. The first is concerned with algebraic and mathematical thinking. It addresses the learning and practice of mathematics from a cognitive science perspective, as well as illustrative applications to some distinctively human concerns (e.g., education, semiotics). The second part focusses on algebraic semigroups and some of their generalizations. It includes a range of applications inspired by the strongly relational character of many important social phenomena (e.g., social networks).

We thank the authors and the reviewers for their expert contributions to this work and extend our gratitude also to Dr. Daya Reddy, Professor Emeritus at the University of Cape Town, for his insightful Foreword.

Cotonou, Benin — Mahouton Norbert Hounkonnou
Windsor, ON, Canada — Dragana Martinovic
Niš, Serbia — Melanija Mitrović
Sydney, NSW, Australia — Philippa Pattison
January 2023

References

Bacon, R., & Burke, R. B. (1928). Mathematics. In *Opus Majus, Volumes 1 and 2* (pp. 116–418). University of Pennsylvania Press. http://www.jstor.org/stable/j.ctt1fzhg1s.8

Senn, P. (2000). Mathematics and the social sciences at the time of the modern beginnings of the social sciences. *Journal of Economic Studies*, *27*(4/5), 271–292.

Introduction

> A detemporalized mathematics cannot tell us what mathematics is, why mathematics is true, why it is beautiful, how it comes to be, or why anybody should care a fig about it. But if one places mathematics squarely within human time and experience, it becomes a warm and rich source of possible meanings and actions. Its ultimate mystery is never dispelled, yet it is exhibited as one of the primary creations of the human intellect. (Davis and Hersh 1986, p. 201)

Regardless of our readers believing that mathematical objects exist independently of humans or are created by human minds, they would probably agree that mathematics is a "language of science." This position was promoted by many scholars, such as Galileo Galilei (1564–1642), Johannes Kepler (1571–1630), René Descartes (1596–1650), and Isaac Newton (1642–1727), who all glorified the superiority of especially geometry, in terms of "intelligibility of geometric proof and ... the manifest applicability of geometry to space, time, and matter" (Gorham et al. 2016, p. 5). Apart from astronomy and physics, which always relied on mathematics to the extent that they were considered its parts, that was not the case for some other sciences. For that reason, some other sciences were considered as less rigorous and more experimental. The "mathematization of science" that Gorham, Hill, and Slowik describe happening in the Western world around the seventeenth century, lasted for about two centuries. The seventeenth and eighteenth centuries are often called the "era of enlightenment," as a testament that during that time sciences (and mathematics) flourished, and many new sciences emerged. This "time of reason" also witnessed a rising interest in understanding and explaining personal and social aspects of the world. Senn (2000) eloquently describes how that happened:

> The current conception of the social sciences as including the disciplines of anthropology, economics, geography, history, political science, psychology, sociology, and their applications ± education, planning, public administration and social work ± is a development less than two centuries old. It only came about after people got the idea that a social science was possible. (p. 273)

While much earlier, various thinkers explored different aspects of human existence (e.g., exchange of resources and accumulation of wealth; organization of society and governance; historical accounts; planning of human dwellings, etc.).

Senn states that for a social science to be established, it was necessary that it uses a scientific method, which started happening only around the eighteenth century. At that time, algebra, statistics, and differential and integral calculus were already founded and the new fields of mathematics and its applications continued to emerge. While not all social scientists are likely to agree on the utility of mathematics in the social sciences, there is clear evidence of progress in using mathematics to model social process, structure, and action (Edling 2002) and the existence of a community of scholars who have successfully applied mathematics to theorizing in social sciences such as sociology (Edling 2007).

The testament to Shenn's (2000) and Edling's (2002, 2007) positions can be also found in the emergence of the terms like "mathematical social sciences," which suggest an orientation toward building connections between mathematics and social sciences.

One example of this trend is the appearance of international peer reviewed journals, such as Mathematical Social Sciences, which since its first issue in September 1980, "emphasizes the unity of *mathematical modelling* in economics, psychology, political sciences, sociology and other *social sciences*" (emphases in original, 2023). The scope of this journal covers fundamental aspects of choice, information, and preferences (decision science) and of interaction (game theory and economic theory), the measurement of utility, welfare and inequality, the formal theories of justice and implementation, voting rules, cooperative games, fair division, cost allocation, bargaining, matching, social networks, and evolutionary and other dynamics models.

A more recent, but also influential development in this domain is a series of annual public lectures titled, *Keyfitz Lecture in Mathematics and the Social Science*. Since 2007, the Fields Institute for Research in Mathematical Sciences in Toronto (Canada) has organized these lectures intended "to inform the public of some of the ways quantitative methods are being used to design solutions to societal problems, and to encourage dialogue between mathematical and social scientists" (Fields Institute, para 1). These lectures have covered a range of questions that our communities (and societies) continue to grapple with, for example, "How many people the Earth can support?" (Cohen 2007) and "How do social and information networks relate and cohabit?" (Kleinberg 2007). The presenters have included sociologists, economists, psychologists, statisticians, epidemiologists, and cognitive scientists, all of whom were either mathematicians working in these fields or social scientists applying mathematics in their work.

Social scientists who are yet to be convinced of the utility of mathematics in their work (Edling 2002) may find developments in qualitative mathematical approaches of particular interest. Dedò (2017) counteracts a public view of mathematics as a discipline that deals only with numbers. In a number of mathematical fields, one can go far in observing relationships between objects without using numbers. Kemeny (1995), for example, writes about visual and conceptual approaches to analyzing social relationships. If one understands a mathematical model as an object stripped of any presently irrelevant information, one could visualize a model of some social situation in which vertices are persons and edges connect people who know each other. The additional symbols on the edges (e.g., "+" or "-") could specify if these people also like or dislike each other. Thus, a social scientist interested in a group dynamic could create a model in the language of graph theory and, after analyzing it there, interpret the results in their social science domain language. Kemeny concludes that this kind of knowledge transfer between the domains also gives opportunity for mathematicians to contribute to social science research, by pointing to some relevant theorems from graph theory or attempting to solve problems identified by the social scientist.

Another of Kemeny's examples of the application of a numberless mathematics is from the group theory of transformations. He uses it to model ways in which a society that does not use written records can avoid marriages between close relatives. These two examples demonstrate how mathematics can be numberless and conceptual. In the third example, this time with numbers, Kemeny uses matrices to analyze communication networks, and in the fourth, nets to organize a consensus ranking of objects. The resourcefulness of mathematical models is evident in these examples, both in terms of their applicability in different target domains, and variability of symbolic and visual representations used.

It is no wonder that many mathematical approaches to representing social phenomena involve concepts from algebra, graph theory, and related areas. Algebra is for Pratt (2022), "strikingly different from other branches of mathematics in both its domain independence and its close affinity to formal logic" (para. 10). Because of its tendency towards "abstraction, simplification and formalism," algebra is found useful in "describing and theorizing social systems" (Squazzoni 2010, p. 201).

However, "mathematization of social systems" brought about a realization that, for example, forecasting in economics or modeling social structures by population dynamics equations may result in Gödel-like undecidable mathematical statements (i.e., they can neither be proven nor refuted within the system; Doria 2017). While that does not mean that creating axiomatic mathematical models of complex social phenomena is not useful, it means that general solutions may not always exist.

The book *Mathematics for Social Sciences and Arts: Algebraic Modeling* addresses several contemporary topics, centered in the broad domains of algebraic structures and semigroups in social sciences and the arts. These encompass some rapidly developing, versatile areas without, of course, well-defined boundaries. They present algebraic structures and modeling in different aspects of human life (e.g., history, education, governance), as well as in diverse applications in day-to-day life, such as in architecture, agriculture, finance and music.

The eleven chapters in this book can address but a small subset of these fields, making use of a cross-section of algebraic methods. Some of the chapters are intended to present reviews of wider areas, while others are more focused. In the latter case, authors have nevertheless provided some general background. The chapters are organized in two sections: the first addressing algebraic thinking and modelling from the standpoints of cognitive science, semiotics, mathematics education, mathematical logic, and social choice theory. The other section covers applications of algebraic structures in social sciences (e.g., social relationships, transport networks).

1. Algebraic Thinking and Modeling

The book begins with a set of seven chapters directly pertinent to problems of algebraic thinking and modeling, while the methods described also may have many other applications. The chapter by Mahouton Norbert Hounkonnou and Melanija Mitrović provides a detailed exposition of key mutually beneficial interactions between mathematics and social sciences and the arts, focusing on the role of algebraic structures in the construction and evolution of these fields.

Marcel Danesi uses semiotic analysis to present algebra (and by extension, mathematics), as a powerful modeling system. This is not a mere "reformulation" of the well-known notions in mathematics from the standpoint of semiotics; rather, it will arguably put us in a better position to grasp the symbolic-formal nature of algebra and what it entails cognitively.

In the third chapter, Jacek Woźny discusses the effectiveness of mathematical modelling, and mathematics in general, from the standpoint of cognitive science. He uses two algebra handbooks' descriptions of functions (mappings) to illustrate how conceptual blending of different interpretations contributes to polysemy of mathematics and affects learning these concepts.

Then, a mathematics educator, Dragana Martinovic argues for early school inclusion of pre-algebraic and algebraic concepts and tasks. To demonstrate how both mathematical fluency and intuition could be enhanced through creation of models and modeling, she uses an example of a (number) line in its multiplicity of curriculum applications.

In the fifth chapter, Zvonimir Šikić gives a brief historical sketch of the clash of dualism vs. naturalism and then analyzes the argument that Gödel's incompleteness theorems support dualism by implying human-machine non-equivalence. He proves that this implication is not valid and puts forward a correct implication.

Aspects of Decision Theory and Social Choice Theory are covered by Branislav Boričić and Marija Srećković who discuss voting systems. They conclude that for each weighted voting system containing some agents with a veto power, there exists an equivalent weighted voting system with no agents with a transparent formal veto power.

2. Semigroups and other Algebraic Structures in Social Sciences

Following these intensive developments, we include five other chapters devoted to semigroups and other algebraic structures in social sciences. In this vein, Melanija Mitrović, Mahouton Norbert Hounkonnou, and Paula Catarino, motivated by previous results on an interactive theorem proving approach of formal verifications, give a review in the seventh chapter of the new algebraic theory – the theory of semigroups with apartness. They outline the constructive co-order theory as the main tool for the development of the theory of constructive semigroups with apartness in the beginning, and later, for the development of some other algebraic structures with apartness. They also provide a critical overview and a comparative analysis between classical and constructive semigroup theories.

Then, the review by Philippa Pattison in the eighth chapter addresses social relationships from an algebraic point of view and emphasizes related successes and shortcomings. Pattison also analyzes the role structure associated with observed social relations using techniques from partially ordered semigroup theory.

In the ninth chapter, J. Antonio Rivero Ostoic introduces the study of multi-level network structures in which relations within and between distinct sets are analyzed. He employs concepts from formal concept analysis and semigroup theory to study the relational system of the transport network and provincial politico-administrative arrangements in ancient Rome.

In the tenth chapter, Lucia Falzon analyzes time and sequence in networks of social interactions. Falzon investigates dynamic relational structures and algebraic structures related to sequential patterns in relational event data, and viable time-ordered paths for dynamic network measures.

In the final chapter, John Levi Martin analyzes different algebraic structures for two-mode asymmetric binary data and discusses their interpretations with respect to social and cognitive processes.

Finally, this book and the conference that engendered it provide an example of a fruitful collaboration, in which the skills and deep knowledge of algebraic structures, modeling, social sciences, and arts brought in by the authors from different countries and continents merge in harmonious ways. Many of the chapters also include substantial bibliographic resources. We expect these expositions to be a rich resource, of interest both to mathematicians and non-mathematicians in the broad sense.

International Chair in Mathematical Physics and Applications (ICMPA-UNESCO Chair), University of Abomey-Calavi, Benin Republic — Mahouton Norbert Hounkonnou
Faculty of Education, University of Windsor, Windsor, ON, Canada — Dragana Martinovic
Faculty of Mechanical Engineering, CAM-FMEN, University of Niš, Niš, Serbia — Melanija Mitrović
University of Sydney, Sydney, NSW, Australia — Philippa Pattison

References

Davis, P., & Hersh, R. (1986). Descartes' dream: The world according to mathematics. San Diego: Harcoun Brace Jovanovich.

Dedò, M. (2017). Q for qualitative mathematics. *Lettera Matematica*, *5*, 167–171.

Doria, F. A. (2017). Axiomatics, the Social Sciences, and the Gödel Phenomenon: A Toolkit. In F. A. Doria (Ed.). Limits of Mathematical Modeling in the Social Sciences – The Significance of Gödel's Incompleteness Phenomenon. World Scientific, London.

Edling, C. (2002). Mathematics in sociology. *Annual Review of Sociology*, *28*, 197–220.

Edling, C. (2007). Interviews with mathematical sociologists. Department of Sociology Working Paper series. University of Stockholm (https://su.figshare.com/articles/preprint/Interviews_with_mathematical_sociologists/14130014).

Fields Institute. (n.d.). Keyfitz Lecture in Mathematics and the Social Sciences. http://www.fields.utoronto.ca/activities/keyfitz-lecture-mathematics-and-social-sciences

Gorham, G., Hill, B., & Slowik, E. (2016). Introduction. In G. Gorham, B. Hill, E. Slowik, & C. K. Waters (Eds.), The Language of Nature: Reassessing the Mathematization of Natural Philosophy in the Seventeenth Century (pp. 1–28). University of Minnesota Press.

Kemeny, J. G. (1959). Mathematics without numbers. *Daedalus*, *88*(4), 577–591.

Pratt, V. (2022). Algebra. In E. N. Zalta and U. Nodelman (Eds.), The Stanford Encyclopedia of Philosophy.

Senn, P. (2000). Mathematics and the social sciences at the time of the modern beginnings of the social sciences. *Journal of Economic Studies*, *27*(4/5), 271–292.

Squazzoni, F. (2010). The impact of agent-based models in the social sciences after 15 years of incursions. *History of Economic Ideas*, *18*(2), 197–233.

Contents

Part I
Algebraic Thinking and Modeling

Problematic of Mathematics, Social Sciences, and Arts: A Ubiquitous Constructive Interaction in Algebraic Modeling

Mahouton Norbert Hounkonnou and Melanija Mitrović

1 Introduction

> Mathematics compares the most diverse phenomena and discovers the secret analogies that unite them. (Joseph Fourier, French mathematician and physicist)

We are certainly not the first who write on extensive, ubiquitous connections between mathematics and sciences, in particular social, and arts. However, we have endeavored to clarify why certain algebraic knowledge and algebraic modeling show themselves useful within the scope of human activities.

At the very beginning, it should be pointed out that we have resisted the natural inclination to go beyond our own areas of expertise. Instead, our main intention is to write this chapter from the position of practicing mathematicians "who are trying to take a very close look through the magnifying glass at our own every day work" (Borovik 2010, p. VII).

Scientific disciplines are in the book *Practicing interdisciplinarity* edited by Weingart and Stehr (2000) described as the eyes through which modern society sees and forms its images about the world, shaping its own future or reconstituting the past. During the most of the twentieth century, the question of knowledge has been shaped by disciplinarity. In recent decades, the growth of scientific and technical knowledge has made scientists to join in addressing complex problems.

It is considered that development of new modes of inquiry can make interdisciplinary research more effective in yielding significant benefits to science and society.

M. N. Hounkonnou (✉)
International Chair in Mathematical Physics and Applications (ICMPA-UNESCO Chair), University of Abomey-Calavi, Benin Republic
e-mail: norbert.hounkonnou@cipma.uac.bj

M. Mitrović
Faculty of Mechanical Engineering, CAM-FMEN, University of Niš, Niš, Serbia
e-mail: melanija.mitrovic@masfak.ni.ac.rs

© The Author(s), under exclusive license to Springer Nature Switzerland AG 2023
M. N. Hounkonnou et al. (eds.), *Mathematics for Social Sciences and Arts*, Mathematics in Mind, https://doi.org/10.1007/978-3-031-37792-1_1

From that point of view, individuals who are familiar with ideas and languages of other fields are of central importance to the scientific progress in the twenty-first century. All this should not be understood as the matter of getting privilege of interdisciplinary research over disciplinary research. Instead, it is preferable to work on the "interdisciplinarity in order to optimize its effectiveness as well as to strengthen both interdisciplinarity and the disciplinary foundations from which it springs" (More in National Academy of Sciences et al. 2005).

Indicated processes and work on their realizations are not an easy kind of job at all. From our point of view demonstrating how mathematical ideas are embodied in the social sciences and arts can "enlighten all who are interested on the topic, in the complex intellectual pursuits, personalities, and cultural settings that connect these vast disciplines". With this in mind, and to respect page limitation, we write this chapter which is far from covering all the topics.

Mathematics floods our world in both its quantitative and qualitative aspects and interacts with all areas of knowledge, know-how, and skills that determine and shape our lives. Today, as explained so well in the Preface of Edward Frenkel's book *Love and Math: The Heart of Hidden Reality* (2013) "there's a secret world out there. A hidden parallel universe of beauty and elegance, intricately intertwined with ours. It's the world of mathematics. And it's invisible for most of us". Paradoxically, "every time we make an online purchase, send a text message, do a search on the Internet, or use GPS device, mathematical formulas and algorithms are in play ... In our world, increasingly driven by science and technology, mathematics is becoming, ever more the source of power, wealth and progress".

There is no single definition of mathematics. For some, mathematics is the language of science; for others, it is just a game of symbols and rules of variables on them, or, simply, mathematics is what mathematicians do (Hounkonnou and Mitrović 2020).

Mathematics studies structures that it creates itself or that originate from other sciences. Paraphrasing Mark Turner's book *The Origin Of Ideas: Blending, Creativity, And The Human Spark* (2014, p. 1), it can be said that mathematicians did not make galaxies, life, viruses, the sun, DNA, or the chemical bond. But they do make new ideas – lots and lots of them.

The physicist Eugene Wigner (1960) says that the miracle of the appropriateness of the language of mathematics for the formulation of the laws of physics is a wonderful gift which we neither understand nor deserve. We should be grateful for it and hope that it will remain valid in future research and that it will extend, for better or for worse, to our pleasure, even though perhaps also to our bafflement, to wide branches of learning.

Arthur B. Powell, in Foreword in Paulus Gerdes *Geometry from Africa* (1999), reported the Egyptian-born mathematician and educational psychologist, Caleb Gattegno, states that the elements of reality upon which mathematics is built are objects, relations among objects, and dynamics linking different relations. The objects, relations, and dynamics may be concrete and contextual or abstract and decontextual. Mathematical ideas may arise from context or apply to contextual situation not necessarily related to the origin of the idea. The power of mathematical ideas often become manifest once they transcend their physical, tangible origins.

One can broadly divide mathematics into "pure", the study of mathematics for its own sake, and "applied", the application of mathematical methods by different fields. Following such reasoning, applied mathematics can be viewed as a combination of mathematical knowledge and specialized knowledge. One must keep in mind, however, that such a division is not a universal model, but rather a consequence of the personal research aims of mathematicians themselves. Mathematicians have a habit of joking as follows: "it was as though applied mathematics was my spouse, and pure mathematics was my secret lover" (Frenkel 2013, p. 139).

It often happens that something so abstract ends up being really useful. About this, Borovik et al. (arXiv:2201.08364v1, p. 11) write that we now have a smaller, but increasingly messy, universe around us: the world of IT. What is more abstract than money (ask Karl Marx), and even more so electronic money? And there is no need even to ask this question about cryptocurrencies. As an old adage goes, money rules the world. In the modern times, electronic money rules the world. Hence mathematical abstraction rules the world.

As it is also written in *Mathematics for Action, Supporting Science-Based Decision-Making* (UNESCO 2022), everything we do is based on some mathematical structure, and although mathematics is often considered abstract, it is fundamental to how we understand nature, the larger universe, with its time and space dimensions and a myriad of uncertainties.

Through this chapter, we use the description (or a definition) of mathematics given in Borovik (2021) as follows: "Mathematics is the study of mental objects and constructions with reproducible properties which imitate the causality structures of the physical world, and are expressed in the human language of social interactions".

Mathematics is a vast discipline. About 40 years ago, it was estimated that between 100 000 and 200 000 new theorems were published every year in mathematical journals around the world – this number could only have increased since then. A mathematical theorem, as a rule, explicitly refers to other theorems and definitions and is integrated into the huge system of mathematical knowledge. This system remains unified, tightly connected, and cohesive; despite the fantastic diversity of mathematics, it also has almost incomprehensible unity. In this regard, it may be interesting to read a document put online by Borovik et al. (arXiv:2201.08364v1, p. 2).

We also agree with our colleague Frenkel (2013, p. 3) in pointing out that mathematical knowledge is different from most other types of knowledge. The way we perceive the world can always be distorted; however our perception of mathematical truths is not liable to such distortion. A mathematical formula or theorem should mean the same thing to anyone anywhere – no matter their gender, religion, or skin color; it should ideally signify the same thing to anyone in a thousand years as well. And it is also amazing that we "own" all of this mathematical knowledge. Mathematics construction as human thought differs in essence from how other sciences are built. Each great mathematician's contribution corrects or extends what came previously.

Albert Einstein, in *The World As I See It – An Essay*, 2006, teaches us that the fairest thing we can experience is the mysterious. It is the fundamental emotion which stands at the cradle of true art and true science. Sciences and mathematics have had a history full of mysteries; many of these are hard to grasp for a modern audience. An important thing to remember here is the long tradition of international cooperation in mathematics. We often think this is a modern tradition, but in fact it dates back to more than half a millennium ago. And influences from outside of Europe, of course, typically relate to an even more distant past. Science is an international endeavor. Its development depends on exchanges of ideas and expertise, which are made possible by people moving from one part of the world to another (Ramakrishnan 2020, p. 148).

Thus, as any science, mathematics emerges from bringing together people and ideas from different cultures around the world, as well as from people migration, and its story cannot focus solely on a region. Its evolution and dissemination follow the same path. This has certainly favored various interpretations of the origin of mathematics, and the fanciful attribution of the first results to such and such a person according to the geographical affinities of each other. So, for example, the legend has ascribed the most interesting and attractive theorem of early geometry, the so-called Pythagorean theorem to Pythagoras of Samos (sixth century BC). Fortunately, the extensive and deep investigations carried out by the Egyptologist, Cheikh Anta Diop (1981, p. 434, p. 479), have definitely proved that Pythagoras has learned the theorem during his long stay in Egypt. Moreover, it is reported that the theorem that is attributed to Thales is illustrated by the figure of problem n°53 of the Papyrus of Rhind, written 1300 years before. Jamblique writes that all line theorems (geometry) come from Egypt (1981, p. 324). A good compilation of all these historical facts can be found in a recent book by O. B. Bagodo (2020, pp. 314–334), prefaced by the eminent archaeologist Augustin F. C. Holl, President of the International Scientific Committee for the General History of Africa, Volumes IX, X and XI (UNESCO). Note that one can obviously discover important mathematical ideas from quantitative, qualitative, and spatial features in a diversity of ancient and modern objects in the multicultural mix of African civilization as well as in other cultural groups in other geographical or planetary regions. A history of mathematics should then be unfettered with nationalistic and ethnocentric bias and acknowledge and valorize multicultural manifestations of mathematical ideas, without any claim of primacy. We must, however, admit the evidence that connects the origin of the mathematics to Africa, the cradle of humanity.

Of course, there is a lot more to the history of mathematics, and, as so nicely described in Frenkel (2013, p. 70, p. 184), one can envisage mathematics as a huge jigsaw puzzle, in which the ultimate image is always a mystery at first. Reaching that image comes as a consequence of a lot of joint work. And people there tend to work in sections: algebraists, number theorists, geometers, etc. Much progress has been made within those sections through history, of course, yet they have rarely teamed up enough to provide a really big picture. Occasionally, however, there have been individuals trying to bridge those gaps. When this happens, important traits of the big picture come out, providing new significance for the individual fields. Every new piece of the puzzle gives us new insights, new tools to uncover the mystery. This in turn further enriches the big picture, which keeps surprising us as we progress along the path of research.

Before getting to the heart of the matter, let us emphasize that the UNESCO, in *Mathematics for Action, Supporting Science-Based Decision-Making* (UNESCO 2022), has devoted a great deal of its work to improving the quality of mathematics education and research but remains something of an enigma to the person in the street. While the UNESCO has put in a lot of effort to broaden the perception of the importance of mathematics worldwide, we still often view it as a discipline detached from daily activities – but for basic numerical operations that we all perform and the somewhat shady awareness that mathematics has a lot to do with the work of various electronic devices we use. Since it is quite difficult for an average person to understand why mathematics is important beyond these limits, structured effort is needed to stress the importance of a "basic" mathematical literacy, which however goes far beyond simple numerical skills. Neither is mathematics a strictly individual activity nor is it totally deductive anymore, which are only two of many facts educators should keep in mind while supporting the need for better mathematics in schools.

In this chapter, we intend to understand what mathematics is, how it functions, what it accomplishes for the world, what it has to offer in itself, and how it serves the physical and social scientists, the philosopher, logician, and the artist. How it satisfies the curiosity of the humankind who surveys the heavens and muses on the sweetness of musical sounds, and so on.

In other words, how does mathematics impact the development of social, intellectual, vocational, moral, cultural, and spiritual abilities and feed education system, economics, the advancement of science and technology, the agriculture, etc.? These questions are in the core of this chapter.

2 Mathematics and Society

In his address to the Prussian Academy of Sciences in Berlin in 1921, Albert Einstein asks the following question: How it can be that mathematics, being after all a product of human thought independent of experience, is so admirably appropriate to the objects of reality? And to Frenkel (2013, p. 27) to specify that Albert Einstein used what looks like the purest and most abstract mathematical knowledge to unlock the deepest secrets of the world around us.

Though Eugene Wigner's unreasonable effectiveness of mathematics mentioned in Sect. 1 has been exploited for centuries, its root is still poorly understood. Mathematical truth seems to exist objectively and independently of both the physical world and the human brain. There is no doubt that the links between the world of mathematical ideas, physical reality, and consciousness are profound and need to be further explored (Frenkel 2013, p. 202).

Thus, to be convinced of the usefulness of mathematics, it suffices to understand the importance of science, technology, and innovation (STI) in the development of our society, as evidenced by the solutions they bring to contemporary world challenges to improve human well-being, to advance sustainability and respect for the environment, to protect biological and cultural diversity of the planet, to promote the sustainable social and economic development, etc.

As mathematics develops, one cannot normally predict in which areas mathematical concepts may later be applied. Still, the complexity of structures and forms analyzed by mathematicians provides invaluable tools for scientists in other fields, often revealing profound connections with nature (Gerdes 2007, p. 163).

Mathematics interacts with other sciences and numerous domains, including life sciences, biology, health and medicine, computer science in general, high-performance computer (HPC), and big data, statistical learning, neural networks, physics at all scales, engineering sciences, chemistry, social sciences in the broadest sense, development of complex systems, modeling, or data analysis. More generally, there are more and more interactions between fundamental mathematics, applied mathematics, modeling, and its applications. These are showing enormous potential but of course again necessitate cooperation among mathematicians from different fields and scientists from disciplines external to mathematics.

Mathematics is also potentially relevant to the growth of a broad community of "citizen scientists". Such projects have made it possible to crucially involve citizens in research. For instance, citizen science in the domain of health may foster health awareness and promote healthy changes in behavior patterns. What mathematicians need to do is to develop algorithms to deal with such real-life, often heterogeneous, data. Nowadays, we luckily have good technology to collect such data en masse: cell phone apps, trackers, and patient records.

In the rest of this section, we briefly consider the impact of mathematics on human development and then describe some of the very productive interactions between mathematics and its applications.

To paraphrase Kline (1967, p. 8, p. 9), to many people, mathematics offers intellectual challenges, and it is well-known that such challenges do engross humans. People do respond to intellectual challenges, and once they get a slight start in mathematics, they encounter in abundance.

In the future, all society will be learning all the time. More and more, education will foster self-esteem, sense of belonging, and advancement. Formal or not, supported by the social system or a consequence of individual effort, education will serve the central role for us to improve on our chances for the future. Further, since education prepare children to earn a living and become independent, mathematicians must prepare students for technical and other vocations where mathematics is applied, e.g., engineering, architecture, accountancy, banking, business, even the agriculture, tailoring, carpentry, surveying, and the office work require the knowledge of mathematics. Learning mathematics arguably supports character development. Mathematics is crucially relevant to the provision of more rigor and precision in multidisciplinary interactions.

The German poet and essayist Hans Magnus Enzensberger, invited speaker at the 1998 International Congress of Mathematicians in Berlin, analyzed the exclusion of mathematics from the cultural sphere by the public. His talk served as the basis of the book: *Drawbridge Up: Mathematics – A Cultural Anathema 1st Edition* (2001). In this book, he compares and contrasts the public's attitude toward mathematics with intellectual achievements in other fields, such as music, paintings, or literature, and poses the question: How does it happen that mathematics has remained as it were a blind spot in our culture – alien territory – in which only the elite, the initiate few, have managed to entrench themselves? He admits that there are problems already in the use of specialized professional jargon, where mathematicians and non-mathematicians often do not even understand one another. Likewise, he discusses the traditional perception of pure mathematics as unprofitable and pointless. However, in the continuation, he proves how effective and unexpectedly useful mathematics can be in solving problems in the real world.

The problem of mathematical illiteracy in the public has been touched upon by many writers, for instance, John Allen Paulos in his book *Innumeracy* (1990). However, one should not think this problem is easily solvable. Indeed, one finds numerous mathematics education reform and anti-reform movements, most of which are related to Enzensberger's point. The problem lies in the fact that resorting to abstract thinking strategies alone may be useless if teachers themselves are not capable of teaching in that fashion. Likewise, curricula that revert to abstract thinking only are bound to produce only number-crunchers, and not really creative, problem-solving individuals, persons who can then start really loving mathematics. What appears certain from such dilemmas is the following: one cannot change the ever-lasting fear and ignorance of mathematics quickly, yet a really good teacher can mitigate this tendency by turning the mathematics class into something intellectually demanding, and thus inspiring. For more information, see https://www.maa.org/press/maa-reviews/drawbridge-up-mathematics-a-cultural-anathema.

Let's refer again to Frenkel (2013, Preface) to say that "mathematics is as much part of our cultural heritage as art, literature, and music. Mathematics is the source of timeless profound knowledge, which goes to the heart of all matter and unites us across cultures, continents, and centuries".

This helps the learner to understand the contribution of mathematics in the development of civilization and culture, as well as its role in fine arts and in beatifying human life.

Let us conclude this development on the following insightful citation: "We have to do mathematics using the brain which evolved 30,000 years ago for survival in the African savanna" (Stanislas Dehaene, cited in Borovik et al. (arXiv:2201.08364v1, p. 17)).

How does mathematics contribute to the development of the main sectors in modern society? Given previous developments, it is an open secret to claim that mathematics is of central importance to all sectors of modern society and provides the vital underpinning of the knowledge of economy.

There is also need for mathematical approach that helps to explain sources of risk and uncertainty, which affect global markets. With the help of mathematical techniques coming from, for example, calculus and higher differential equations, it is possible to build a complicated machine, a skyscraper, a bridge, a subway, a railroad, a large ocean-going ship, an electric lighting plant, a telephone, a telegraph cable, or any other manifestations of modern ingenuity. Of course, the main questions remain: Who pays for the costs of all this modernity? Who benefits and who loses from such induced economic growth?

Machine learning and artificial intelligence were quite recently emerging fields. Now they underpin much of modern society. Scientific and engineering computing advances, among other things, strongly depend on new research on data analysis, machine learning tools, and data-centric architectures. It is obviously a good thing that today mathematics is generally accepted as crucial for dealing with big scientific, technological, and societal challenges. Yet one should note that its role should not be limited to giving the non-mathematician some tools to use, ideally supported by a good theoretical framework. The case is quite the opposite nowadays: one finds more and more research across the fields based on strictly mathematical issues. Thus one expects mathematics to keep developing as an applied field itself as well. The applied nature of research in other disciplines can piggyback on new fundamental research in mathematics. So it is always a two-way relationship, with a constant need to combine fundamental research with the work on a particular application problem. Of course, in the beginning, one cannot really predict where these proposed fundamental contributions will lead. Yet constant development is required to be a successful event today and certainly in the future. Additional information can be found in *Mathematics for Europe* (European Commission 2016).

All of the above points out that the mathematical sciences, in addition to being a key discipline in their own right, are crucial for the advancement of all areas of science and technology (see, for example, Hounkonnou et al. 2022). This underpins a wide range of discoveries across the research spectrum from health and security to the environment, agriculture, ecology, epidemiology, tumor and cardiac modeling, DNA sequencing, and gene technology.

Mathematical models provide invaluable tools for public health decision-making, both by forecasting the likely impact of an epidemic and by predicting the effectiveness of measures of disease containment and prevention (see *Mathematics for Action, Supporting Science-Based Decision-Making UNESCO* 2022, p. 6).

Mathematical models are indispensable in reaching decisions in matters involving the public. And this goes both ways: they can help predict the likely strength and consequences of an epidemic and also assess how effective the measures to contain it will be. There is an increasing need for big data and artificial intelligence in the SARS-CoV-2 pandemic. These tools and techniques can give policymakers more accurate, timely, and locally nuanced analysis. In turn, this would facilitate health decision-making. Indeed, to prevent or better respond to future outbreaks, researchers from multiple disciplines would need to combine their knowledge and insights. This would help both Africa and the whole world better prepare for such challenging times.

Mathematical modeling can be used to foster the betterment of vaccines: this can be done through the analysis of complex immunological dynamics and through the assessment of which parts of a pathogen will most likely induce an immunogenic response.

Mathematical and statistical models can assist in creating clinical trials and research of the efficiency of vaccines in search of more safety and efficacy. Complex mathematical algorithms are used to quickly break down sequences of pathogen genomes and predict or track changes to these sequences that may reduce vaccine effectiveness. Such analyses are then used to decide on new vaccine designs. Mathematical tools can help investigate and resolve issues across a range of complex phenomena, for example, the game-theoretical aspects of climate change and vaccine uptake. Cross-disciplinary game-theoretical models may prove indispensable matters of public health. Once people get closer to the so-called herd immunity, policymakers can expect that new individuals should be less prone to vaccination. In such circumstances, programs for global eradication and local elimination should work hand in hand to address such lowered vaccination levels. Likewise, campaigns should be developed to reach remote populations but also to talk to those vaccine-hesitant persons who are still open to the idea of vaccination.

We also know that a food system is a complex network, with unlimited actors and relationships, that is deeply associated with health, society, and the environment. A food system is a complex network that is deeply associated with health, society, and the environment. Mathematics faces a huge challenge in trying to understand how such a network works. Food systems, for instance, are complex networks in which actors and relationships are practically infinite in scope. Thus researchers find such complicated networks quite hard to model. It turns out that precisely mathematics may hold the key to a better understanding of food systems and hence the well-being of future generations. Optimal solutions for management objectives may be investigated within mathematical models, which would strive to minimize costs, capture market feedback, or simulate policies. Multiple risk scenarios can equally be tested mathematically. They are, of course, important for the development of mitigation and adaptation strategies. Useful details can be also obtained from *Mathematics for Action, Supporting Science-Based Decision-Making* UNESCO, 2022 (p. 3, p. 6, p. 12).

In fact our society as a whole is a network of structures: they transport people and provide goods, information, energy, and social communication. Such large-scale, multiple-tier networking is both a challenge and a danger. Whichever aspect we focus on, as mankind we need to well understand, control, and optimize such networks.

Cyberspace systems and processes are fundamentally controlled by mathematical principles. Think of the encryption or decryption of file transmissions, modeling of networks, performance of data analyses, quantification of uncertainty, measurement of risk, and reaching decisions. And this is just a small sample of activities highlighted by mathematics. One thus needs to focus on a greater awareness of the role of mathematics in cyber systems and processes in current and future research.

We are limited in time and cannot address here all these aspects otherwise well described in Goethals et al. (2022).

All this invasion of mathematics in our life incite to ask the following questions: Why to mathematize social sciences? What are the story and dynamics of mathematics and social sciences co-routing? Do such scientific interactions induce new developments and benefit to these fields? Are semigroups relevant in social science algebraic modeling? What about contemporary challenges in this epoch of global science?

Theoretical modeling in the social sciences can entail the specification and analysis of parameterized systems of nonlinear differential equations. Many critical insights have been obtained by social scientists using this powerful classical mathematics approach. This leads us to question what is the "right" mathematics for the social sciences? And what can the social sciences and mathematics learn from each other? These questions are of great interest in elucidating the interactions between social and sciences and mathematics.

3 Algebra, Algebraic Modeling, and Sciences

> Till algebra, that great instrument and instance of human sagacity, was discovered, men, with amazement, looked on several of the demonstrations of ancient mathematicians, and could scarce forbear to think the finding several of those proofs to be something more than human. (John Locke, 1690)

In the words of John Derbyshire, in our own time, algebra has become the most rarefied and demanding of all mental disciplines, whose objects are abstractions of abstractions of abstractions, yet whose results have a power and beauty that are all too little known outside the world of professional mathematicians. Most amazing, most mysterious of all, these ethereal mental objects seem to contain, within their nested abstractions, the deepest, most fundamental secrets of the physical world (Derbyshire 2001, p. 6).

It is widely considered that description of the development of algebraic thinking through history presents a challenge. For example, John Derbyshire (2001) wrote that the development of algebra was irregular and haphazard. More precisely, as written by Dov Tamari (1978) "[We] have no definition of Algebra, no central idea or unifying principle, not even a well centered picture of the whole, but rather an enumeration of algebraic topics in the literature of the epoch, even though this enumeration is in some sense natural but not uniquely determined order . . . "

Over the course of the nineteenth century, algebra made a transition from a subject concerned entirely with the solution of mostly polynomial equations to a discipline that deals with general structures within mathematics. Van der Waerden in his book *A History of Algebra-From al-Khwarizmi to Emmy Noether* (1985, p. 6) wrote: "Modem algebra begins with Evariste Galois. With Galois, the character of algebra changed radically. Before Galois, the efforts of algebraists were mainly directed towards the solution of algebraic equations . . . After Galois, the efforts of the leading algebraists were mainly directed towards the structure of rings, fields, algebras, and the like".

"Algebra is beautiful. It is so beautiful that many people forget that algebra can be very useful as well" (Lidl and Pilz 1998). Paraphrasing the Nobel laureate in physics, Steven Weinberg's wording (1983, p. 125), we, algebraists, "led by our sense of mathematical beauty, develop formal structures that scientists from other disciplines only later find useful, even where we had no such goal in mind". Abstract algebra is the highest level of abstraction. Understanding it means, among other things, that one can think more clearly, more efficiently. Oliver Wendell Holmes, an American jurist and legal scholar, who served as an associate justice of the Supreme Court of the United States from 1902 to 1932, wrote in *The Autocrat of the Breakfast Table*: "I was just going to say, when I was interrupted, that one of the many ways of classifying minds is under the heads of arithmetical and algebraical intellects. [...] We are mere operatives, empirics, and egotists until we learn to think in letters instead of figures".

An algebraic system, roughly speaking, is a set with operations and relations defined on it. Graph theory is the study of graphs, that is, a pictorial representation of a set of objects where some pairs of objects are connected by links. Close connections between graphs and algebraic structures have been widely used in different kind of applications. The structural approach to algebra has provided opportunities for:

- already solved as well as open problems to give solutions in a more efficient and elegant way;
- appearance of new directions in research in the area in particular, and mathematics in general.

Algebraic structures such as semigroups, groups, rings, fields, lattices, and vector spaces, began to be used by mathematicians in their work, as tools for solving problems in other areas of mathematics: geometry, topology, number theory, theory of functions, probability theory, statistics, etc. All in all, abstract algebra has grown into one of the most highly developed and pervasive domains of contemporary mathematics.

Can the whole of modern algebra be described in a couple of sentences? Paraphrasing Woźny (2018), yes it can; it has been designed to be elegantly simple: the story starts with sets (collections of objects) and relations on them and proceeds to the concept of a semigroup; each new concept is based on the previous ones, and, ultimately, the whole multistory edifice rests on the sparse foundation of sets and relations. Semigroups serve as the building blocks for the structures comprising the subject which is today called modern algebra. In the wording of B. M. Schein (1997) "Semigroups aren't a barren, sterile flower on the tree of algebra, they are a natural algebraic approach to some of the most fundamental concepts of algebra (and mathematics in general), this is why they have been in existence for more then half a century, and this is why they are here to stay".

A semigroup is an algebraic structure consisting of a set with an associative binary operation defined on it. Boyd wrote in his book *Social Semigroups: A Unified*

Theory of Scaling and Blockmodelling as Applied to Social Networks (1991) that "semigroups have such a simple definition that it is sometimes hard to believe that anything can be proven about them… A semigroup is a simplest possible mathematical object since it has only one set, one operation, one axiom and many common examples". Sets of relations closed under composition are examples of semigroups, giving rise to interesting nontrivial ideas, while providing links, both, in results and methodology with other mathematical disciplines as well as other sciences.

In the history of mathematics, the algebraic theory of semigroups is a relative newcomer, with the theory proper developing only in the second half of the twentieth century. Historically, it can be viewed as an algebraic abstraction of the properties of the composition of transformations on a set. Nowadays, semigroup theory is an enormously broad topic and has advanced on a very broad front. It is considered that a huge variety of structures studied by mathematicians are sets endowed with associative binary operation. If, in addition, one considers the many possibilities of constructing new semigroups starting from a few reference semigroups (vectors, matrices, polynomials, formal series, etc.), an almost unlimited source of examples can be found. So, it appears that semigroup theory provides a convenient general framework for unifying and clarifying a number of topics in fields that are seen, at first sight, unrelated (See Mitrović et al. 2021).

Let us start our short journey through the applications of semigroups with the connections to the algebra of relations. The theory of semigroups is one of the main algebraic tools used in the theory of automata as well as in the theory of formal languages. Theory of automata, formal languages, and codes are, according to Pilz et al. (2002), among important and best known (to the date) areas where the semigroup-theoretic approach is quite substantial. Applications of semigroups within these areas are often called *combinatorial application*. Some investigations on transformation semigroups of synchronizing automata show up interesting implications for various applications for robotics or more precisely, robotic manipulation.

Piecewise deterministic Markov processes, PDMPs, are among advanced probabilistic tool used for the study of environmental and molecular processes in modern biology. On the other hand, the theory of semigroups of linear operators provides the primary tools in the study of continuous-time Markov processes, an important type of PDMPs connected to certain biological phenomena. More about PDMPs and biological models and their applications can be found in Rudnicki and Tyran-Kamińska's book (2017). Other areas of natural sciences also make use of semigroups. Let us mention a few of existing applications. Tiling semigroups can be applied in solid state physics (Pilz et al. 2002). Galilei-type semigroups found their applications within the dynamics of open quantum systems. In the framework of dynamics-generating approach, DGA, they can be used for deriving the dynamics of decohering or dissipating systems (Mensky 2013). Semigroup models as a natural setup to treat the emergence of autocatalytic biochemical reaction networks are described in Loutchko (arXiv:1908.04642v2).

Following Boorman and White (1976) and Lidl and Pilz (1998), the sociology includes the study of human interactive behavior in group situations, in particular in underlying structures of societies. The study of such relations can be elegantly formulated in the language of semigroups. The book *Social Semigroups: A Unified Theory of Scaling and Blockmodelling as Applied to Social Networks* (Boyd 1991) is written for social scientists with the main aim to help readers to apply "interesting and powerful concepts" of semigroup theory to their own fields of expertise. The book uses the algebraic theory of semigroups to study social networks. Established link between semigroups and social sciences is an important interface between semigroups and the "outside world". In Schein's review (1997) of Boyd's book, it is written: "semigroups do appear very naturally in various studies known under the common umbrella name of social sciences".

Philippa Pattison in her book *Algebraic Models for Social Networks* (1993) presents a unified approach to the algebraic analysis of both complete and local networks. Social networks are collections of social or interpersonal relationships linking individuals in a social grouping. Social networks have been used to explain various characteristics and behaviors of individuals as well as to account for social processes occurring in both small and large group of individuals.

Social network data can be described as a set of social units, such as individuals, and a collection of pairs of units who are linked by a social relationship of some kind. The starting point for the models is the characterization of social networks in the terms of blockmodels and the subsequent construction of semigroup models for role. Developed models vary in complexity from single indices summarizing a particular structural feature of a network to quite complex algebraic and geometric representations.

Philippa Pattison argues "that the algebraic description of structure is natural from the perspective of social theory and the perspective of data analysis. In particular, it allows for a more general means of analyzing network representations into simpler components, a property that greatly enhances the descriptive power of the representations".

Informatics "provides the scientist's workbench of the 21st century" – it is "the 21st century's means of expression". In what follows till the end of this section, we intend to list some of the applications of algebra in general, and semigroups in particular, within areas of informatics.

Let us list some of the most prominent examples. Boolean algebra is hardly recognized by the mathematical community when it was developed; nowadays it is known as the foundation for most of computer science. Linear algebra is useful in all kinds of applications and situations, such as the feature-based classification techniques in machine learning and the method for face recognition. Quotient structures have many applications within informatics – in particular in modeling and in automated program analysis. More examples of applications of different algebraic contents within informatics can be found in Steffen et al. (2018). As claimed by the authors, *inductive approaches and algebraic thinking are combined in order to illustrate the art of perfect modeling.*

An interesting application of algebra is related to data analysis and multivariate analysis, i.e., the statistical study of data where multiple measurements are made on each experimental unit and where the relationships among multivariate measurements and their structure are important.

Traditionally, algebraic statistics, which is the use of algebra to advance statistics, is associated with the design of experiments and multivariate analysis (especially time series).

The use of algebra in statistics can be traced to the beginning of the twentieth century. The work on algebraic statistics has, in turn, resulted in the development of new themes to be discussed in algebra and combinatorics, e.g., association schemas. These schemas deal with relations between ordered pairs of elements of a given set. Equivalent definitions are given in terms of partitions, graphs, and matrices. Association schemas were originally used in statistics, where they provided the basic structure for numerous experimental designs. Aside from statistics, association schemas have found a place in the theory of permutation groups.

The "unreasonable effectiveness" of algebra has shown up all over the place. With the development of computing in the last several decades, applications that involve algebraic structures have become increasingly important. Semigroup-theoretic approach becomes quite substantial into the study of more complex branches of mathematics and computer science.

In processes of finding models of computing the power of a specific model can be described by the complexity of the language it generates or accepts. For more information, see the book by Mateescu and Salomaa (1997).

Regular languages and finite automata are considered to be among the oldest topics in formal language theory. The theory of finite semigroups and monoids represent powerful framework for study of regular languages. On the other hand, the language approach showed to be very useful in the further development of finite semigroups and monoids. There are numerous examples of applications of regular languages and finite automata.

Following Paperman et al. (date. ffhal-03831752f) vectorial programming, the combination of Single Instruction Multiple Data, SIMD, instructions with usual processor instructions, is known to speed up many standard algorithms. The main result is the construction of compilation procedures that turns syntactic semigroups into vectorial circuits. Obtained circuits are small in that they improve known upper-bounds on representations of automata within the logical formalisms. All complexity measures of circuits are provided in terms of the semigroup size. Results presented in (date. ffhal-03831752f) belong to the active and important interdisciplinary research area: machine learning, text processing, vectorial circuits, regular languages and finite automata, mathematical logic, semigroups, and monoids.

At the very end of this section, we give two examples of applications of the research and problem-solving connected to the testing for the safety validation in railway and highway traffic. Both projects made use of machine learning and computer vision.

Engineers from the Faculty of Mechanical Engineering Niš took part in the SMART2 project, whose objective was to research, innovate, implement, and validate advanced holistic obstacle detection and track intrusion detection system, or, shortly, OD&TID system, for railways. A prototype of a novel holistic OD&TID system for railways was developed, consisting of three sub-systems, on-board and trackside/unmanned aerial vehicle (UAV)-based (i.e., drone-based), which were demonstrated in a realistic environment under realistic test conditions. In the frame of the project, the railway OD&TID dataset was generated representing the one of the very first datasets reflecting real-world challenges of object detection in railway applications such as long-range object detection. See more at https://smart2rail-project.net/.

Krajewski et al. (2018) wrote in the Abstract of their paper that "scenario-based testing for the safety validation of highly automated vehicles is a promising approach that is being examined in research and industry. This approach heavily relies on data from real-world scenarios to derive the necessary scenario information for testing. Measurement data should be collected at a reasonable effort, contain naturalistic behavior of road users and include all data relevant for a description of the identified scenarios in sufficient quality. However, the current measurement methods fail to meet at least one of the requirements. Thus, we propose a novel method to measure data from an aerial perspective for scenario-based validation fulfilling the mentioned requirements. Furthermore, we provide a large-scale naturalistic vehicle trajectory dataset from German highways called highD. We evaluate the data in terms of quantity, variety and contained scenarios. Our dataset consists of 16.5 hours of measurements from six locations with 110 000 vehicles, a total driven distance of 45 000 km and 5600 recorded complete lane changes".

4 Mathematics and Arts

> The scientist does not study nature because it is useful, he studies it because he delights in it, and he delights in it because it is beautiful. If nature were not beautiful, it would not be worth knowing, and if nature were not worth knowing, life would not be worth living. Of course, I do not here speak of that beauty that strikes the senses, the beauty of qualities and appearances, not that I undervalue such beauty, far from it, but it has nothing to do with science, I mean that profounder beauty which comes from the harmonious order of the parts, and which a pure intelligence can grasp. (Henri Poincaré, cited in Lafollette, P. S., Jr., (1999))

Talking about mathematics and art, let us paraphrase wording of Thomas McEvilley, professor of art history, Rice University, "what's hard for people to accept is that issues of art are just as difficult as issues of mathematics; you cannot expect to open up a page on mathematics and understand it".

On the other hand, Neil deGrasse Tyson wrote in Foreword to the Lynn Gamwell's book *Mathematics and Art: A Cultural History* (2016): "The value of mathematics to the scientist is clear and present. Given its potency and ubiquity in granting access to the operations of nature, should we be surprised that mathematics has served (and continues to serve) as an irresistible muse for philosophers and artists alike? Isn't it one of the jobs of the artists to help non-artists to interpret world around us, and the world within us? As long as mathematics is what shapes those world, the observant artist cannot help but embrace and express this influence on us all".

Within the scope of this section, we will demonstrate some of the interactions between mathematics and arts. Once again, we have to emphasize that in writing this section (as the chapter as whole), we do that from the position of practicing mathematicians "who are trying to take a very close look through the magnifying glass at our own every day work". In addition, respecting title of the book, and the scientific event which result is this book (http://mathsocart.masfak.ni.ac.rs/), we limited ourselves mostly to algebra and its connections to art.

In the announcement of Racois' book *Propos sur la beauté des maths* (2021), several important questions are posted: What is this beauty mathematicians try to achieve? Where is it hiding? How to perceive it? In the geometric form? In language? In a feeling of attraction coming from the number? Feel real emotion while building reasoning? Appreciate the beauty of a demonstration? Or, as it is sometimes considered (Starikova 2018) that mathematical beauty primarily involves mathematician's sensitivity to aesthetic to abstract?

Aesthetics may be defined narrowly as the theory of beauty, or more broadly as that together with the philosophy of art (https://iep.utm.edu/aesthetics/). The term aesthetics can be, also, described as theoretical principles "underlying art for the purpose of historical comparison, e.g., art movements from different periods, composition changes in music, etc". (Sriraman and Lee 2021, p. 3). Together with art and taste, beauty is one of the main subjects of aesthetics. Going through literature on the topics, it can be concluded that the nature of beauty is one of the most long-lasting and controversial themes considered in Western philosophy. What is beauty? What makes, for example, a beautiful face, appealing painting, pleasing design, or charming scenery? This question has been debated for at least 2500 years and has been given a wide variety of answers – for more information, see Reber et al. (2004). Anyway, beauty has traditionally been counted among the ultimate values, with goodness, truth, and justice (https://plato.stanford.edu/entries/beauty/). It seems quite natural to connect art with aesthetic. On the other hand, the connection of mathematics with aesthetics is not so obvious and, sometimes, considered as even unusual. Can one discuss mathematics per se and aesthetics without using art as a mediator? This is the question Sriraman and Lee (2021, p. 3) posted concluding that the relationship between mathematics and aesthetics is more pervasive than one imagines.

The question posted at the very beginning of this section What is mathematical beauty? implies new one: Is there anything which distinguishes it from other kinds of beauty? Following, Breitenbach and Rizza (2018a) "mathematicians often appreciate the beauty and elegance of particular theorems, proofs, and definitions, attaching importance not only to the truth but also to the aesthetic merit of their work". Proof can be elegant, clumsy, enlightening, explanatory, deep, and simple – let us mention some of numerous attributes a proof might have, as stated by Starikova (2018). Anyway, the tendency among mathematicians to judge mathematical work according to aesthetic standards raises a number of additional questions. As for existing answers, let us cite Breitenbach and Rizza (2018a) again: "a number of authors in aesthetics and the philosophy of mathematics have tried to shed light on mathematical beauty by highlighting its relation to such factors as order, harmony, unity, symmetry, and simplicity ... Others have argued that judgments about the beauty of mathematics are related to the understanding or enlightenment that the mathematics affords ... Some philosophers of mathematics and theoreticians in the field of mathematics education have furthermore stressed the need to take seriously the aesthetic dimension of mathematical practices". What is certain, mathematics and its knowledge offer a possibility to appreciate art, while "art continually presents possibilities for mathematics to develop a lens through which one might enlarge our aesthetic understanding". (Sriraman and Lee 2021, p. 5). Pointing out that mathematics deserves a more attention from aestheticians than it has so far had, Rieger (2018) concludes that there are plenty of interesting avenues for further exploration in the area. For more reading, see *Handbook of the Mathematics of the Arts and Sciences* (Sriraman 2021) and *Aesthetics in Mathematics* (Breitenbach and Rizza 2018b).

Although it might look like at first glance that the arts are independent of mathematics, it was mathematics which has fashioned major styles of painting and architecture, for example, the rich, symbiotic relationship that exists between architecture and mathematics.

- From the construction of primitive tribal huts to the design of technologically advanced high-rise buildings, architects have relied on and been inspired by mathematics.
- From ancient Egypt to the modern era, architects have used mathematics to solve practical problems and been inspired by the beauty and poetry of numbers and geometry (Ostwald 2021, p. 1132).

Mario Salvadori (1996), the author of both well-respected textbooks on architectural structures and applied mathematics, wrote: "The relationships between mathematics and architecture are so many and so important that, if mathematics had not been invented, architects would have had to invent it themselves".

On the other hand, some mathematicians have used architectural examples to explain mathematical concepts. Let us mention a few: Naum Vilenkin's Mathematical Art Museum and Henri Poincaré's Gallery of Monsters point to the ways mathematical knowledge can be ordered and conceptualized architecturally; mathematicians talk about constructing a theorem, laying its foundations, and testing the strength of its principles. More about mutual relationships between mathematics and architecture can be found in the *Handbook of the Mathematics of the Arts and Sciences*, Part III in Sriraman (2021).

In addition to the case with architecture, the whole development of visual arts was strongly related to the development of mathematics. Through literature it is often considered that artistic beauty follows some order and simplicity, harmony, and symmetry. From this point of view, Euclidean geometry (and geometries in general) and group theory are among useful mathematical theories guided with the same principles. Fractals are related to complexity and sometimes, to ugliness; Euclidean geometry is related to simplicity and classical beauty, see Marcus (1999). Marcus mentioned, as an example, the chiral icosahedral symmetry group, which is, by George W. Hart, an American sculptor and geometer (or, as it is written on his web, a freelance mathematical sculptor/designer), pronounced to be a source of creativity in sculpture. For Hart *geometry is married to algebra*, due to the essential involvement of a group (source Marcus 1999). The works by P. Gerdes are also relevant regarding arts and geometry. See, for instance, Gerdes (2007).

It is considered that in the field of textile art and design, mathematics seems to have advanced the design of textiles, particularly in the field of technical textiles. Mathematical contents are used in studying the nature craft knots as well as in stimulating new ideas, which may not have occurred otherwise. On the other hand, Harris (1988, 1997) studied the mathematical basis of various textile activities so as to suggest that one can obtain mathematical knowledge by learning textile crafts. She used textiles to visualize mathematical concepts, such as symmetry, pairs, patterns, sets, lattices, tension, nets and solids, visual, tactile, and three-dimensional. Through this process, she developed a method for teaching mathematics through the observation of domestic textile craft objects. More about the relationship between mathematics and textile knot practice can be found in Nimkulrat and Nurmi (2021).

Cellular automata provide a natural meeting place for mathematics and art. Many fiber arts make a use of it. Recall, a cellular automaton is a mathematical construct which models a system evolving in time. System in question could be physical, chemical, biological, social, computational, or purely mathematical. The fiber artwork based on cellular automata among others includes techniques of knitting, crochet, weaving, braiding, cross stitch, needlepoint, and bead weaving. For more information, we refer to Holden and Holden (2021).

As Kline (1967, p. 8) wrote, the service of mathematics renders to music has not only enabled man to understand it but has spread its enjoyment to all corners of our globe. Following book written by Montiel and Peck (2019, Introduction), mathematical models can be found for almost all levels of musical activities, from theoretical analysis to actual composition and sound production. During the last decades, it is noticed that there is an intention of modern music theory to rely more and more on mathematical content such as algebra and algebraic combinatorics, theory of graphs, and topology.

Abstract algebra can provide a great framework for analyzing music and abstracting the relationships found in modern (Western) music theory to uncover other possible music. The connection between algebraic structures and music theory has been widely studied. Andreatta (2004) gives the group-theoretical approach to music theory and composition focusing on a family of the non-Hajós groups. Described algebraic approaches to music compositions are particularly interesting from a mathematical perspective too. Amiot (2005) connected the study of rhythmic canons with Galois theory. Can you hear an action of a group? Or a centralizer? Crans et al. (2009) present how music may be interpreted in terms of the group structure of the dihedral group of order 24 and its centralizer by explaining two musical actions.

Semigroups also found their place within music. The science of acoustics describes how harmonics of a given fundamental tone appear together with the fundamental tone when this one is played by a mechanical musical instrument. The way harmonics arise in each instrument describes its timbre. Bras-Amorós (2019) considers the algebraic structure of the sequence of harmonics when combined with equal temperaments. The sequence of physical harmonics is an increasingly enumerable submonoid of (R, +) whose pairs of consecutive terms get arbitrarily close as they grow. Obtained properties lead to new type of monoids author called tempered monoids. In continuation, Bras-Amorós (2020) proceeded studying classes of increasingly enumerable additive submonoids of R.

It is considered that the theory of words is connected to numerous different fields of mathematics and applications. Being a natural environment of a word, a finitely generated free monoid in particular, and algebra in general, combinatorics, theory of automata, and to a certain extent probability theory and topology are among them. Combinatorial word theory and musical scale theory are connected by Clampitt et al. (2008) who identified the study of scale patterns with the algebraic theory of words.

Mathematics has often been compared with music. Gottfried Leibniz in a Letter to Christian Goldbach, April 17, 1712, wrote: "Music is a hidden arithmetic exercise of the soul, which does not know that it is counting". And, James Joseph Sylvester: "May not music be described as the mathematic of sense, mathematic as music of the reason? the soul of each the same! Thus the musician feels mathematics, the mathematician thinks music, music the dream, mathematic the working life each to receive its consummation from the other... " (source: Bell 2015).

Mathematics and literature, or more broadly sciences and arts, are too often presented as separate, or opposites. But when they fuse and combine, the results can be beautifully challenging, or perhaps excitingly absurd. From the historical point of view, the area of Southwest Asia is of special importance for the development of mathematics. The various poems and fragments of poems written in ancient Mesopotamia allow us to understand the origins of mathematics in its historical, social, and cultural context. One of ancient Mesopotamia's most interesting personalities was Enheduanna, the daughter of Sargon the Great, who lived in the twenty-third century BCE. It is widely considered she is the first mathematician and writer to be named or have a mathematical and literary work attributed to her. That is, she is the first recorded author in world history. Part from one of Enheduanna's hymns dedicated to Nisaba's temple in Ereš:

The true woman who possesses exceeding wisdom,
She consults a tablet of lapis lazuli,
She gives advice to all lands,
She measures off the heavens,
she places the measuring-cords on the earth.
Nisaba, praise!

Recall, Nisaba, the grain goddess, was patron of writing arts and mathematical calculations. (Source: Glaz and Growney 2008, Introduction -xii)

Karaali and Sriraman (2021, p. 965) wrote that mathematics and language are intricately connected, and together they draw for us the boundaries and expanses of our humanity. They claim: "if poetry is concise linguistic expression of emotion and experience, and if as many people believe, mathematics is not experiential and should be free of emotion, then there should be no overlap! But emotions do not leave the building when math makes its entrance" (Karaali and Lesser 2021, p. 968). On the contrary, there are mathematician who wrote poetry. Jakob Bernoulli in *Treatise on Infinite Series* wrote (Glaz and Growney 2008, p. 130):

Even as the finite encloses an infinite series
And in the unlimited limits appear,
So the soul of immensity dwells in minutia
And in narrowest limits no limits inhere.
What joy to discern the minute in infinity!
The vast to perceive in the small, what divinity!

On the other hand, there are poets who incorporated mathematical concepts in their work. Emily Dickinson is nowadays recognized as one of the most important figures in American poetry. Given below is her song *We Shall Find the Cube of the Rainbow*:

We shall find the Cube of the Rainbow.
Of that, there is no doubt.
But the Arc of a Lover's conjecture
Eludes the finding out.

J. L. Brouwer claimed that mathematics is an essentially languageless activity and that language can only give descriptions of mathematical activity after the fact. (https://plato.stanford.edu/entries/brouwer/)

Bell (2015) claims just on contrary. Following him, "it seems evident that mathematics is language-based, both as a formal/symbolic practice and in its mode of transmission (through textbooks, lectures, etc.) It has been observed that mathematics resembles literary fiction in its systematic introduction of concepts such as numbers, circles, sets, etc. which, while lacking concrete existence, are then reified, that is, treated as if they really exiested-this is as true of constructive as of classical mathematics, by the way. In fiction, characters and events are treated [...] as if they were real".

Commenting Jorge Luis Borges' citation "*Art is fire plus algebra*", award-winning author of historical fiction Susanne Dunlap wrote: "The fire in Borges's quote is the emotional content, the passion of creativity that makes it possible for a writer to get in the 'zone' and let the words flow, spending hours, days, weeks, years crafting a book. The algebra is the structural underpinnings, the shape of your story and the details of the plot. The framework that the fire brings to life". (https://writersinprogress.com/2020/04/08/behold-the-inside-outline/).

An interesting application of the theory of matrices appears in the mathematical theatrology, which was born in 1966 when Solomon Marcus was engaged with his collaborators to teach a class of mathematical theatrology at the Faculty of Letters of Bucharest University. As it is written in Marcus (1998b), their starting point was the idea to associate to any theatrical play A a Boolean matrix, where the columns are associated to the scenes of A, in their chronological order, while the horizontal lines are associated to the characters of the play. At the intersection of the line of the character a with the column of the scene b, the digit 1 is put if a appears in b and the digit 0 if a does not appear in b. "The mathematical and computational processing of this apparently very primitive information leads to a lot of results with high theatrical relevance. In some further steps, the dramatic dialogue is also considered and various mathematical methods coming from game theory, graph theory, formal grammar are used". Since then, as already stated in Marcus (1998b), mathematical methods were also applied to the great old Greek tragedies (such as those of Aeschylus), to Shakespeare, to Comeille, Racine, and Moliere, and to some more recent theatrical plays.

Zawislak and Kopec (2019) analyze the famous Anton Chekhov play *Uncle Vanya* by means of graph theory. They believe that the idea of inter-field or multidisciplinary knowledge transfer (graph theory – drama) could be an additional tool for drama analysis.

We cite some quotes by mathematicians on their positing on the topics and questions considered within this section.

Freeman Dyson (2009) wrote that "mathematics is both great art and important science, because it combines generality of concepts with depth of structures".

Robert P. Langlands (2010) pointed out: "I appreciate, as do many, that there is bad architecture, good architecture and great architecture just as there is bad, good, and great music or bad, good and great literature but neither my education, nor my experience nor, above all, my innate abilities allow me to distinguish with any certainty one from the other. Besides the boundaries are fluid and uncertain. With mathematics, my topic in this lecture, the world at large is less aware of these distinctions and, even among mathematicians, there are widely different perceptions of the merits of this or that achievement, this or that contribution". Even more, "There is also a question to what extent it is possible to appreciate serious mathematics without understanding much of mathematics itself".

Following John Lane Bell (2015), the beauty of a mathematical concept often rests on the contrast between the simplicity and elegance of the concept itself and the richness and variety of the mathematical structures embodying it. This kind of mathematical beauty – let us term it conceptual beauty, in which variety emerges from simplicity – was elegantly encapsulated by Poincaré in his characterization of mathematics as the art of calling different things by the same name. (In that case, one might ask parenthetically, isn't poetry the art of calling the same thing by different names?) *Abstract algebra* abounds in instances of conceptual beauty. The algebraic concepts of semigroup, group, ring, field, monoid, vector space, module, and lattice are simple and elegant, and instances are encountered everywhere in mathematics.

Following Frenkel (2013), the process of creating new mathematics is a passionate search, a deep personal experience, just like creating art and music. It requires love and dedication, a struggle with the unknown and oneself, which often elicits strong emotions.

As a matter of conclusion, is mathematics an art? According to the belief of many great mathematicians, mathematics is both art and science. On the other hand, there are notable differences between them. As it is pointed by Bell (2015) "art can be negated and yet remain art, while mathematics cannot be negated and remain mathematics, even though it has much in common with art". That is, as Bell further proceeds, "the artist can choose to create works of art that are surpassingly beautiful, or shockingly ugly – artists have been liberated from the constraints of the Beautiful!" On the other hand, this is not the case with mathematicians, "the mathematician has no such freedom (except in the production of expository works) for in the end he or she is constrained by the dictates of mathematical truth and proof".

So, there are notable differences between the world of art and mathematics. Let us also point out that one of the most advanced ideas related to mathematics is that it has long been a catalyst of a great number of transfers of ideas or results from one field to another. At the end, paraphrasing Solomon Marcus (1998a, 1999), it would be realized that most oppositions between mathematics and art take place within the framework of some similarities and give rise, in their turn, to new possible similarities; differences and similarities alternate in an endless succession; the right way to approach the mathematics-art relation is to observe how some discrepancies emerge from some similarities and conversely, i.e., the discrepancies between art and mathematics can be well understood only if we look at them in the framework of the similarities between art and mathematics.

Mathematics and arts with all their similarities and discrepancies, discrepancies within similarities, and similarities within discrepancies have a common task: going all together in favorable directions they should make this world to be a better place for all of us.

5 Concluding Remarks

Scientific disciplines are in the book *Practicing Interdisciplinarity* (Weingart and Stehr 2000) described as the eyes through which modern society sees and forms its images about the world, frames its experience, and learns, shaping its own future or reconstituting the past. Thus, disciplines are the intellectual structures in which the transfer of knowledge from one generation to the next is cast; that is, they shape the entire system of education. During the most of the twentieth century, the question of knowledge has been shaped by disciplinarity.

In recent decades, the growth of scientific and technical knowledge has made scientists, engineers, social scientists, and humanists to join in addressing complex problems. Human society in its natural setting contends with enormously complex systems that are influenced by numerous forces.

Going through literature, there is no unique definition of interdisciplinary research. Description given below is taken from National Academy of Sciences et al. (2005, p. 26). Interdisciplinary research (IDR) is a mode of research by teams or individuals that integrates information, data, techniques, tools, perspectives, concepts, and/or theories from two or more disciplines or bodies of specialized knowledge to advance fundamental understanding or to solve problems whose solutions are beyond the scope of a single discipline or field of research practice.

Research is truly interdisciplinary when it is not just pasting two disciplines together to create one product but rather is an integration and synthesis of ideas and methods.

Social sciences, in the broadest sense, is an important field for investment by mathematics, via the development of complex systems or modeling and data analysis.

It is considered that in today's mathematics, it would be artificial and counter-productive to draw a boundary between theory and application. It is preferable to ensure a continuum between theoretical and applied mathematics and overcome the siloed approach of traditional mathematics. To this aim, mathematicians need training, incitation and incentives, and the creation of interdisciplinary and inter-mathematical teams to stimulate new ways of collaborating.

Borovik et al. (arXiv:2201.08364v1, p. 2) pointed out that mathematics continues to grow, and if you look around, you see that mathematical results and concepts involved in practical applications are much deeper, more abstract, and more difficult than ever before. And we have to accept that the mathematics hardwired and coded, say, in a smartphone, is beyond the reach of the vast majority of graduates from mathematics departments in many universities in the world. The cutting edge of mathematical research moves further and further away from the stagnating mathematics education. Due to Frenkel, the importance of having access to the mathematical knowledge is in protecting us from arbitrary decisions made by the powerful few in an increasingly math-driven world. "Where there is no mathematics, there is no freedom" he concludes (Frenkel 2013, p. 5).

To sum up, the value of mathematics to society is advanced by the intertwined development of mathematics as a discipline and the application of mathematics across the many fields of human endeavor. The benefits are both practical and beautiful. Increasingly solving problems at the intersection of multiple fields such as mathematics, computer science, and the social science will be key to addressing major societal challenges – hence the focus of the book. These developments though need to avoid a divide between those who can develop and use such solutions and those who cannot; hence an important educational imperative comes with this direction of future fruitful research.

As matter of conclusion, let us point out two citations.

> Mathematics is not a book confined within a cover and bound between brazen clasps, whose contents it needs only patience to ransack; it is not a mine, whose treasures may take long to reduce into possession, but which fill only a limited number of veins and lodes; it is not a soil, whose fertility can be exhausted by the yield of successive harvests; it is not a continent or an ocean, whose area can be mapped out and its contour defined: it is limitless as that space which it finds too narrow for its aspirations; its possibilities are as infinite as the worlds which are forever crowding in and multiplying upon the astronomer's gaze; it is as incapable of being restricted within assigned boundaries or being reduced to definitions of permanent validity, as the consciousness of life, which seems to slumber in each monad, in every atom of matter, in each leaf and bud cell, and is forever ready to burst forth into new forms of vegetable and animal existence. (Sylvester 1909)

> [Mathematics] is security. Certainty. Truth. Beauty. Insight. Structure. Architecture. I see mathematics, the part of human knowledge that I call mathematics, as one thing-one great, glorious thing. Whether it is differential topology, or functional analysis, or homological

algebra, it is all one thing. They are intimately interconnected, they are all facets of the same thing. That interconnection, that architecture, is secure truth and is beauty. (Halmos 1991)

That is the content and meaning of mathematics to us.

Acknowledgements The authors are grateful to referees for careful reading of the manuscript and helpful comments. M. N. H. is supported by the ICMPA-UNESCO Chair, which is in partnership with Daniel Iagolnitzer Foundation (DIF), France, and the Association pour la Promotion Scientifique de l'Afrique (APSA), supporting the development of mathematical physics in Africa. M. M. was financially supported by the Ministry of Science, Technological Development and Innovation of the Republic of Serbia (Contract No. 451-03-47/2023-01/ 200109).

References

Amiot, E. (2005). Rhythmic canons and Galois theory. *Grazer Math. Ber., Vol. 347*, 1–25.

Andreatta, M. (2004). On group-theoretical methods applied to music: some compositional and implementational aspects. Pp. 122–162 in Mazzola, G., Noll, T., & Puebla, E. (Eds.), *Perspectives in Mathematical and Computational Music Theory*. Osnabrück: Electronic Publishing Osnabrück.

Bagodo, O. B. (2020). Quatre millénaires de techniques, arts, cultes et métiers des métaux dans la baie du Bénin et son arrière-pays, Place et rôle de la divinité Goun/Gu au *XXI*^{e} siècle, Ecole du Patrimoine, ISBN 978-99982-0-627-4.

Bell, J. L. (2015). Reflections on Mathematics and Aesthetics. *Aisthesis. Pratiche, linguaggi e saperi dell'estetico 8 (1)*, 159–179.

Boorman, S. A., & White, H. C. (1976). Social structures from multiple networks. II. Role structures. *The American Journal of Sociology, Vol. 81, No. 6*, 1384–1446.

Borovik, A. V. (2010). *Mathematics under the Microscope, Notes on cognitive aspects of mathematical practice.* AMS, 2010.

Borovik, A. (2021). Mathematician's view of the unreasonable ineffectiveness of mathematics in biology. *Biosystems, Volume 205*, 104410.

Borovik, A., Kocsis, Z., & Kondratiev, V., Mathematics and mathematics education in 21st century. arXiv:2201.08364v1

Boyd, J. P. (1991). *Social semigroups a unified theory of scaling and blockmodelling as applied to social networks*. George Mason University Press, Fairfax, Virginia.

Bras-Amorós, M. (2019). Tempered Monoids of Real Numbers, the Golden Fractal Monoid, and the Well-Tempered Harmonic Semigroup. *Semigroup Forum, Springer, 99(2)*, 496–516.

Bras-Amorós, M. (2020). Increasingly Enumerable Submonoids of R: Music Theory as a Unifying Theme. *The American Mathematical Monthly, Mathematical Association of America, 127(1)*, 33–44.

Breitenbach, A., & Rizza, D. (2018a). Introduction to Special Issue: Aesthetics in Mathematics, *Philosophia Mathematica, Volume 26, Issue 2*, 153–160, https://doi.org/10.1093/philmat/nkx014

Breitenbach, A., & Rizza, D. (2018b). Guest Eds., Philosophia Mathematica, Series III, *Special Issue: Aesthetics in Mathematics*, Volume 26, Issue 2, Oxford University Press.

Clampitt, D., Domínguez, M., & Noll, T. (2008). Well-formed scales, maximally even sets, and Christoffel words. In Klouche, T., Noll, T. (Eds.), *MCM 2007. CCIS., Vol. 37*. Heidelberg: Springer, 2008.

Cheikh, A-D. (1981). *Civilisation ou Barbarie, Anthropologie sans Complaisance*. Présence Africaine, Paris.

Crans, A. S., Fiore, T. M., & Satyendra, R. (2009). Musical actions of dihedral groups. *Amer. Math. Monthly. 116 (6)*, 479–495. https://doi.org/10.4169/193009709X470399.

Dayson, F. (2009). Frogs and birds. *Notices of the AMS, Vol. 56*, 212–223.

Derbyshire, J. (2001). *Unknown Quantity: A Real And Imaginary History of Algebra*. Joseph Henry Press, 2006. Enzensberger H. M., Drawbridge Up: Mathematics – A Cultural Anathema 1st Edition, Taylor & Francis, Apr 27.

Einstein, A. (1921). Geometry and Experience. *Address to the Prussian Academy of Sciences in Berlin, January 27*.

Einstein, A., The World As I See It – An Essay, 2006, https://history.aip.org/exhibits/einstein/essay.htm

Enzensberger, H. M. (2001). *Drawbridge Up: Mathematics–A Cultural Anathema 1st Edition*, Taylor & Francis, Apr 27, 2001.

Frenkel, E. (2013). *Love and Math: The Hearth of Hidden Reality*. Basic Books, New York, 2013.

Gamwel, L. (2016). *Mathematics and Art: A Cultural History*. Princeton University Press.

Gerdes, P. (2007). *Adventures in the world of matrices*. Nova Science Publishers, Inc. New York.

Gerdes, P. (1999). *Geometry from Africa, Mathematical and Educational Explorations*. Mathematical Association of America, ISBN 0-88385-715-4, USA.

Glaz, S., & Growney, J. (2008). *Strange Attractors: Poems of Love and Mathematics*. CRC Press, Taylor & Francis Group.

Goethals, P. L., Scala, N. M., & Bennett, D. T. (2022). *Mathematics in Cyber Research*. Taylor & Francis Group.

Halmos, P. R. (1991). From interview with Donald J. Albers. In John H. Ewing and Frederick W. Gehring, Paul Halmos Celebrating 50 Years of Mathematics, 13.

Harris, M. (1988). Common threads: mathematics and textiles. *Math Sch 17(4)*, 24–28.

Harris, M. (1997). *Common threads: women, mathematics and work*. Trentham Books, Stoke- on-Trent. ISBN-13: ? 978-1858560151

Holden, J., & Holden, L. (2021). A Survey of Cellular Automata in Fiber Arts. Pp. 443–466 in Sriraman B. (Ed.), *Handbook of the Mathematics of the Arts and Sciences, Volume 1*. Springer, 2021.

Hounkonnou, M. N., Landalidji, M. J., & Mitrović, M. (2022). Hamiltonian Dynamics of a spaceship in Alcubierre and Godel metrics: Recursion operators and underlying master symmetries. *Theoretical and Mathematical Physics 212*(1), 1001–1018. https://link.springer.com/article/10.1134/S0040577922070091

Hounkonnou, M. N., & M. Mitrović, M. (2020). Mathematics for human flourishing in the time of COVID-19 and post COVID-19. Pp. 387–390 *in the Proceedings of the Fifth International Conference "Mechanical Engineering in XXI Century", MASING, December 9-10, 2020, Niš*.

Karaali G., & Lesser L. M. (2021). Mathematics and Poetry: Arts of the Heart. Pp. 967–980 in Sriraman B. (Ed.), *Handbook of the Mathematics of the Arts and Sciences, Volume 1*. Springer.

Karaali, G., & Sriraman, B. (2021). Mathematics, Humanities, and the Language Arts: An Introduction. Pp. 963–966 in Sriraman B. (Ed.), *Handbook of the Mathematics of the Arts and Sciences, Volume 1*. Springer.

Kline, M. (1967). *Mathematics for nonmathematicians*. Dover Publications, INC.

Krajewski, R., Bock, J., Kloeker, L., & Eckstein, L. (2018). The highD Dataset: A Drone Dataset of Naturalistic Vehicle Trajectories on German Highways for Validation of Highly Automated Driving Systems. Pp. 2118–2125. *21st International Conference on Intelligent Transportation Systems (ITSC)*. https://doi.org/10.1109/ITSC.2018.8569552.

Lafollette, P. S., Jr., (1999). Aesthetics, Beauty, and Computer Science. Pp. 185–190 in Sarhangi R. (Ed.), *Bridges – Mathematical Connections in Art, Music, and Science*. Bridges Conference

Langlands, R. P. (2010). *Is there beauty in mathematical theories?* University of Notre Dame, January 2010.

Lidl, R., Pilz, G. (1998). *Applied abstract algebra*. Springer.

Loutchko, D., Semigroup models for biochemical reaction networks, July 2022, https://arxiv.org/pdf/1908.04642.pdf

Marcus, S. (1998a). Mathematics and Poetry: Discrepancies within Similarities. Pp. 175–180 in Sarhangi R. (Ed.), *Bridges – Mathematical Connections in Art, Music, and Science*. Bridges Conference.

Marcus, S. (1998b). The Theater of Mathematics and the Mathematics of the Theater. Pp. 293–294 in Sarhangi R. (Ed.), *Bridges – Mathematical Connections in Art, Music, and Science*. Bridges Conference.

Marcus, S. (1999). The Art-Science Marriage, From Quarrel to Understanding. Pp. 279–290 in Sarhangi R. (Ed.), *Bridges – Mathematical Connections in Art, Music, and Science*. Bridges Conference.

Mateescu, A., & Salomaa, A. (1997). Formal Languages: an Introduction and Synopsis. Pp. 1–40 in Rozenberg G., & Salomaa A., (Eds.), *Handbook of Formal Languages, Volume 1: Word, Language, Grammar*. Springer.

Mathematics for Europe, European Commission, 2016. https://ec.europa.eu/futurium/en/system/files/ged/finalreportmaths.pdf

Mathematics for Action, Supporting Science-Based Decision-Making UNESCO, 2022. https://unesdoc.unesco.org/ark:/48223/pf0000380883.locale=en

Mensky, M. B. (2013). Dynamics-generating semigroups and phenomenology of decoherence. Pp. 46–58 in Kamalov T. F. (Ed.), *Theoretical Physics and its new Applications, Selected contributions at the 3rd International Conference on Theoretical Physics. June 24–28, 2013*. Moscow, Russia: MIPT, 2014. ISBN 978-5-7417 0519-3; 46-58.

Mitrović, M., Hounkonnou, M. N., & Baroni, M.A. (2021). Theory of constructive semigroups with apartness – foundations, development and practice. *Fundamenta Informaticae 184(3)*, 233–271.

Montiel, M., & Peck, R. W. (2019). *Mathematical Music Theory: Algebraic, Geometric, Combinatorial, Topological and Applied Approaches to Understanding Musical Phenomena*. World Scientific.

National Academy of Sciences, National Academy of Engineering, and Institute of Medicine, Facilitating interdisciplinary research, The National Academies Press Washington, D. C., 2005.

Nimkulrat, N., & Nurmi, T. (2021). Mathematical Design for Knotted Textiles. Pp. 381–409 in Sriraman B. (Ed.), *Handbook of the Mathematics of the Arts and Sciences, Volume 1*. Springer.

Ostwald, M. J. (2021). Architecture and Mathematics: An Ancient Symbiosis. Pp. 1117–1134 in Sriraman B. (Ed.), *Handbook of the Mathematics of the Arts and Sciences*. Springer.

Paperman, C., Salvati, S., & Soyez-Martin, C., An algebraic approach to vectorial programs. date. ffhal-03831752f

Pattison, P. (1993). *Algebraic models for social networks*. Cambridge University Press.

Paulos, J. A. (1990). *Innumeracy*. Vintage Books A Division of Random House, Inc. New York.

Pilz, G. F., Shevrin, L. N., & Trotter, P. G. (2002). Applications of Semigroups. Pp. 66–70 in Mikhalev A. V., & Pilz G. F. (Eds.), *The Concise Handbook of Algebra*. Springer-Science+Business Media, B.V.

Racois, M. (2021). *Propos sur la beauté des maths*, l'Art-dit, ISBN: 978-2-919221-50-9.

Ramakrishnan, V. (2020). Migration in Science. Pp. 148–166 in Knolle, J., & Poskett, J. (Eds.), *The Darwin College Lectures*. Published online by Cambridge University Press.

Reber, R., Schwarz, N., & Winkielman, P. (2004). Processing fluency and aesthetic pleasure: Is beauty in the perceiver's processing experience? *Personality and Social Psychology Review. 8 (4)*, 364–382.

Rieger, A. (2018). The Beautiful Art of Mathematics. *Philosophia Mathematica, Volume 26, Issue 2*, 234–250, https://doi.org/10.1093/philmat/nkx006

Rudnicki, R., & Tyran-Kamińska, M. (2017). *Piecewise Deterministic Processes in Biological Models*. Springer, 2017.

Salvadori, M. (1996). Can there be any relationships between Mathematics and Architecture. Nexus: Architecture and Mathematics, Kim Williams, ed. (Fucecchio, Florence: Edizioni deB'Erba, 1996), 12.

Starikova, I. (2018). Aesthetic Preferences in Mathematics: A Case Study. *Philosophia Mathematica, Volume 26, Issue 2*, 161–183, https://doi.org/10.1093/philmat/nkx01

Schein, B. M. (1997). Book Review – Social semigroups a unified theory of scaling and blockmodelling as applied to social networks (by J. P. Boug). *Semigroup Forum, 54*, 264–268.
Sriraman, B. (Ed.). (2021). *Handbook of the Mathematics of the Arts and Sciences*, Springer, 2021.
Sriraman, B., Lee, K.(2021). Mathematics, Art, and Aesthetics: An Introduction. Pp. 3–6 in Sriraman B. (Ed.), *Handbook of the Mathematics of the Arts and Sciences*. Springer.
Steffen, B., Rüthing, O., Huth, M. (2018). *Mathematical Foundations of Advanced Informatics Volume I Inductive Approach*, Springer.
Sylvester, J. J. (1909). From Commemoration Day Address (22 Feb 1877) at Johns Hopkins University, Baltimore, collected in The Collected Mathematical Papers: (1870-1883) (1909), 77–78.
Tamari, D. (1978). Algebra: Its Place in Mathematics and Its Role, Past and Present for Mankind. A Historical and Critical Essay, *Eleuteria (September 1978)*, 187–201.
Turner, M. (2014). *The Origin Of Ideas: Blending, Creativity And The Human Spark.* Oxford & New York: Oxford University Press.
Waerden, B. L. van der. (1985). *A History of Algebra-From al-Khwarizmi to Emmy Noether*. Berlin/New York, Springer.
Weinberg, S. (1983). *Dreams of a Final Theory*. London, Vintage.
Weingart, P., & Stehr, N. (Eds.). (2000). *Practicing Interdisciplinarity*. University of Toronto Press, Toronto Buffalo London.
Wigner, E. (1960). The Unreasonable Effectiveness of Mathematics in the Natural Sciences. *Communications in Pure and Applied Mathematics, Issue 13(I)*, 1–14.
Woźny, J. (2018). *How We Understand Mathematics*. Springer, 2018.
Zawislak, S., Kopec, J. (2019). A Graph-Based Analysis of Anton Chekhov's Uncle Vanya, *Journal of Humanistic Mathematics, Volume 9, 2 (July 2019)*, 157–186.

Algebra as a Semiotic Modeling System

Marcel Danesi

1 Introduction

The publication of Lakoff and Núñez's book, *Where Mathematics Comes From* (2000), which established links between metaphorical cognition and mathematics, led immediately thereafter to a broadening interest among mathematicians, cognitive scientists, and semioticians in the relation between sign structures and mathematical ideas and methods (for example, Rotman 2006; Turner 2012; Danesi and Bockarova 2014). The main area of interest has been mathematics education (Radford 2010; Presmeg et al. 2016; Sáenz-Ludlow and Kadunz 2016). Research on the Lakoff-Núñez perspective of mathematics has continued in all areas, not just the pedagogical one. However, as far as can be told, the theory has rarely been applied to examining the semiotic roots of algebraic method, despite the fact that it presents itself as a system of signs and structures that are utilized not only to do mathematics at an abstract level, but to model its inner workings. The purpose of this essay is to provide such an application in terms of a specific version of semiotic analysis called *modeling systems theory* (MST) (Lotman 1991; Sebeok and Danesi 2000; Nöth 2018).

Algebra is a modeling system in the semiotic sense (Dantzig 1930; Thom 1975, 2010; Gershenfeld 1998). Even a simple equation can be described in terms of its *form*, *structure*, and *modeling* dimensions. For example, the Pythagorean equation, $c^2 = a^2 + b^2$, has been constructed with a specific form (as opposed to other types of equations); it has a recognizable *structure*, with the variables within it assembled in such a way that shows how they are related to each other; and it functions as a *model* standing, initially, for the practical observation that "the square on the hypotenuse

M. Danesi (✉)
Department of Anthropology, Victoria College, University of Toronto, Toronto, ON, Canada
e-mail: marcel.danesi@utoronto.ca

© The Author(s), under exclusive license to Springer Nature Switzerland AG 2023
M. N. Hounkonnou et al. (eds.), *Mathematics for Social Sciences and Arts*, Mathematics in Mind, https://doi.org/10.1007/978-3-031-37792-1_2

is equal to the sum of the squares on the other two sides." The equation thus turns a specific type of information into a compact form via a specific structure, becoming a model of that information. In so doing, the power of models is that they can be unpacked to extract further, previously non-obvious, information (Turner 2012)—for example, identifying the infinite set of integer triples that satisfy the equation; what they might entail theoretically, such as Fermat's Last Theorem, which states that no three positive integers a, b, and c satisfy the equation $a^n + b^n = c^n$ for any integer value of n greater than 2. "It is impossible for a cube to be the sum of two cubes, a fourth power to be the sum of two fourth powers, or in general for any number that is a power greater than the second to be the sum of two like powers. I have discovered a truly marvelous demonstration of this proposition that this margin is too narrow to contain" (Fermat, translated from Latin, in Nagell 1951: 252). The argument will be made here that semiotic analysis presents a useful and insightful way of grasping the *raison d'être* of algebra as a powerful modeling system.

This is not a mere "reformulation" of well-known notions in mathematics from the standpoint of semiotics; rather, it will arguably put us in a better position to grasp the symbolic-formal nature of algebra and what it entails cognitively. As Bellos (2010: 123) has aptly put it:

> Replacing words with letters and symbols was more than convenient shorthand. The symbol *x* may have started as an abbreviation for "unknown quantity," but once invented, it became a powerful tool for thought. A word or an abbreviation cannot be subjected to mathematical operation in the way that a symbol like *x* can be. Numbers made counting possible; but letter symbols took mathematics into a domain far beyond language.

In an in-depth study, Keith Devlin (2012) identified what he called the "symbol barrier" as the biggest obstacle to a mastery of mathematics. Ordinary people, Devlin asserted, can do practical mathematics (counting, measuring, comparing quantities, etc.). But they have more difficulty doing more complex mathematics without possessing the symbolism used to represent complicated ideas. As the mathematics becomes more complex and abstract, so too does the reliance on symbolism, which at the most abstract levels supersedes practical experience. It is in deconstructing the "abstract levels" as Devlin calls them that MST can be especially useful—to learners, educators, and mathematicians themselves, as will be argued in this essay.

2 Modeling Systems Theory: An Outline

Modeling in semiotics can be defined simply as the use of sign forms (letters, various symbols, numbers, etc.), which, when assembled in a specific way, are used to represent something. A model, such as the Pythagorean equation, provides a form (with structural features) for explaining something. It can be simulative (iconic), relational (indexical), or based on conventions (symbolic). An example of an iconic model would be Pascal's Triangle, an arrangement of numbers that resembles a triangle wherein each number within it is the sum of two numbers directly above it (except for the edges, which are all "1"). It is designed to show the various expansions of the binomial $(a + b)^n$ in terms of an implied triangular layout. An example of an indexical model would be the number line, which is a linear model relating numbers to each other in a vectorial (left-right) way. An example of a symbolic model would be the Argand Diagram for complex numbers, which relates imaginary numbers to real ones in terms of a coordinate system; another example is the notion of field, which provides a way of conceptualizing a set of objects on which addition, subtraction, multiplication, and division are defined, thus modeling the organization of a set of numbers in a symbolic way.

Now, once the model has been constructed, it leads to further ideas that are inherent in it and which would not have been obvious without having been given a specific form and structure. The numerical coefficients in the expansion of the binomial above form the shape of a triangle with infinite dimensions. Significantly, Pascal's Triangle has been used in other areas of mathematics, such as in the calculation of probabilities. It also turns up in combinatorial analysis. A myriad of patterns within it have been discovered over time—in effect, the initial modeling function of the triangle has been unpacked in an infinitude of ways, making the model a multi-functional one. The same analysis can be applied to any model.

Sebeok and Danesi (2000) proposed a useful typology of four form types within MST, each with its structural features and modeling functions: *singularized*, *composite*, *cohesive*, and *connective*. The notion of *forms* is more or less isomorphic to the traditional notion of *signs*; the difference is that *form* is used in this case in relation to the conventional philosophical one, where it is seen as separate from content, which emerges as the form is used.

A *singularized form* (a numeral, a variable, and so on) is one that has been made to represent a singular (unitary) referent or referential domain. Numerals are singularized forms. For example, the digit 3 is equivalent to Roman numeral III, even though they differ structurally—the former assumes its value via its position in a numeral, while the latter does not (it enfolds a value independently of its position, constituting an iconic form). Both, however, when used constituted singularized models of "threeness." *Composite forms* stand for referents via the combinatory structuring of elements. An equation is a composite form. As discussed, the Pythagorean equation, $c^2 = a^2 + b^2$, for instance, stands for sets of three integers that are related to each other in the exact way stipulated by the form of the equation. It is a composite model of the relationship that the sides of a right triangle exhibit. A *cohesive form* indicates how various parts (or sub-forms) are related to each other structurally. The mathematical notion of set is an example of a cohesive model, since it defines elements in terms of shared properties—hence the set of integers, rational numbers, complex numbers, and so on. As such, each set is a cohesive model of some aspect of mathematical structure. Another example is the notion of group, which involves the consideration of algebraic structures (rings, fields, and vector spaces) that display specific properties. Finally, a *connective form* is one that results from fusing or blending other forms ontologically to produce a model of something previously unknown. Mathematical metaphors, for example, are connective models, linking one type of referent to another. The Lakoff and Núñez book mentioned at the start of this essay is based on examining connective modeling within mathematics (to be discussed below).

These four types of forms are not mutually exclusive. They are referentially interdependent—singularized forms go into the make-up of composite ones which, in turn, are dependent upon the forms that cohesive systems make available, and so on. Now, the key notion in MST is that each form type subserves a particular *modeling function*. For instance, the function of singularized forms is to represent single, unitary referents (as discussed). Roman numerals such as I, II, and III are iconic models because they stand for their referents in a simulative way (one stroke = one unit, two strokes = two units, three strokes = three units). The use of letters to represent unknowns in algebra is another example of singularized modeling—a usage traced to René Descartes' *La géométrie* (1637), which established the convention of using the lowercase letters at the beginning of the alphabet for known quantities (a, b, and c) and those at the end of the alphabet for unknown quantities (x, y, and z).

A composite form allows for the modeling of some complex referent or as a textual model of some mathematical procedure, such as proof. Without going into any of the historical debates and definitions of proof here, it can be said, for the sake of convenience, that proof is any set of statements *composed* (literally) in such a way that demonstrate that something is necessarily so, much like literary texts are intended to model some aspect of reality in a chronological way (Ernest 2008). A proof is a form based on putting parts together to show that something holds logically. The composition is not random; it is based on sequential structure, much like narrative texts, with each statement depending on the one before. As Stewart (2008: 34) puts it:

> What is a proof? It is a kind of mathematical story, in which each step is a logical consequence of the previous steps. Every statement has to be justified by referring it back to previous statements and showing that it is a logical consequence of them.

In effect, the composite form of a proof models how the different forms of logic itself unfold. Diagrams and equations are also examples of composite forms used to model specific referents. Eulerian graph theory, for instance, is based on iconic diagrammatic forms that model the structure of specific figures or systems. The model has been extended to provide a symbolic pattern among the sides of figures. In the case of three-dimensional figures, called polyhedra, Euler found that if we subtract the number of edges (e) from the number of vertices (v) and then add the number of faces (f), we will always get 2 as a result: $\text{v} - \text{e} + \text{f} = 2$.

Sets, as mentioned, are examples of cohesive forms. Among various other things, these forms have allowed for the modeling of otherwise intractable notions such as infinity, as provided by Cantor's (1874) proofs—which were based on considering the integers as a cohesive system. As is well known, but worth reiterating here for the sake of illustration, the first insinuation of the cardinals as cohesive forms for investigating (and thus modeling) infinity is found in Galileo's *Two New Sciences* (1638), where he noted that the set of square integers can be compared, one-by-one, with all the whole numbers (positive integers), leading to the apparent paradox that there are as many square integers as there are integers (even though the square integers are themselves only a part or subset of the entire set of integers). This suggested to Galileo that the "number" of elements in the set of all positive integers and the "number" of those in one of its proper subsets is the same. As such, the possibility that the cardinal numbers constituted a modeling system of numbers generally was proved later by Cantor.

Finally, connective modeling is a means of linking the referents of forms in terms of their bidirectional implications, with one referential domain implying the other—an area of research initiated, as mentioned, by Lakoff and Núñez (2000), and developed further by Fauconnier and Turner (2002), among others (for example, Gibbs and Colston 2012). Actually, in his 1987 book, *Women, Fire, and Dangerous Things*, Lakoff cites a 1981 paper by Saunders Mac Lane which is consistent with the view of connective modeling, expressed from the viewpoint of a mathematician. Mac Lane (1981: 471) summarizes this kind of approach to mathematics as follows, predating Lakoff and Núñez: "mathematical development uses experience and intuitive insights to discover appropriate formal structures, to make deductive analyses of these structures, and to establish formal interconnections between them. In other words, mathematics studies interlocking structures."

According to Lakoff and Núñez, connective modeling is based on conceptual metaphors—*change is motion*, *sets are collections in containers*, *continuity is gapless*, *functions are sets of ordered pairs*, *geometric figures are objects in space*, *numbers are object collections*, *recurrence is circular*, etc. Each of these underlies a specific mathematical conceptualization such as the calculus (change is motion), infinity (continuity is gapless), set theory (numbers are object collections), and so on. In this framework, metonymy (the part for the whole) is seen separately as the connective mechanism that allows for generalizations from particular instances. Suffice it to say here that, unlike metaphor, metonymy does not function to create knowledge through connective reasoning, but rather it allows us to cast specific kinds of light on certain situations, so as to be able to grasp their mathematical implications. It is an indexical modeling process that allows relations to be made explicit. Generally defined, and index is a sign that puts referents into some kind of existential relation. A simple example is the figure of an arrow, which indicates the direction of something in relation to a point of departure. So, indexical modeling is, in essence, a way of putting referents into an existential relation pattern. Metonymy is an example of indexicality, whereby a part stands for the whole in a relational way. Marcus (2012: 146) thus sees metonymy as an adjunct to metaphor within mathematics:

> Complementary to metaphorical thinking is metonymical thinking. The former is related to iconic thinking, the latter, to indexical thinking. Metonymy is everywhere in mathematics, either as *pars pro toto* or as in *if-then* thinking. The whole mathematical enterprise is metonymical, since mathematics is looking for a suitable representation of infinity by countable forms, then to reduce the latter to a finite representation and after that to reduce the large finite to the small finite. There is the claim that mathematics is the science of approximations; but approximation is a metonymical notion. Most real numbers have

> essentially infinite representations (decimal or by continuous fractions) and we try to capture finite parts of them, as large as possible. This process never stops. A famous example is the constant effort to capture the decimals of the number π. This began with Archimedes and is continued today in computer programs and by clever procedures such as those found in the notebooks of Ramanujan. The basic metonymy, *if-then*, is at the root of the deductive thinking essential in the final presentation of mathematical proofs. It is the main tool to validate a mathematical theorem.

While various critiques of the Lakoff-Núñez approach have come forth, it is fair to say that it has been adopted broadly by mathematicians and mathematics educators, as a means to penetrate the metaphorical-metonymic sources of many ideas within mathematics (Danesi 2020). In a 2011 lecture given at the Fields Institute for Research in Mathematical Sciences, Lakoff gave a specific example of the power of connective modeling, demonstrating how it undergirded one of the indeterminacy proofs of Kurt Gödel (1931) (see Danesi 2011 for a summary). The preexistent composite form that influenced Gödel was Cantor's (1874) famous diagonal proof, whereby he arranged the set of all rational numbers in a diagonal array, called Cantor's sieve. Since the singularized numeral form of the numbers is p/q, Cantor constructed each row with successive denominators (q)—{1, 2, 3, 4, 5, 6, ...}, making the numerator (p) of the numbers in the first row equal 1, of those in the second row equal 2, of those in the third row equal 3, and so on. In this way, all numbers of the form p/q are covered in the sieve. Cantor then highlighted every fraction that is a multiple of another fraction. If these fractions are skipped, then every rational number appears once and only once. Cantor was thus able to set up a one-to-one correspondence between the integers and the numbers in the array to show that the matching goes on ad infinitum, moving through the sieve in a diagonal fashion: the cardinal number 1 corresponds to 1/1 at the top left-hand corner; the cardinal number 2 corresponds to the number below (2/1); following the zig-zagging path, the cardinal number 3 corresponds to 1/2; the cardinal number 4 corresponds to 1/3; and so on, throughout the sieve. The diagonal path thus allowed Cantor to establish a one-to-one correspondence between the cardinal numbers and all the rational numbers.

Gödel's proof showed that within any formal logical system, there are results that are undecidable, and he did so by using the same type of diagonalized argument. Without going into details here, Lakoff demonstrated how Gödel was guided by Cantor's model—a model that is consistent with what Lakoff and Núñez (2000) had called the Basic Metaphor of Infinity (BMI). As Rafael Núñez (2005: 1717) wrote several years after the publication of *Where Mathematics Comes From*:

> [Cantor's] analysis is based on the Basic Metaphor of Infinity (BMI). The BMI is a human everyday conceptual mechanism, originally outside of mathematics, hypothesized to be responsible for the creation of all kinds of mathematical actual infinities, from points at infinity in projective geometry to infinite sets, to infinitesimal numbers, to least upper bounds. Under this view "BMI" becomes the Basic Mapping of Infinity.

The BMI, in simplified terms, implies the perception that mathematical processes which go on indefinitely are conceptualized as having an ultimate result. Winter and Yoshimi (2020) have observed, cogently, that the Lakoff-Núñez approach may not be as all-encompassing as the two cognitive scientists suggest. However, the history of discoveries in mathematics typically reveals connective modeling. The discovery of imaginary numbers is a case-in-point and worth revisiting schematically here for illustrative purposes.

As is well known, it was in his treatise on equations, *Ars magna*, that Gerolamo Cardano (1545) came unexpectedly across the square root of negative numbers. This appeared while he was considering the solution to the following two simultaneous equations in his book: $x + y = 10$ and $xy = 40$, whose solutions are $x = (5 + \sqrt{-15})$ or $(5 - \sqrt{-15})$. What possible number is $\sqrt{-15}$, Cardano asked himself? Unable to literally connect it to anything known at the time, he decided to put it aside, calling it simply a *fictitious* number. It was Cardano's contemporary Rafaello Bombelli who made the first connections between such numbers and the *real* ones in his *L'algebra* of 1572; they were later called *imaginary*, leading eventually to the concept of *complex number*. The point here is that this discovery came about via connective modeling—that is, by connecting imaginary and real numbers into a composite form (complex numbers). After this discovery, complex numbers became themselves models of other referents and referential domains.

The importance of connective modeling to mathematics was emphasized by Charles Peirce, who emphasized that diagrams in mathematics do not simply portray information but also mirror how connective thinking about the relevant information occurs in actu (Peirce, vol. 4: 6). As Kiryushchenko (2012: 122) has aptly put it, for Peirce "graphic language allows us to experience a meaning visually as a set of transitional states, where the meaning is accessible in its entirety at any given 'here and now' during its transformation."

It is relevant to note at this point that mathematicians have used the notion of modeling, to indicate that specific mathematical constructs can be used to model various aspects of physical reality. The calculus, for instance, is seen as a modeling system for portraying change and flux, both encoding it and predicting it via its composite and cohesive forms (such as differential and integral equations). Crilly (2011: 84) describes the meaning of modeling in mathematics as essentially the language of physics:

> The field of differential equations is huge, and besides mathematicians it attracts physicists involved with physical theories, chemists interested in diseases and how fast they are spread. These are studied within the framework of *mathematical modeling* [emphasis ours], where simplifying assumptions are made in order to understand a process. Many areas where the Calculus is applied involve quantities with more than one variable, such as space *and* time.

Crilly (2011: 152) goes on to state that a "mathematical model is a way of describing a real-life situation in mathematical language, turning it into the vocabulary of variables and equations." In MST, modeling is seen as a formalization of structural patterns, allowing for an understanding of how these relate to reality. The classic example is that of a formula or equation, which contains structural patterns that refer to some aspect of mathematical reality. First, an actual experience or occurrence leads to the unwitting discovery of something unknown (as did Cardano with $\sqrt{-15}$). By giving this discovery a *form* ($i = \sqrt{-1}$), the basis was laid for an exploration of the structural properties that the form entails, which in this case involved composite structures (complex numbers). Once these have been established as cohesive structures of their own, then they assume modeling functions (as discussed above). This semiotic "flow" of ideas can be represented as follows:

$$\text{Experience} \rightarrow \text{Form} \rightarrow \text{Structure} \rightarrow \text{Modeling Functions} \left(\text{singularized, composite, cohesive, connective}\right)$$

If the discovery of numbers such as $\sqrt{-15}$ had remained at the level of a practical problem, each time they surfaced from solving quadratic equations, it would have remained meaningless as a tool of mathematical cognition. By giving it a form and connecting it structurally to existing forms, it became possible to expand the modeling capacities of mathematics.

3 Principles

The overriding principle in MST is the *standing for principle* (SFP), whereby all forms, no matter their size or function, display similar patterns of representation via iconic, indexical, and symbolic modeling systems. Simply put, an iconic system is one that aims to reproduce by some form of resemblance some referent. In arithmetic the classic example is the Roman numeral system in which, for example, one stroke (I) stands for the number one and two strokes (II) for the number two, among other iconic forms. An indexical system (as mentioned) is one that aims to relate referents to each other in some existential way. In mathematics, equations are general indexes since they relate variables in some way to each other. Symbolic systems are those derived conventionally, including symbols such as π or e. The late semiotician Thomas Sebeok derived various corollary principles from this main purview (in Sebeok and Danesi 2000).

1. Modeling is the end-result of assigning different forms (singularized, composite, cohesive, connective) to concrete observations, experiences, etc. (the *modeling principle*).
2. Knowledge is indistinguishable from the models derived from the different forms (the *knowledge principle*).
3. Modeling unfolds in terms of iconic, indexical, and symbolic forms (the *dimensionality principle*).
4. Complex (abstract) forms are derivatives of simpler (more concrete) ones (the *extensionality principle*).
5. Models and their referential domains are interconnected to each other (the *interconnectedness principle*).
6. All models display the same pattern of structural properties (the *structuralist principle*).

The *modeling principle* asserts that modeling is fundamentally a meaning-making process, implying that in order for something to be known and remembered, it must be assigned meaning in form-based ways. In this case, "meaning" is equivalent to function, in the sense that the meaning of the Pythagorean theorem is its function in describing relationships among integers of a specific kind. The *dimensionality principle* maintains that there are three dimensions that forms assume—further research in this area might show that there is an evolutionary flow here as well, from iconic to indexical and symbolic forms. The *extensionality principle* posits that abstract models are derivatives of more concrete, sense-based forms, while connective models in particular result from the process of extensionality. The *interconnectedness principle* asserts that a specific model is interconnected to other models, as became obvious with the discovery of imaginary numbers. The *structuralist principle* claims that certain elemental structural properties characterize all modeling systems, given that they are derivates of the same form types (singularized, composite, cohesive, and connective).

These are all well-known notions within semiotics. But the reworking of these notions in the context of MST gives them a much more concrete applicative modality, since they can be seen to occur across form sizes (signs, texts, codes, and figurative assemblages); that is, they underlie the structural formation of all kinds of models (Sebeok and Danesi 2000). MST is thus different from the traditional semiotic modes of analysis, which separates signs from texts, for example; it inheres in uniting all forms into a comprehensive understanding of how modeling works at all levels of meaning-making.

One of the more important theoretical functions of modeling in mathematics is to provide a way to envisage the general mathematical structure of specific forms. For example, Giuseppe Peano's aim in 1889 to formalize the operations of arithmetic, by breaking them down into their simple logical components, recalling Euclid's axioms for geometry, was intended to constitute such a generic modeling system. It started by establishing the first natural number (no matter what numeral system is used to represent it), which is zero. The other axioms (composite forms) in the model are successor ones showing that they apply to every successive natural number:

1. Zero is a number.
2. If x is a number, the successor of x is a number.
3. Zero is not the successor of a number.
4. Two numbers of which the successors are equal are themselves equal.
5. (Induction axiom) If a set S of numbers contains zero and also the successor of every number in S, then every number is in S.

Peano's axioms may seem self-evident, but the goal of modeling systems is to shed light on their structural essence and why it allows us to envisage further modeling functions.

4 Algebra as a Modeling System

Definitions or characterizations of algebra have one thing in common: it is perceived as a way of generalizing arithmetic, which itself is a generalization of counting patterns. If we consider these to be "orders" of modeling, then algebra can be defined semiotically as a fourth-order modeling system.

The first order is the instinctive ability itself to count. Using numerical signs—singularized forms—to stand for the objects or referents of counting constitutes a second order, at which counting concepts are modeled in terms of singularized forms (numerals). The third order is the arithmetical one. This is the level at which numerals are organized into composite and cohesive forms. At this level, modeling systems emerge, such as the associative and commutative laws. What is relevant here is that the associative law of addition—$(a + b) + c = a + (b + c)$—and of multiplication, $(a \times b) \times c = a \times (b \times c)$, zeroes in on a particular modeling domain, showing that not all operations have the associative property (subtraction and division are not associative). Finally, algebra constitutes a fourth-order modeling system that has the capacity to generalize the arithmetic by giving it abstract form:

Fourth order	*Algebra* = Modeling system of arithmetic
Third order	*Arithmetic* = Modeling of numerical forms
Second order	*Numeration* = Numeral forms as modeling counting processes
First order	*Counting* = the instinctive ability to separate referents into distinctive quantities

The concept of *modeling order* thus allows us to grasp the connection between levels in mathematics. The emergence of algebraic competence is the end result of following a "natural semiotic order," as it can be called, and it might be the basis solving the symbol barrier problem identified by Devlin (2012). Marcus (2012: 184) writes on this theme insightfully as follows:

> When mathematics is involved in a cognitive modeling process, both analogical and indexical operations are used. But the conflict is unavoidable, because the model M of a situation A should be concomitantly as near as possible to A (to increase the chance of the statements about M to be relevant for A too), but, on the other hand, M should be as far as possible from A (to increase the chance of M to be investigated by some method which is not compatible with the nature of A). No mathematical *construction* [read: model] can be constrained to have a unique interpretation, its semantic freedom is infinite.

One of the hardest problems of mathematics has been to determine if there are higher orders than the algebraic one, and if there is one overarching modeling system, or "metalanguage," that would encompass all of mathematics. Russell and Whitehead (1913) were probably the first to tackle this very question in a systematic way. But it became transparently obvious right after publication of their work that there might be a limit to the orders, which they called types. The Polish mathematician Alfred Tarski (1933) attempted to rectify some of the problems related to the notion of types by naming each increasingly higher order a metalanguage. A metalanguage, Tarski argued, is essentially a statement about another statement. At the bottom of the hierarchy are straightforward statements about things such as "Earth has one moon." Now, if we say "The statement that Earth has one moon is true," we are using a different type of language, because it constitutes a statement about a previous statement. It is a metalanguage. The problem with this whole approach is, of course, that more and more abstract metalanguages are needed to evaluate lower-level statements. And this can go on ad infinitum. In effect, Tarski's system only postpones making final decisions. So, it can be posited that algebra is the upper limit of modeling orders. Now, this does not mean that algebraic models are static. Indeed, there are countless such models—linear algebra, matrix algebra, abstract algebra, etc. It can be called, putatively, a meta-modeling system.

This suggests that there is a mental modeling system that the biologist-semiotician Jakob von Uexküll (1909) called the *Bauplan* that reaches its limits at a certain order of abstraction. This in no way denies the infinite potential of creativity in the human species. But creativity is not random. A simple example might be used. A musical composer such as Mozart was highly creative, but within established structures that defined the symphony, the concerto, etc. So too mathematicians are highly creative, but within the modeling systems that they have acquired from history. This modeling constraint may actually be the reason why some problems are undecidable and why some may not have a solution, since they might actually exist at a higher order to which, at least presently, we have no access. Problems such as the Collatz conjecture may be unsolvable because they may transcend a fourth-order algebraic modeling system. In highly condensed form, the conjecture stipulates that we always end up with the number one (or more specifically the sequence 4–2–1) if we apply the following rule: If a number, n, is even, we make it half, or n/2; if it is odd, we triple it and add one, or 3n + 1; and if we keep repeating this rule, we always end up with the 4–2–1 sequence. Several mathematicians have proved that the conjecture is almost always true but have not demonstrated that it is always true. As Paul Erdös said about the Collatz conjecture (cited in Guy 2004: 336–337): "Mathematics may not be ready for such problems," which translated in MST terms implies that for now the fourth order is the limit of mathematical representation.

Another classic example of a seemingly simple problem that has resisted a proof is, of course, the Goldbach conjecture—namely, that every even integer greater than 2 is the sum of two primes ($4 = 2 + 2$, $6 = 3 + 3$, etc.). By exploration, one can always find two primes with which to write an even number. But all we can do is assume this to be true. So far, no one has been able to prove this, even though it has been shown to hold for all integers less than 4×10^{18}. The conjecture above is the *strong conjecture*; the *weak conjecture* is that any number greater than 5 could be written as the sum of three primes ($6 = 2 + 2 + 2$, $8 = 2 + 3 + 3$, etc.). Actually, the weak conjecture was proved by Harald Helfgott in 2013, although it is still being discussed by mathematicians. The strong conjecture remains elusive to proof so far.

There certainly may come a time when both these conjectures will be provided with a coherent composite proof. But it could be that, in some cases, proofs may

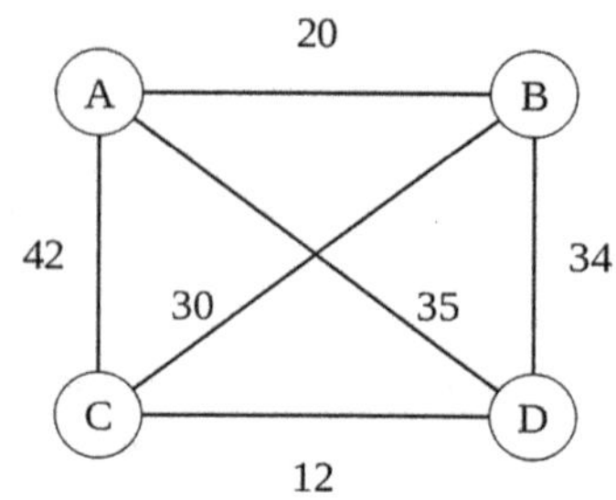

Fig. 1 Symmetric model of the traveling salesman problem

not be possible, and this is due to the upper boundary of four orders of modeling. Today, modeling by computer has led to a compromise solution in devising putative solutions to intractable problems. The most famous of these is the four-color problem (Appel and Haken 1986). Another one is the so-called traveling salesman problem (TSP):

> A salesman wishes to make a round-trip that visits a certain number of cities. He knows the distance between all pairs of cities. If he is to visit each city exactly once, then what is the minimum total distance of such a round trip?

If we represent cities as vertices and trips as paths, then we can transform the problem into one involving graph theory. The TSP requires that we find the most efficient path that the salesman can take through each of the vertices—starting and finishing at a specified vertex after having visited each other vertex exactly once. There are two types of TSPs: symmetric and asymmetric. The one above is symmetric, involving four odd vertices (cities), A, B, C, and D, each of a degree 3 (having three paths leading in and out) (Fig. 1):

In this case, there is no way to complete the circuit without doubling back at a vertex. In an asymmetric TSP, paths may not exist in both directions or the distances might be different, as illustrated in the path below (Fig. 2):

The TSP belongs to the class of combinatorial optimization problems known as P versus NP. The TSP is classified as NP-hard because it has no "quick" solution and the complexity of calculating the best route will increase when we add more destinations to the problem. More to the point of the present discussion, it may be beyond solution because we do not have the metalanguage (an order beyond the fourth order) available at the present time. The origins of the TSP are unclear. A handbook for traveling salesmen from 1832 mentions the problem and includes example tours through Germany and Switzerland but contains no mathematical treatment. As a graph theory problem, it was formulated in the nineteenth century by the Irish mathematician William Rowan Hamilton and by the British mathematician Thomas Kirkman. Hamilton's Icosian game is a recreational version of the TSP. The general form of the TSP was first studied in the 1930s by Austrian mathematician Karl Menger, who originally called it the *messenger problem*. Now, the point here is that the problem can be modeled on computer, by translating the physical aspects of the problem (distances, cities, and so on) into symbolic constructs, such as paths, vertices, and so on, that apply to graph systems.

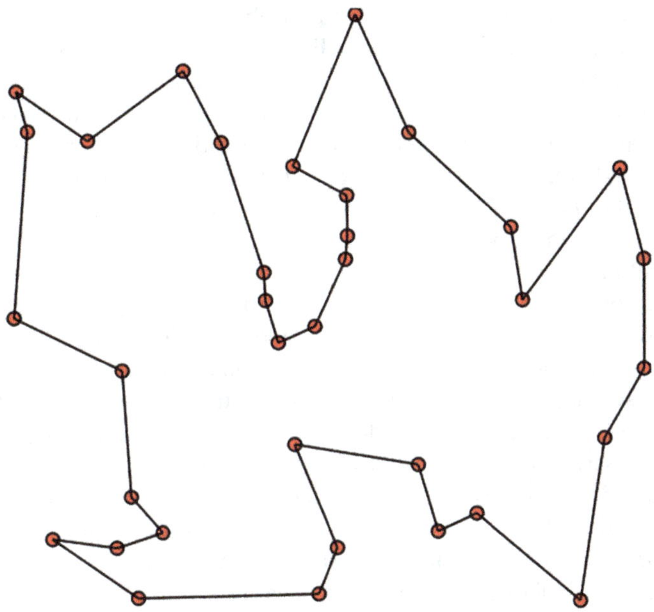

Fig. 2 Asymmetric model of the traveling salesman problem

Interestingly, a new branch of semiotics, called *algebraic semiotics*—the brainchild of computer scientist Joseph Goguen (1999)—involves using semiotic theory, and especially the connective forms of modeling (although it is not named as such within it), to come up with interface design that builds on semiotic principles. One of the main premises of this branch is that of *morphisms* (i.e., mappings between sign systems): If some class of objects is interesting, then structure preserving morphisms of those objects are also interesting. For semiotics, these morphisms are representations or in MST terms, models. Algebraic semiotics is based on the idea that we can compare morphisms by how well they preserve structure. Hence semiotic morphisms map forms to forms, structures to structures, and models to models.

5 Concluding Remarks

The purpose of this essay has been to argue and illustrate concretely that algebraic notions (and mathematical notions in general) can be deconstructed to reveal a core set of semiotic representational processes and principles encompassed by modeling systems theory. This implies that, by giving specific mathematical information a certain *form*, the result is a *structural* representation that packs the information in such a way that it can be utilized to subserve various modeling functions. Now, as the model is used more and more, it lends itself to further unpacking, providing ideas that transcend the impetus for formally representing the original information. In other words, the model becomes a discovery tool on its own. To reiterate here, the Pythagorean theorem was devised to model a structural pattern among the sides of a right triangle via a composite form (the equation $c^2 = a^2 + b^2$). As this was used by mathematicians, it was unpacked further to reveal hidden information within it (as discussed). The same analysis can be applied to any notion. The determinant, for example, was devised as a composite form for representing various computational processes—via a square arrangement of singularized forms with an equal number of rows and columns. The determinant, more specifically, involved iconic (the vertical layout of symbols) and indexical (the relation of the symbols to each other) properties. Now, many discoveries and applications became possible after it was used, leading to the emergence of matrix algebra as a new modeling system. In sum, MST reveals how ideas, intuitions, experiences, inferences, etc. are given expressive form and how this then subserves modeling functions within mathematics and without (as in science), which may lead to further knowledge.

As mentioned at the start of this essay, the impetus for establishing a semiotic approach to the study of mathematical representation was laid, indirectly, by George Lakoff and Rafael Núñez in their book *Where Mathematics Comes From* (2000), in which they discussed how mathematicians come to use and invent their ideas, proofs, and theorems through what has been called here connective modeling. If they are correct, then the same neural circuits are involved in mathematics and other expressive-representational enterprises, and this would open up new suggestive investigative avenues for connecting mathematics, language, the arts, science, and so on. Whether or not a neural interconnection can be established empirically, the point is that it is plausible and highly interesting and, thus, needs to be explored seriously if we are ever to come to an understanding of what mathematics is.

It is not happenstance that algebra emerged as a modeling system to generalize arithmetical forms and structures. The ancient Egyptians and Babylonians used a proto-form of algebra, as did the Greeks, Chinese, and people of India later. Diophantus devised what we now call quadratic equations and symbols for unknown quantities. But it was between 813 and 833 that al-Khwarizmi, a teacher in the mathematical school in Baghdad, finally showed how algebra worked as a modeling system of arithmetical information. Since then, as Crilly (2011: 104) aptly observes, the "desire to find a formula [has become] a driving force in science and mathematics." Constructing a formula is, in effect, devising a specific structural form that can then be used as a model of something.

One could ask, however: Does this add anything to understanding what algebra (and mathematics more generally) is? Admittedly, the application of MST to algebra may be nothing more than an interesting perspective of this branch of mathematics. But then, interesting ideas may bear unexpected insights. In this case, it can be argued that the main one is that mathematics and other faculties, such as language, are based on the same semiotic principles. Since at least the 1960s (e.g., Jakobson 1961; Hockett 1967; Harris 1968), mathematical notions have influenced the evolution of various research paradigms within linguistics, both intrinsically and contrastively; at the same time, mathematicians started to look at questions explored within linguistics, such as the nature of grammar and linguistic rules of word and sentence formation. As rigid disciplinary territories started breaking down in the 1990s, and with interdisciplinarity emerging as a powerful investigative mindset, the boundaries between research paradigms in linguistics and mathematics have actually been steadily crumbling ever since.

The underlying premise in MST (and semiotics more generally) is that all human systems of representation and communication are grounded on the same kinds of forms and structures (Sebeok and Danesi 2000). When these assume specific modeling functions, they become themselves powerful tools of discoveries. So, arguably, MST begs one of the oldest questions in philosophy and mathematics: Is mathematics invented or is it discovered? As modern physics attempts to develop a "theory of everything" from increasingly abstract mathematics, it does not seem far-fetched to imagine that mathematics may indeed hold the key to the universe—a prospect contemplated initially by Pythagoras. Does the cosmos make mathematics, or does mathematics make the cosmos? The answer that MST putatively provides is that it is moot to make such a distinction. Differences in singularized forms of numeration (Roman, decimal, binary) are, of course, culture-based and invented, but the similarity of the modeling functions of all such systems transcends cultural specificity. So, the question of which came first constitutes a mathematical version of the chicken-and-the-egg conundrum. All we can do is document how this conundrum works at the level of expression and formalization.

Modeling leads to knowledge, not because it is designed by instinct to be "knowledge-productive," but because it packs information in creative ways from its contexts of occurrence by giving it a structural form, which in turn evolves into a model of something. The exploratory power of the "modeling instinct" (as it can be called) suggests that we are probably programmed to first represent and then discover things through our semiotic strategies. As such, these strategies might be mirrors of how the mind works. As literary critic John William Navin Sullivan (1925) so aptly put it: "The significance of mathematics resides precisely in the fact that it is an art; by informing us of the nature of our own minds it informs us of much that depends on our minds."

Glossary

Basic Metaphor of Infinity a conceptual mechanism responsible for the creation of all kinds of mathematical infinities

Bauplan the innate mental modeling system that imposes limits on the order of models, but not on creativity

Cohesive form any form that contains sets of forms related to each other in some cohesive way

Composite form any form that is made up of simpler forms

Conceptual metaphor a metaphorical formula, such as *numbers are objects in a container* that undergirds abstract conceptualization
Connective form any form that emerges by connectivity to other forms
Dimensionality principle of modeling systems theory whereby models are seen as displaying specific modalities—iconic, indexical, and symbolic
Extensionality principle of modeling systems theory whereby complex forms are seen as derivative of simpler forms
Form any unit that can be used to represent something
Iconic a form that resembles its referent in some way, concrete or abstract
Indexical a form that puts signs in existential relation to each other
Interconnectedness principle of modeling systems theory whereby models are seen as connected to each other in specific ways
Metalanguage a language or statement that refers to a lower order language or statement
Model any form that has structure and meaning
Modeling order a level at which a model exists, from simple to abstract (e.g., counting is a first-order modeling system while algebra is a fourth-order modeling system)
Modeling system the use of models to portray something in a consistent manner
Sign anything that stands for something other than itself
Singularized form a form that stands for some unitary referent
Structuralist principle of modeling systems theory whereby all models are seen as exhibiting particular types of structure
Structure the actual shape of a form, which thus possesses meaning
Symbolic form a form derived from conventions

References

Appel, K., & Haken, W. (1986). The four color proof suffices. *The Mathematical Intelligencer, 8*, 10-20.

Bellos, A. (2010). *Here's Looking at Euclid: A Surprising Excursion through the Astonishing World of Math*. Princeton: Princeton University Press.

Bombelli, R. (1572). *L'algebra*. Bologna: Giovanni Rolsi.

Cantor, G. (1874). Über eine Eigneschaft des Inbegriffes aller reelen algebraischen Zahlen. *Journal für die Reine und Angewandte Mathematik, 77*, 258-262.

Cardano, G. (1545). *Ars magna*. New York: Dover.

Crilly, T. (2011). *Mathematics*. London: Quercus.

Danesi, M. (2011). George Lakoff on the cognitive and neural foundation of mathematics. *Fields Notes, 11*, 14-20.

Danesi, M. (2020). Math cognition and the arts. *Journal of Humanistic Mathematics, 10*, 1-20.

Danesi, M., & Bockarova, M. (2014). *Mathematics as a Modeling System: A Semiotic Approach*. Tartu: Tartu University Press.

Dantzig, T. (1930). *Number: The Language of Science*. London: Allen & Unwin.

Descartes, R. (1637). *La géometrie*. Paris: Presses Universitaires de France.

Devlin, K. (2012). The symbol barrier to mathematics learning. In M. Bockarova, M. Danesi, and R. Núñez (Eds.), *Semiotic and Cognitive Science Essays on the Nature of Mathematics*, pp. 54-60. Munich: Lincom Europa.

Ernest, P. (2008). Towards a semiotics of mathematical text (Part 2). *For the Learning of Mathematics, 28*, 39-47.

Fauconnier, G., & Turner, M. (2002). *The Way We Think: Conceptual Blending and the Mind's Hidden Complexities*. New York: Basic.

Galilei, Galileo (1638) *Dialogue Concerning the Two Chief World Systems*, trans. by S. Drake. New York: Modern Library.

Gershenfeld, N. (1998). *The Nature of Mathematical Modeling*. Cambridge: Cambridge University Press.

Gibbs, R. W & Colston, H. L. (2012). *Interpreting Figurative Meaning*. Cambridge: Cambridge University Press.

Gödel, K. (1931). Über formal unentscheidbare Sätze der Principia Mathematica und verwandter Systeme, Teil I. *Monatshefte für Mathematik und Physik, 38*, 173-189.

Goguen, J. (1999). An introduction to algebraic semiotics, with application to user interface design. In C. Nehaniv, (Ed.), *Computation for Metaphors, Analogy, and Agents*, pp. 242-291. Heidelberg: Springer.

Guy, R. K. (2004). *E17: Permutation Sequences: Unsolved Problems in Number Theory*. Berlin: Springer-Verlag.

Harris, Z. (1968). *Mathematical Structures of Language*. New York: John Wiley.

Helfgott, H. A. (2013). The ternary Goldbach conjecture is true. *arXiv, 1312*, 7748.

Hockett, C. F. (1967). *Language, Mathematics and Linguistics*. The Hague: Mouton.

Jakobson, R. (Ed.) (1961). *Structure of Language and Its Mathematical Aspects*. New York: American Mathematical Association.

Kiryushchenko, V. (2012). The Visual and the Virtual in Theory, Life and Scientific Practice: The Case of Peirce's Quincuncial Map Projection. In M. Bockarova, M. Danesi, & R. Núñez (Eds.), *Semiotic and Cognitive Science Essays on the Nature of Mathematics*, pp. 61–70. Munich: Lincom Europa.

Lakoff, G. (1987). *Women, Fire, and Dangerous Things: What Categories Reveal about the Mind*. Chicago: University of Chicago Press.

Lakoff, G., & Núñez, R. E. (2000). *Where Mathematics Comes From: How the Embodied Mind Brings Mathematics into Being*. New York: Basic Books.

Lotman, Y. (1991). *Universe of the Mind: A Semiotic Theory of Culture*. Bloomington: Indiana University Press.

Mac Lane, S. (1981). Mathematical models: A sketch for the philosophy of mathematics. *American Mathematical Monthly, 88*, 462-472.

Marcus, S. (2012). Mathematics between semiosis and cognition. In M. Bockarova, M. Danesi, & R. Núñez (Eds.), *Semiotic and Cognitive Science Essays on the Nature of Mathematics*, pp. 98-182. Munich: Lincom Europa.

Nagell, T. (1951). *Introduction to Number Theory*. New York: Wiley.

Nöth, W. (2018). The semiotics of models. *Sign Systems Studies, 46*, 7-43.

Núñez, R. (2005). Creating mathematical infinities: Metaphor, blending, and the beauty of transfinite cardinals. *Journal of Pragmatics, 37*, 1717-1741.

Peano, G. (1889). The Principles of Arithmetic, Presented by a New Method. In J. van Heijenoort (Ed.), *A Source Book in Mathematical Logic* (1967), pp. 83-97. Cambridge, Mass.: Harvard University Press.

Presmeg, N., Radford, L., Roth, W-M., & Kadunz, G. (2016). *Semiotics in Mathematics Education*. New York: Springer.

Radford, L. (2010). Algebraic thinking from a cultural semiotic perspective. *Research in Mathematics Education, 12*, 1-19.

Rotman, B. (2006). Towards a semiotics of mathematics. In R. Hersh (Ed.), *18 Unconventional Essays in Mathematics*, pp. 97-127. New York: Springer.

Russell, B., & Whitehead, A. N. (1913). *Principia mathematica*. Cambridge: Cambridge University Press.

Sáenz-Ludlow, A. & Kadunz, G. (Eds.) (2016). *Semiotics as a Tool for Learning Mathematics: How to Describe the Construction, Visualisation, and Communication of Mathematical Concepts*. New York: Springer.

Sebeok, T. A. & Danesi, M. (2000). *The Forms of Meaning: Modeling Systems Theory and Semiotics*. Berlin: Mouton de Gruyter.

Stewart, I. (2008). *Taming the Infinite*. London: Quercus.

Sullivan, J. (1925). *Aspects of Science*. New York: Knopf.

Tarski, A. (1933). Logic, semantics, metamathematics. In *Papers from 1923 to 1938,* J. Corcoran (Ed.). Indianapolis: Hackett Publishing Company.

Thom, R. (1975). *Structural Stability and Morphogenesis: An Outline of a General Theory of Models*. Reading: Benjamin.

Thom, R. (2010). Mathematics. In T. A. Sebeok & M. Danesi (Eds.), *Encyclopedic Dictionary of Semiotics*, 3rd edition. Berlin: Mouton de Gruyter.

Turner, M. (2012). Mental Packing and Unpacking in Mathematics. In M. Bockarova, M. Danesi, & R. Núñez (Eds.), pp. 248-294. *Semiotic and Cognitive Science Essays on the Nature of Mathematics*. Munich: Lincom Europa.

Uexküll, J. von (1909). *Umwelt und Innenwelt der Tierre*. Berlin: Springer.

Winter, B. & Yoshimi, J. (2020). Metaphor and the philosophical implications of embodied mathematics. *Frontiers in Psychology*. doi:https://doi.org/10.3389/fpsyg.2020.569487.

The "Unreasonable" Effectiveness of Mathematical Modeling

Jacek Woźny

1 Introduction

On July 20, 1969, the lunar module of Apollo 11 landed on the Moon. The trajectory of this historic space flight has been calculated by hand by a group of the so-called human computers. It is just one example of the effectiveness of mathematics in modeling (and changing) the world around us. Mathematics continues to be productively applied in engineering, medicine, chemistry, biology, physics, social sciences, communication, and computer science, to name but a few. This paper examines the manner in which simple dynamic scenarios allow, through the process of conceptual integration, for multiple ways of constructing the meaning of mathematical mapping – "probably the single most important and universal notion that runs through all of mathematics" (Herstein 1975, p. 10). The analysis of selected fragments of two popular academic handbooks of mathematics shows that the authors prompt for the use of a number of simple dynamic scenarios to avoid the problem of circularity of the static, formal definition of mathematical mapping. The results of the study indicate that the crucial notion of mathematical mapping is extensively polysemous, which may account for the flexibility of mathematics and its "unreasonable" (Wigner 1960) effectiveness in modeling the world around us.

As Hohol (2011, p. 143) points out the effectiveness of mathematical modeling is often treated by philosophers as an argument for mathematical realism of the Platonian or Aristotelian variety. It is from this perspective that Quine-Putnam "Indispensability Argument," Heller's "Hypothesis of the Mathematical Rationality of the World," or Tegmark's "Mathematical Universe Hypothesis" have been discussed. Eugene Wigner, a Nobel Prize laureate in physics, finished his paper

J. Woźny (✉)
Institute of English Studies, University of Wrocław, Wrocław, Poland
e-mail: Jacek.wozny@uwr.edu.pl

© The Author(s), under exclusive license to Springer Nature Switzerland AG 2023

M. N. Hounkonnou et al. (eds.), *Mathematics for Social Sciences and Arts*, Mathematics in Mind, https://doi.org/10.1007/978-3-031-37792-1_3

titled "The Unreasonable Effectiveness of Mathematics in the Natural Sciences" in the following way:

> The miracle of the appropriateness of the language of mathematics for the formulation of the laws of physics is a wonderful gift which we neither understand nor deserve. We should be grateful for it and hope that it will remain valid in future research and that it will extend, for better or for worse, to our pleasure, even though perhaps also to our bafflement, to wide branches of learning. (1960, p. 14)

James C. Alexander, a professor of mathematics, also sees the "unreasonable effectiveness" of mathematics as of a mystery but offers the following explanation for it:

> It is a mystery to be explored that mathematics, in one sense a formal game based on a sparse foundation, does not become barren, but is ever more fecund. I posit, [. . .] that mathematics incorporates blending (and other cognitive processes) into its formal structure as a manifestation of human creativity melding into the disciplinary culture, and that features of blending, in particular emergent structure, are vital for the fecundity. (Alexander 2011, p. 3)

Over the last two decades conceptual blending and other mental processes in mathematics have been extensively researched by, among others, Lakoff and Núñez (2000), Fauconnier and Turner (2002), Turner (2005, 2012), Núñez (2006), Alexander (2011), Danesi (2016), and Woźny (2018a, b). As an illustration, let us just have two quotations, starting with the ground-breaking *Where Mathematics Comes From – How the Embodied Mind Brings Mathematics Into Being* by George Lakoff and Raphael Nunez:

> Blends, metaphorical and nonmetaphorical, occur throughout mathematics. Many of the most important ideas in mathematics are metaphorical conceptual blends. (2000, p. 48)

Mark Turner adds the concept of "small spatial story" as a vital component of conceptual blending in mathematics:

> Our advanced abilities for mathematics are based in part on our prior cognitive ability for story [. . .] – understanding the world and our agency in it through certain kinds of human-scale conceptual organizations involving agents and actions in space. Another basic human cognitive operation that makes it possible for us to invent mathematical concepts [. . .] is "conceptual integration," also called "blending." Story and blending work as a team. (2005, p. 4)

Small spatial stories are dynamic scenarios which constitute one of the inputs of the conceptual integration network, always involving agents/actors moving and manipulating (interacting with) objects. For example, a person moving objects from one place to another. The main claim of this article is twofold. Firstly, the crucial notion of the mathematical mapping is "understood through" a number of selected small spatial stories, like the one above. Or – more precisely – the algebra handbooks prompt the reader to understand the mathematical mappings (functions) in this way. Secondly, incorporating small spatial stories – through conceptual blending – into the structure of mathematics is responsible for the effectiveness and productivity of the latter. To paraphrase Wigner from the above quotation – conceptual blending makes the effectiveness of mathematics "reasonable." The next section contains a short introduction to conceptual integration (blending) theory.

2 Conceptual Integration (Blending) Theory: The Basic Architecture

Three theories feature prominently in the cognitive semantics: Cognitive Metaphor Theory,[1] Mental Spaces Theory,[2] and Conceptual Integration Theory,[3] the latter related to the previous two and often described as an extension of them.

> Blending Theory is most closely related to Mental Spaces Theory, and some cognitive semanticists explicitly refer to it as an extension of this approach. This is due to its central concern with dynamic aspects of meaning construction and its dependence upon mental spaces and mental space construction as part of its architecture. However, Blending Theory is a distinct theory that has been developed to account for phenomena that Mental Spaces Theory and Conceptual Metaphor Theory cannot adequately account for. Moreover, Blending Theory adds significant theoretical sophistication of its own. The crucial insight of Blending Theory is that meaning construction typically involves integration of structure that gives rise to more than the sum of its parts. Blending theorists argue that this process of conceptual integration or blending is a general and basic cognitive operation which is central to the way we think. (Evans and Green 2006, p. 400)

Mark Turner (2014) begins his book, titled *The Origin of Ideas: Blending, Creativity, and the Human Spark*, with the following statement:

> The human contribution to the miracle of life around us is obvious: We hit upon new ideas, on the fly, all the time, and we have been performing this magic for, at the very least, 50,000 years. We did not make galaxies. We did not make life. We did not make viruses, the sun, DNA, or the chemical bond. But we do make new ideas—lots and lots of them. [...] Each of us is born with this spark for creating and understanding new ideas. But where exactly do new ideas come from? The claim of this book is that the human spark comes from our advanced ability to blend ideas to make new ideas. Blending is the origin of ideas. (1)

Blending then is the way we construct meaning and create new ideas, but what is it exactly? James Alexander (2011) begins his explanation of conceptual blending in the following way:

> Blending is a common but sophisticated and subtle mode of human thought, somewhat, but not exactly, analogous to analogy, with its own set of constitutive principles, explicated, for example, in Fauconnier and Turner's book *The Way We Think: Conceptual Blending and the Mind's Hidden Complexities*. (Alexander 2011, p. 2)

---

[1] cf. Lakoff and Johnson (1980, 1999), Lakoff and Turner (1989), Lakoff (1993), Gibbs and Steen (1999).

[2] cf. Fauconnier (1994, 1997), Fauconnier and Sweetser (1996).

[3] cf. Fauconnier and Turner (1998, 2002), Coulson and Oakley (2000).

Blending, as we learn, is "somewhat, but not exactly, analogous to analogy" – does not sound very precise, does it? But James Alexander is perfectly right – let us take a closer look at "analogy". For example – solving a problem is analogous to cracking a hard-boiled egg (or a walnut if it is a tough one). And we can easily see this analogy (or metaphor). In both cases (a problem, a walnut), prolonged effort, applying pressure, is involved. In both cases, we are trying to get inside, to uncover something that is hidden and – if successful – we are rewarded. This analogy, or metaphor, can be described as a mapping from the domain of cracking walnuts to the domain of solving complex theoretical problems (say, solving a differential equation). And we are now "somewhat, but not exactly" there. Let me remind the reader – we are trying to explain what conceptual blending is. So far we have established a set of analogies:

the walnut cracker (person) – the mathematician
walnut shell – the mathematical difficulty
physical effort, pressure – mental effort
peeling the walnut – constructing the solution
the content of the shell – the satisfaction of solving the equation
the nutcracker (tool) – the Calculus

In conceptual metaphor theory, the above would be called *the metaphorical mapping*. But human imagination is capable of more than that, more than just mapping the existing elements. For example, we can now imagine a person who uses advanced mathematics to find the best methods of cracking the walnut shell. This brilliant mathematician/walnut enthusiast one day invents a perfect nut-cracking machine, sells the patent to Kellogg's, becomes immensely rich, gets bored with life, drinks herself to death, etc. We are capable of integrating, merging, compressing the input elements (the walnut cracker, the mathematician), importing new elements (Kellog's, patent office, money, drinking habit), and then imaginatively running the story, inventing a whole new scenario. And then we may look back at the mathematician and the walnut cracker in the new light – in blending theory it is called "projecting back from the blended space to the input spaces." The process of blending is also referred to as building a conceptual integration network. This is how Fauconnier and Turner (2002), the creators of conceptual blending theory, describe it:

> Building an integration network involves setting up mental spaces, locating shared structures, projecting backwards to inputs, recruiting new structure to the inputs or the blend, and running various operations in the blend itself. (44)

The four mental spaces mentioned above are represented schematically in Fig. 1.

In our "nutcracker example," one of the input spaces is the small spatial story of a person trying to crack a nut, and the other represents the mathematician trying to solve an equation. The generic space contains the shared features – the analogies between the two, and the blend (or "blended") space is where the action of compressing the two stories takes place. The operations taking place in the blend space are the already exemplified compression, completion, and elaboration

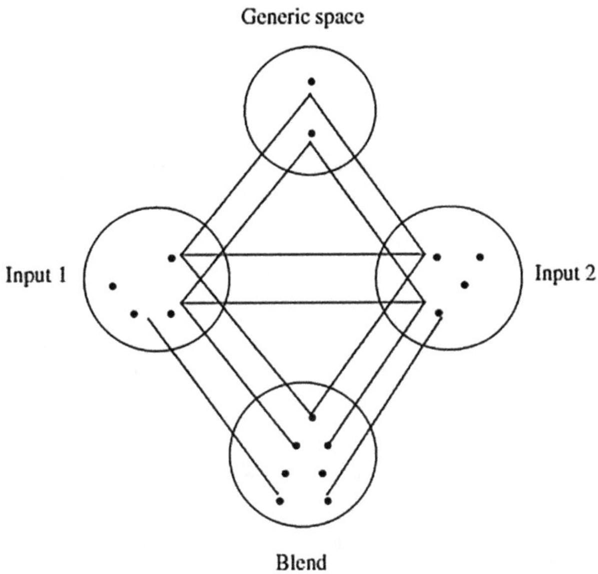

Fig. 1 Schematic representation of a conceptual integration network

(imagining the new scenario, also called "running the blend"). The lines represent selective mappings between the spaces.

In the next section, we will analyze fragments of two popular academic level handbooks of mathematics to see how the mapping (function) is defined and described there. The choice is not accidental – we were looking for standard, yet "chatty" ones. The first of them is Nathan Herstein's (1975) excellent *Topics in Algebra* – a classic handbook[4] addressed to "the most gifted sophomores in mathematics at Cornell" (8). The second is *First Semester Calculus for Students of Mathematics and Related Disciplines* by Michael M. Dougherty and John Gieringer (2012). Probably the only advanced calculus handbook which "is meant to be read, perhaps even curled up with and read" (iii).

3 The Concept of Mathematical Mapping and Small Spatial Stories

Nathan Herstein in his popular *Topics in Algebra* introduces the mapping as "probably the single most important and universal notion that runs through all of mathematics" (1975, p. 10) Quite typically,[5] he defines it as follows:

[4] Undergraduate modern algebra courses are sometimes referred to as "Herstein level courses."

[5] This definition can be found in all advanced level algebra handbooks and is usually referred to as the definition "by graph" or "Peano's definition."

> If S and T are nonempty sets, then a mapping from S to T is a subset, M, of SxT such that for every s in S there is a unique t in T such that the ordered pair (s, t) is in M. (1975, p. 10)

And then adds:

> This definition serves to make the concept of a mapping precise for us *but we shall almost never use it in this form. Instead we do prefer to think of a mapping as a rule which associates* (emphasis mine) with any element s in S some element t in T. (ibid.)

A "precise" definition, yet – for some reason – to be avoided. Dougherty and Gieringer (2012) in another popular handbook, speaking of the same definition, express similar reluctance:

> while we mention it, *we will not use it within the rest of the text because it de-emphasizes functions as actions or processes* (emphasis mine) taking inputs and deterministically returning outputs. (124)

Both handbooks recommend understanding functions in terms of small spatial stories – dynamic scenarios in which objects are manipulated by agents. In the first, we have "a rule" which performs the action of tying objects into pairs, in the second there is a black box which takes arguments and returns values. Herstein prompts for yet another small spatial story which can be used to construct the meaning of a mathematical mapping:

> It is hardly a new thing to any of us, for we have been considering mappings from the very earliest days of our mathematical training. When we were asked to plot the relation $y = x^2$ we were simply being asked to study the particular *mapping which takes every real number onto its square* (emphasis mine). (10)

In this scenario the function becomes an actor who carries objects (numbers) from one place to another. So far we have had as many as three small spatial stories which are to be used as the basis of our understanding of mathematical mapping. Let us call them *the matchmaker*, *the black box*, and *the carrier*, respectively. The matchmaker associates clients into pairs, the black box grinds the rough material of the arguments into the finished product of the function values, and the carrier moves cargo from the function domain to the target set. In Fig. 1 the black box scenario is diagrammed on the left. The diagram on the right represents both the matchmaker and the carrier.

It may be surprising to see that the narrative of modern algebra prompts the reader to understand the crucial concept of a mapping in so many ways. It seems even to undermine the status of mathematics as a paragon of formal rigor and precision. However, the polysemy we have just detected is exactly the reason for the fecundity of mathematics – it prevents mathematics from becoming barren (cf. Alexander 2011, p. 3, quoted above).

Further to explain it, let us return to the static definition of the mapping as a set of ordered pairs. We will try to see more clearly why both handbooks mentioned above[6] avoid using it and instead prefer the dynamic scenarios of the black box,

[6] Herstein 1975 and Dougherty and Gieringer 2012.

the matchmaker, and the carrier. The reason is certainly not didactic because the static definition is simple enough: a mapping from set S to set T is a set of ordered pairs (s,t) such that every element s in S is paired with some element t in T. The "ordered pair" is one of the undefined primary concepts (just like set and element, for example). Let us concentrate on the notion of an ordered pair for a moment. What makes a pair "ordered"? Well – we have to know which element comes first (or left) and which is second (or right). In other words – we need a set of two indexes like {1,2} or {left, right} or {first, second}, etc., and then we have to associate each of the indexes with the elements of the pair. So, for example, "first" can be associated with x and "second" with y and thus an ordered pair of (x,y) is created. An ordered pair, therefore, is a mapping from the set of indexes to the set of the elements of the pair. It would certainly be a good definition but then of course the static definition of a mapping (a set of ordered pairs) would be circular – mapping would be defined as a set of mappings. The circularity is avoided by treating the ordered pair as a primary, undefined concept. Formally – the circularity is avoided but certainly not conceptually. To refer to Alexander (2011, p. 3) again – this definition of a mapping is "barren," implicitly circular.

If that is the case – perhaps all those other ways of understanding the mapping (the black box, the matchmaker, the carrier) are also implicitly circular? The carrier, for example, carries x onto x^2 and in this way also creates ordered pairs of (2,4), (3,9), (4,16), etc. How do we know those pairs are ordered? Well, we do not need any primary concepts for that – the order is built into the dynamic scenario. The number on the left is the one that is being carried and the number on the right indicates the spot on which the carried number is placed. The pairs are ordered because their left and right elements occupy different roles in the scenario. The same applies to the two other scenarios – the black box and the matchmaker. First we put x into the black box and then it returns f(x). In the matchmaker scenario – x is the client for which a match y has to be found. The matchmaker has to know x first to find a suitable match y later. Because the scenarios are dynamic, the pairs (x,y) are ordered not only according to their roles but also temporarily – x comes sooner, y comes later.

So far, we have mentioned as many as four ways of constructing the meaning of mapping:

1. *the set of ordered pairs* (static, circular)
2. *the matchmaker* (dynamic scenario)
3. *the black box* (dynamic scenario)
4. *the carrier* (dynamic scenario)

But in fact there are many more and this is where conceptual blending comes to play:

> There are many ways to describe functions. To name a couple, we can look at them as mappings (which "map" the independent variable values to their respective dependent variable outputs), and they can be described as processes or "machines." We will include

> the abstract definition[7] later, to be complete. *Ideally it is best to consider functions in all these ways* (emphasis mine). However for our purposes we will concentrate on the notion of functions as machines. (Dougherty and Gieringer, p. 124)

The reader of an algebra handbook is invited to understand mathematical mapping "in all these ways." In other words – we are invited to blend (we tend to do it even without invitation, all the time). Each of the four construals above can be blended with the other three, which gives us as many as six[8] additional ways of constructing the meaning of this "most important and universal notion that runs through all of mathematics" (Herstein 1975, p. 10). If we add those new six to the previous ("unblended") four, we get ten ways of constructing the meaning of the mapping in mathematics. For example, in the blend for 1 and 2, we can imagine a set of pairs that were created by the matchmaker. In the blend of 2 and 3, we can imagine a matchmaker that has a machine[9] for finding a suitable match, etc.

Of course, ten ways of understanding functions (mappings) is certainly an underestimate because we would have to allow for the secondary blends (blends becoming inputs in other blends) and also for other dynamic scenarios, we have not discussed yet that can be found in the algebra handbooks. For example, the graphical representation, which is also in fact a dynamic scenario because we know how a graph "works": if you want to know what the value of the function (mapping) for x is, draw a line perpendicular to the x-axis until it crosses the graph and then from that point, another line, parallel to the x-axis until it crosses the y-axis – and this is where your y is. We can call this scenario *the hiker* (Fig. 2) as it involves motion (trajectories) determined by signposts (points of the function graph), which indicate where to turn (Fig. 3).

Yet another way of representing functions dynamically is with an expression, like $f(x) = x^2 + 1$, which "describes *the action* (emphasis JW) of the function [. . .] the function in the above example can be described as a process, by which the input is first squared, and the result is added to 1" (Dougherty and Gieringer 2012, p. 126). Let us call this dynamic scenario *the chef* (the function formula as a recipe). If we added those two ways of "understanding"[10] functions to our initial four, the extended list of input spaces would be as follows:

1. *the set of ordered pairs* (static, circular)
2. *the matchmaker* (dynamic scenario)
3. *the black box* (dynamic scenario)
4. *the carrier* (dynamic scenario)
5. *the hiker* (dynamic scenario)
6. *the chef* (dynamic scenario)

---

[7] The static one, a set of ordered pairs

[8] (4*3)/2 = 6 (because the order of the input spaces does not matter).

[9] A computer with professional matchmaking software

[10] That is, constructing conceptual integration networks with those dynamic scenarios as inputs

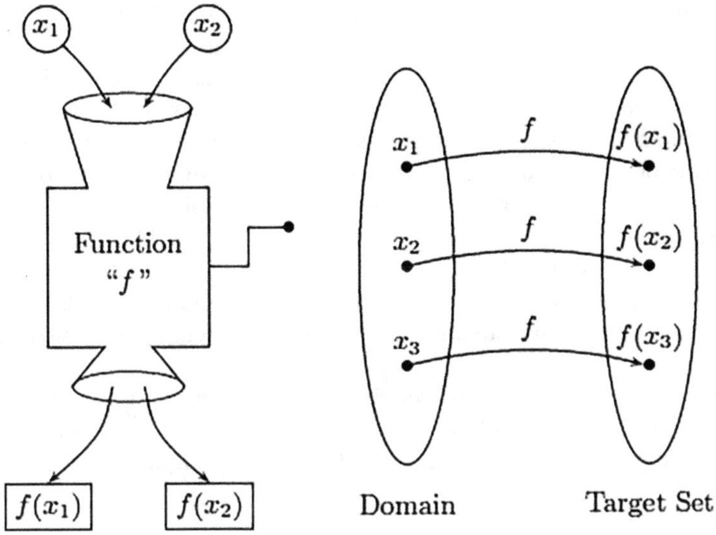

Fig. 2 Functions (mappings) as actions. The diagram on the left represents *the grinder* scenario – the function as an actor transforming elements into values. The diagram on the right represents both *the carrier* and *the matchmaker* scenarios, in which the function either moves the elements or ties them together

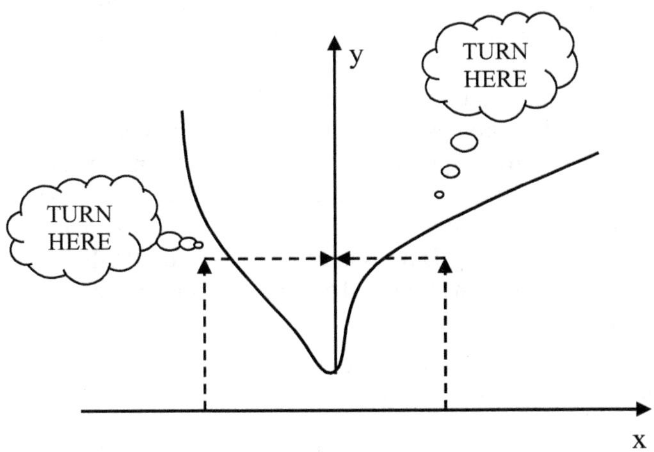

Fig. 3 Graph of a function – the dynamic scenario of *the hiker*. The function is represented as an actor moving from place x to place y

And again, each of the now six inputs can be blended with the other five, which increases the number of possible (first-order) blends to 15.[11] If we add it to the original "unblended" six from the above list, we will get 21, which justifies the title of this article. For example, the blend of 3 (the black box) and 6 (the processor) could result in a "transparent meat grinder" – a black box that we can peek into to see the whole process of turning input into output. Blending 1 (the set of pairs) and 6 (the chef) would result in a set of pairs in which the right element is obtained by processing the left element. In the blended space of 3 (the black box) and 5 (the hiker), the hiker has to find her way from the input to the output of the black box, and so on. Again, we have to point out that 21 is an underestimate because of the iterative nature of the conceptual integration process (blends can be used as inputs again).

The teaching methodology is not the primary focus of this chapter, but in the next section we will briefly consider the possible practical application of the "mathematical polysemy" described above for teaching purposes.

3.1 Small Spatial Stories and the Mathematical Classroom

Naturally, the fact that crucial mathematical concepts like mapping are understood through the process of blending small spatial stories should have a significant influence on the teaching of mathematics. As an example, let us use the perennially thorny and "unsolvable" issue of introducing the concept of zero and negative numbers to children. The traditional method of teaching arithmetic is to start with addition and then gradually introduce the "more complex" concepts of zero, subtraction and negative numbers. The following examples come from a handbook for teachers called "Practical Approaches to Developing Mental Maths Strategies for Addition and Subtraction,"[12] based on the insights of a well-known mathematics educator – John Van de Walle[13] and his 6th edition of " Elementary and Middle School Mathematics Teaching Developmentally" (Van de Walle 2007). The book contains the relevant quotations followed by practical tips for teachers and exercises that can be used in the classroom.

> Occasionally pupils feel that 6 + 0 must be more than 6 because 'adding makes numbers bigger' or that 12 – 0 must be 11 because 'subtracting makes numbers smaller'. Instead of making arbitrary-sounding rules about adding and subtracting zero, build opportunities for discussing zero into the problem-solving routine. (Van de Walle 2007, p. 154)

[11] Each of the six inputs can be paired with the other five, which gives us 6*5 = 30 combinations but, since the order of blending does not matter, we have to divide 30 by 2, with the final result of 15.

[12] http://www.pdst.ie/sites/default/files/Mental%20Maths%20Workshop%201%20Handbook.pdf, accessed 2017–01–03

[13] https://www.nctm.org/Grants-and-Awards/Supporters/John-A_-Van-de-Walle-Biography/, accessed 2017–01–04

The ensuing practical advice for the math teacher is to pose problems involving zero. For example, "Robert had eaten 8 grapes. He was too full to eat any more. How many did Robert eat altogether? In discussion of the problem, use drawings/counters to illustrate the empty set (zero)" (ibid.). The "unnatural and complex" concept of zero is to be taught separately from addition/subtraction, with the use of a rich variety of teaching aids and completely redundant questions about Robert. Subtraction, according to Van de Walle, must also be tackled with care and only after the pupils have got the hang of addition.

> Evidence suggests that children learn very few, if any, subtraction facts without first mastering the corresponding addition facts. In other words, mastery of $3 + 5$ can be thought of as prerequisite knowledge for learning the facts $8 - 3$ and $8 - 5$. Without opportunities to learn and use reasoning strategies, students may continue to rely on counting strategies to come up with subtraction facts, a slow and often inaccurate approach. (Van de Walle 2007, p. 175)

Adding negative numbers, like $2 + (-2) = 0$, is even more terrifying and "unnatural" for pupils than subtraction. After all, it does not usually happen that when we place one object next to another, they both magically disappear in a puff of smoke. And children are usually taught addition and subtraction with the use of manipulating collections of small physical objects. The solution to this problem would be to switch the "small spatial story" – handling pebbles, buttons and Lego bricks is not the only scenario available for a group binary operation like addition:

> When groups first arose in mathematics they usually came from some specific source and in some very concrete form. Very often it was in the form of a set of transformations of some particular mathematical object. In fact, most finite groups appeared as groups of permutations, that is, as subgroups' of Sn. (Sn = A(S) when S is a finite set with n elements.) The English mathematician Cayley first noted that every group could be realized as a group of A(S) for some S. (Herstein 1975, p. 71)

Arthur Cayley, a Lucasian[14] professor at Cambridge University, was one of the founders of what we call today "modern school of pure mathematics." And that means he helped to bring mathematics from the domain of "concrete and specific" into the realm of "purely abstract". And in 1854 he formulated the theorem, mentioned above, which was later named in his honor. And "the theorem enables us to exhibit any abstract group as a more concrete object, namely, as a group of mappings" (Herstein 1975, p. 72). A mapping, as we explained above, may be understood in many ways – one of them is the "carrier" story of moving objects from place to place. Let us look again at the "unnatural" equation of $2 + (-2) = 0$ but not in terms of collecting small objects but in terms of motion. "2" is a move now from place A to place B and "-2" is a reverse move, coming back from place B to place A. What happens when we "add" a move with a reverse move? – We land back at the starting position. There is nothing strange about it – nothing suddenly

[14] The so-called Lucasian Chair of Mathematics is considered to be one of the most prestigious professorships in the world. Its occupiers over the years were, among others, Sir Isaac Newton, Paul Dirac, and most recently – Stephen Hawking.

disappears, no magic tricks this time. Of course, teaching negative numbers and subtraction with the use of the scenario of motion – moving back along the number line – is nothing new; however, it would certainly help the teacher if she were aware that there always is a collection of "small spatial stories" to choose from because – as we tried to demonstrate in the previous sections – mathematical concepts are polysemous.

I strongly support Raphael Núñez, the co-author of *Where Mathematics Comes From: How the Embodied Mind Brings Mathematics into Being* (2000), when he postulates that "mathematics education should demystify truth, proof, definitions, and formalisms" and that "new generations of mathematics teachers, not only should have a good background in education, history, and philosophy, but they should also have some knowledge of cognitive science".[15] Conceptual blending (responsible for the polysemy of the core mathematical concepts like mapping) is a constantly ongoing process that is part of the so-called backstage cognition – we are not able to control it but it would certainly be beneficial for the teachers of mathematics to be aware of it.

4 Concluding Remarks

We have demonstrated how simple dynamic scenarios (small spatial stories) allow, through the process of conceptual integration, for multiple ways of constructing the meaning of mathematical mapping. We analyzed fragments of two popular academic handbooks of mathematics to find that the authors prompt for the use of small spatial stories we chose to call *the matchmaker*, *the black box*, *the carrier*, *the hiker*, and *the chef*. We have also shown how those dynamic scenarios help to avoid the problem of circularity ("barrenness") of the static, formal definition of the mapping as a set of ordered pairs (aka "definition by graph" or "definition of Peano"). Both handbooks we quoted (Herstein 1975; Dougherty and Gieringer 2012) include this definition and describe it as "precise" or "abstract," respectively, but at the same time advise the reader to understand the mapping in terms of the dynamic scenarios ("small spatial stories") in which actors manipulate objects. One of the reasons for selecting this strategy may be the implicit circularity of the formal definition, which is based on the undefined notion of an "ordered pair." The dynamic scenarios do not require the artificial introduction of "order" as a primary concept because they are "naturally ordered" in two ways. Firstly, the function (mapping) argument and value occupy different roles in the scenarios. And secondly, the $\{x, y\}$ sets are ordered according to the timeline – in the scenarios x always comes before y.

A more general conclusion is that the flexibility of mathematics, its ability to keep up with the fast developing technology and natural sciences, may, at least in part,

[15] http://www.cogsci.ucsd.edu/ nunez/web/PME24_Plenary.pdf, accessed 12.12.2016

stem from the polysemy of crucial mathematical concepts. Polysemy is based on the inventory of "small spatial stories," which can be selected as inputs for multiple conceptual integration networks.

Of course, the use of dynamic scenarios does not set "mathematical thinking" apart. On the contrary – it demonstrates that mathematicians think just like the rest of us, ordinary humans because:

> We are very good at thinking in terms of small spatial stories. We are built for it, and we are built to use small spatial stories as inputs to conceptual blends. In small spatial stories, we separate events from objects and think of some of those objects as actors who perform physical and spatial actions. We routinely understand our worlds by constructing a conceptual integration network in which one of the inputs is a small spatial story. (Turner 2005, p. 6)

As a final conclusion of the article, we would like to repeat after the co-author of *Where Mathematics Comes From*[16] that:

> There is nothing easy or automatic or magical about the success of mathematics in empirical domains. It arises from [. . .] understanding of the phenomena in ordinary, everyday terms, which are then translated into corresponding mathematical terms. It is the human capacity to understand experience in terms of basic cognitive concepts that is at the heart of the success of mathematics. (Lakoff 1987, p. 364)

Originally (as the phrase "final conclusion" above would suggest), I intended to finish the article here but the Editors (to whom I am eternally grateful for their help, patience, and insight and of course for including my chapter in this splendid tome) have asked me several extremely relevant questions that perhaps other readers would like to have answered as well. One of them was about the educational implications of conceptual blending in mathematics and I have already attempted to answer this question by adding Sect. 3.1 above. One of the other questions refers to the evidence that the "mathematical polysemy" actually helps brilliant mathematicians like Albert Einstein or Emmy Noether to solve problems. The answer is a resounding "yes" – mathematicians use conceptual blending (and hence the polysemy) because they are *Homo sapiens*[17] and human mind blends ideas, it does it all the time; however, we have to remember most of them (and most of us, non-mathematicians) would not do it consciously. All the same, it would certainly be beneficial for the problem solvers and teachers alike to be more aware of this constantly ongoing "backstage cognition" process. Blending small spatial stories is not just a way of illustrating core mathematical notions – it is IT, not an exception but a universal feature of human cognition, residing at the heart of mathematical (and non-mathematical) thinking.

[16] Lakoff, G. & R. Núñez (2000). *Where Mathematics Comes From: How the Embodied Mind Brings Mathematics into Being*. New York: Basic Books.

[17] "Mathematicians [. . .] – all members of the species Homo Sapiens" (Lakoff and Nuñez 2000, p. 1); Lakoff and Nunez were certainly not the first ones to notice it but the fact that they even had to make this statement speaks volumes of the awe ordinary people, like the author of this chapter, feel for mathematicians.

Appendix: Glossary of Notation and Definitions

Conceptual integration (blending). A universal feature of human thought, "somewhat but not exactly analogous to analogy" (Alexander 2011, p. 2), a more general sister notion to conceptual metaphor. The foundational source text which best explains it is *The Way We Think: Conceptual Blending and the Mind's Hidden Complexities* (Fauconnier & Turner 2002).

Mapping. A mapping from set A to set B is a set of *ordered pairs* in which every element of A is paired with an element of B. This is the definition Herstein calls "rigorous" and then adds that he will almost never use it, preferring a different way of thinking about mapping (Herstein 1975, p. 10).

Metaphor. Analogy, similarity between two entities, typically involving several "common" features, usually expressed with the verb "to be," as in "Achilles is a lion" (with respect to speed, strength, courage, invincibility, etc.). George Lakoff and Mark Johnson were not the first to notice the proliferation of metaphors in human speech and thought but their book *Metaphors We Live By* (1980) is considered by many to be the foundation of cognitive linguistics. Cognitive linguists believe that language is the window into the human mind.

Ordered pair. A set of two elements which are ordered (one element is first, and the other is second), marked with (,). For example, (a,b) reads "an ordered pair of a and b."; a primary notion (not defined). An ordered pair can be thought of as a mapping from the set of indexes {left, right} into {a,b} but we can't say it out loud because that would make the "rigorous" definition of mapping circular – a mapping would be defined as a set of mappings; however, even if we did say it out loud, the edifice of modern algebra would remain standing (instead of crashing down in a spectacular way) because it is safely buttressed with many other ways of understanding a mapping. In other words, the concept of mathematical mapping is polysemous and the polysemy is based on a collection of dynamic scenarios, called *small spatial stories*.

Polysemy. Having more than one meaning. For example, a mapping in mathematics may be understood as a set of ordered pairs but also as an agent that carries elements from place to place, or ties them together or transforms one element into another, or shows the way from one element to another, etc.

Set. Any collection of objects ("elements of a set"), a set of integers but also three bricks in a suitcase. Primary notion (not defined).

Small spatial story. A scenario involving an actor manipulating physical objects in space. For example, a person moving objects from one place to another. According to Mark Turner (2005) "we routinely understand our worlds" through small spatial stories.

References

Alexander, J. (2011). "Blending in mathematics". *Semiotica*, Issue 187, 1–48.

Coulson, S. & Oakley, T. (eds.). (2000). "Special issue on conceptual blending". *Cognitive Linguistics*, Issue11(3/4), 175–360.
Danesi, M. (2016). *Language and Mathematics: An Interdisciplinary Guide*. New York: Mouton de Gruyter.
Dougherty, M. & Gieringer, J. (2012). *First Year Calculus For Students of Mathematics and Related Disciplines*. http://faculty.swosu.edu/michael.dougherty/book/book.pdf, accessed 2017–12–11.
Evans, V., & Green, M. (2006). *Cognitive Linguistics: An Introduction*. Edinburgh: Edinburgh University Press.
Fauconnier, G. (1994) *Mental Spaces*. Cambridge: Cambridge University Press.
Fauconnier, G., & Sweetser, E. (eds.). (1996). *Spaces, Worlds and Grammar*. Chicago: University of Chicago Press.
Fauconnier, G. (1997) *Mappings in Thought and Language*. Cambridge: Cambridge University Press.
Fauconnier, G., & Turner, M. (1998). Conceptual integration networks. *Cognitive Science*, Issue 22(2), 33–187.
Fauconnier, M., & Turner M. (2002). *The Way We Think: Conceptual Blending And The Mind's Hidden Complexities*. New York: Basic Books.
Gibbs, R. W., & Steen, G. (1999). *Metaphor in Cognitive Linguistics*. Amsterdam: John Benjamins.
Herstein, I. (1975). *Topics in Algebra*. New York: John Wiley & Sons.
Hohol, M. (2011). "Matematyczność ucieleśniona". In B. Brożek, J. Mączka, W.P. Grygiel, M. Hohol (eds.), *Oblicza racjonalności: Wokół myśli Michała Hellera*. Kraków: Copernicus Center Press, 143–166.
Lakoff, G. (1987). *Women, Fire and Dangerous Things. What categories reveal about the mind.* Chicago: Chicago University Press.
Lakoff, G., & Johnson, M. (1980). *Metaphors We Live By*. Chicago: University of Chicago Press.
Lakoff, G., & Turner, M. (1989). *More Than Cool Reason: A Field Guide to Poetic Metaphor*. Chicago: University of Chicago Press.
Lakoff, G. (1993). "The contemporary theory of metaphor", in A. Ortony (ed.), *Metaphor and Thoughtx*. Cambridge: Cambridge University Press, 143-166.
Lakoff, G., & Johnson, M. (1999). *Philosophy in the Flesh: The Embodied Mind and Its Challenge to Western Thought*. New York: Basic Books.
Lakoff, G., & Núñez, R. (2000). *Where Mathematics Comes From: How the Embodied Mind Brings Mathematics into Being*. New York: Basic Books.
Núñez, R. (2006). "Do Real Numbers Really Move?". In R. Hersh (ed.), *18 Unconventional Essays on the Nature of Mathematics*. New York: Springer, 160-181.
Turner, M. (2005). "Mathematics and Narrative". Paper presented at the International Conference on Mathematics and Narrative, Mykonos, Greece, 12–15 July 2005. http://thalesandfriends.org/wp-content/uploads/2012/03/turner_paper.pdf, accessed Nov. 11, 2016.
Turner, M. (2012). "Mental Packing and Unpacking in Mathematics". In Mariana Bockarova, Marcel Danesi, and Rafael Núñez (eds.), *Semiotic and Cognitive Science Articles on the Nature of Mathematics*. Munich: Lincom Europa, 248-267.
Turner, M. (2014). *The Origin Of Ideas: Blending, Creativity And The Human Spark*. Oxford & New York: Oxford University Press.
Van de Walle, J. (2007). *Elementary and Middle School Mathematics Teaching Developmentally*. Boston: Allyn and Bacon (Pearson).
Wigner, E. (1960). "The Unreasonable Effectiveness of Mathematics in the Natural Sciences". *Communications in Pure and Applied Mathematics*, Issue 13(I), 1–14.
Woźny, J. (2018a). *How We Understand Mathematics – Conceptual Integration in the Language of Mathematical Description.* In *Mathematics in Mind* Series, Marcel Danesi (ed.). New York: Springer.
Woźny, J. (2018b). "The role of conceptual integration and simple dynamic scenarios in the meaning construction of the mapping in mathematics". *Cognitive Studies | Études cognitives*, 2018(18).

Algebra and Modeling in Mathematics School Curricula

Dragana Martinovic

> *Mathematics epistemology begins with experience, abstracts from it and arrives at the unexperienced (and even the unexperientable).*
>
> *(Resnik 1981, p. 531)*

> I vividly remember several breakthroughs in my learning of mathematics. One of the first was that mathematics expressions also use letters in addition to numbers. I was in early elementary school when I realized that the order in which numbers are added or multiplied does not matter, that the order could be expressed as $a + b = b + a$ and $a * b = b * a$. That is how my love for algebra was born. I remember how excited I was when I first studied analytic geometry, realizing that algebra and geometry are connected. This fascination with mathematics and my admiration of the great mathematical explorers and inventors has accompanied me throughout my studies and continues to inspire me as an educator to this day.(Author's personal note 2020)

1 Introduction

The quote from Resnik (1981) succinctly describes one of the biggest challenges in mathematics education, which is to introduce students with mathematics as rooted, limited only by imagination. As we will see in this chapter, students (and in many cases, teachers) of mathematics have difficulties with mathematics curriculum, since for the most part, they cannot connect its content to their lives and thus perceive its future utility. It may be that much of the opposition to learning mathematics could be diminished by experiencing mathematics as Resnik suggested and helping learners to gradually familiarize themselves with its beauty, language, and logic.

D. Martinovic (✉)
Faculty of Education, University of Windsor, Windsor, ON, Canada
e-mail: dragana@uwindsor.ca

© The Author(s), under exclusive license to Springer Nature Switzerland AG 2023
M. N. Hounkonnou et al. (eds.), *Mathematics for Social Sciences and Arts*, Mathematics in Mind, https://doi.org/10.1007/978-3-031-37792-1_4

The author's personal note is a testament that algebra may have a fundamental role in attracting students to, rather than deterring them from, learning mathematics.

In this chapter, we will address algebra and modeling from the standpoint of a mathematics educator. For cognitive scientists, "[a]lgebra is about essence" (Lakoff and Núñez 2000, p. 110); for mathematicians and mathematics educators, "algebraic reasoning... is among the most powerful intellectual tools that our civilization has developed in schools" (Kaput 2000, p. 4). For students, however, algebra often serves as a gatekeeper, blocking them from progressing further in mathematics or related disciplines. Mathematics educators must therefore reconcile these multifaceted and often opposing notions that people have of algebra. Infusing algebra throughout the mathematics curriculum from early on is important, but educators must also pay attention to the developmental readiness of students and use activities, tools (including technology), and tasks to engage them.

In school, models and modeling are crucial in supporting students in developing mathematical intuition and algebraic fluency. To investigate modeling and algebra as currently presented in the Ontario school curriculum, we will use an example of a number line as a model widely applicable throughout elementary and secondary grades. Furthermore, a number line connects geometry and algebra, which will allow us to highlight the aspects that create stumbling blocks for students learning these subjects.

Since in writing this essay we have used the literature from mathematics education, psychology, cognitive science, and philosophy of mathematics, we expect that, in addition to educators, it will interest people from different walks of life.

2 Algebra in the Curriculum

> Algebra seems to be taken late, with teachers referring to algebra as abstract and therefore difficult to teach and difficult for students to learn. (Grønmo 2018, p. 183)

Teachers of mathematics and curriculum designers often consider algebra too difficult for students to learn and thus delay it in the curriculum (Grønmo 2018). For students, algebra is a gatekeeper; more than with any other domain of mathematics, the extent to which one is good in algebra determines the highest school level one will achieve and the profession one will choose (Jupri, Drijvers and van den Heuvel-Panhuizen 2014).

While numeracy is widely recognized as an important life skill, using algebra outside of school is perceived as rare. When the editor of the *American Educator* journal asked Zalman Usiskin, professor of education at the University of Chicago, to explain why it is important to learn algebra, Usiskin's response was that, above all, "[a]lgebra is the language of generalization" (1995, p. 31). It can describe relations, rules, and patterns in a concise form. It "enables a person to answer all the questions of a particular type at one time" (p. 32).

John Mason, a mathematics educator at the Open University and at Oxford University, asserts it is never too early to learn to think algebraically. Using a "sensitively directed generalization and abstraction" with young learners means that teachers must cater to the children's abilities. However, Mason is adamant that "[a]rithmetic without generality is a purely clerical activity; arithmetic which calls upon children to become aware of generality is mathematics" (2018, p. 330). If students do not use mathematical generalizations, they are not experiencing and learning mathematics (Mason, Graham and Johnston-Wilder 2005). An example of mathematical generalization may be in noticing a general property when multiplying 29 by 31 (i.e., $(30 - 1) \times (30 + 1)$), or that, $29 + 31 = 21 + 39$. Choosing and discussing such examples can help students to develop their algebraic thinking, even in the early grades.

For Lakoff and Nunez (2000), "[a]lgebra is about [mathematical] essence ... [and] Essence Is Form." (p. 110). The idea that algebra is about form (i.e., structure) makes it central to all mathematics. Any structure consists of building blocks—rules for producing more complex configurations and relations between substructures—and one who understands the basics of this idea will start to appreciate the elegance, beauty, logic, and language of mathematics.

However, algebra as currently taught in school seems to be more about manipulating expressions than understanding the objects students manipulate with (Mason 2016). As a result, many people end up considering mathematics, and especially algebra, as a set of symbols void of meaning and a set of rules to memorize. Kaput's (2000) response to such views is that:

> ... algebraic reasoning in its many forms, and the use of algebraic representations such as graphs, tables, spreadsheets and traditional formulas, are among the most powerful intellectual tools that our civilization has developed. Without some form of symbolic algebra, there could be no higher mathematics and no quantitative science, hence no technology and modern life as we know them. (pp. 3–4)

Kaput is also one of many who fought to infuse algebra throughout the mathematics curriculum from the very beginning of elementary school. Like Mason, Kaput suggested that students should be exposed to generalizations while learning mathematics, regardless of whether it is arithmetic, geometry, or modeling.

As they progress through school, students become accustomed to using the increasingly formal language of mathematics, but algebraic thinking should not be equated with the use of symbolism. Radford claims that "the use of letters in algebra is neither a necessary nor a sufficient condition for thinking algebraically" (2015, p. 211). His projects with both adolescents and elementary school students revealed that "[p]erception, speech, gesture, and imagination develop in an interrelated manner" (p. 219) and thus must be considered when trying to decipher student's mathematical (algebraic) thinking. Radford explains that both the student's eye and hand should be domesticated/accustomed to recognizing and following the spatial-numerical characteristics first seen in patterns. While students' algebraic thinking can be expressed in various forms, teachers should be there to recognize its emergence and to scaffold it through further generalizations and abstractions.

For example, the principle of abstraction is used in early grades to introduce students to natural numbers as cardinal numbers of different sets. In mid-grades, a reduced fraction is introduced as a representative of infinitely many equivalent fractions. In geometry, a parallelogram or an equilateral triangle could be introduced through both examples and non-examples (e.g., "these are examples of equilateral triangles, and these are not"), allowing its definition to emerge. Any new concept (e.g., parallel lines; quadratic function) can be introduced by paying attention to common features of its representatives and distinctions in comparison with previously learned concepts.

However, most teachers rarely touch on the idea that in mathematics there are families of objects with some properties in common, allowing us to reduce them all to a single representative. When abstraction of that kind is missing, students get stuck with trying to learn an object's multiple instances by heart and not seeing what unites them. For example, in the Ontario mathematics curriculum, students in Grade 9 (first year of secondary school) learn about linear relationships; in Grade 10, about quadratics; and in Grade 11, about polynomial functions. When the official curriculum presents mathematical objects and processes in such a piecemeal fashion, it is incumbent upon teachers to bring clarity and the big picture to their students.

Mathematics educators highlight different aspects of school algebra. For Dossey (1998), there are four sides to algebra: "structural, linguistic, functional, and modeling, none of [which] can survive or prosper alone" (p. 19). The structural side of algebra (e.g., the study of sets of objects with their characteristics, operations, and relationships) has been described in the previous references to Usiskin (1995), Lakoff and Nunez (2000), and Mason (2018). Under the linguistic side of algebra, Dossey (1998) includes graphical, tabular, symbolic, and verbal representations, to which Radford (2015) adds gestures, and Kaput (2000) adds spreadsheets and other semiotic representations created through technology. The functional side includes the study of functions and relations (mathematicians would call it analysis), while the modeling side "involves looking at and building models to represent real-world contexts and problems within those contexts" (Dossey 1998, p. 19). Kaput (2000), however, singles out modeling as "the primary reason for studying algebra" (p. 23).

3 Algebra and Modeling

Sebeok and Danesi (2000) state that "sophisticated, ingenious, [and] resourceful" model-making "distinguishes human beings from other species [, typifying] all aspects of human intellectual and social life" (p. 6). Mathematical modeling as described in the Ontario Grade 1–8 mathematics curriculum from 2020—that is, understanding a problem, analyzing the situation, creating a mathematical model, and analyzing and assessing the model—can be applied in various contexts and various grades.

According to Thomas et al. (2015), teachers should make a clear distinction between using modeling to learn/teach mathematics and learning/teaching mathematics for modeling. Not making this distinction clear presents a problem both in the delivery of lessons and in presenting the importance of modeling to the students. The authors claim that "upper secondary students have little experience working with real situations and modeling problems [and] that teachers [of mathematics] tend not to make many real-world connections in teaching" (p. 277). Accordingly, they suggest a more prominent role for modeling in the secondary school curriculum; more specifically, they recommend "a system-wide focus emphasizing an applications and modeling approach to teaching and assessing mathematical subjects in the last two years of school and interdisciplinary project work from primary through secondary school, with mathematics as the anchor subject" (p. 278).

Defining modeling in broad terms may result in confusing it with other pedagogical approaches. In a book, *Modeling Students' Mathematical Modeling Competencies*, edited by Lesh et al. (2010), several authors (e.g., Zawojewski; Højgaard) make a case that modeling is different from solving word problems or, indeed, problem-solving in general. According to Højgaard (2010), mathematical problems can be pure or applied, open or closed, but essentially, they require investigation and application of a small number of routine skills. On the other hand:

> [b]y virtue of the "underdetermined" nature of the initial parts of the mathematical modeling process, the crux of this challenge is to learn to handle the many often equally sensible choices that needs [sic] to be made before mathematical concepts and techniques can be of any use, and the lack of a clearly defined strategy to use when making these choices. (p. 259)

Modeling competency also involves use of knowledge that is not solely of a mathematical nature (e.g., understanding the context, choosing between strategies and options). Højgaard hypothesizes that problem-solving is often presented as modeling because it takes less time and skill to engage in. Furthermore, the so-called "pseudo extra-mathematical" tasks (e.g., "The total length of The Loch Ness monster is 40 meters plus half its own length. How long is the monster?" p. 261) that require neither problem-solving nor modeling competency are used on tests and national examinations as easiest to mark. Different tasks involve different competencies and modeling tasks are characterized by their openness.

Carmona and Greenstein (2010) state that modeling is like a spiral curriculum in which the strong themes are continuously revisited. The authors found confirmation for this claim after they posed the same problem (i.e., the team ranking problem) to two groups of participants: one under-skilled and the other over-skilled (in terms of their level of schooling). In Carmona and Greenstein's study, Grade 3 students and post-university students had to compare the performance of 12 teams whose wins and losses were presented in the unitless Cartesian coordinate system. Based only on the relative arrangement of 12 points, each labeled with a distinct letter, the two groups of students had to develop a strategy for ranking the top five teams. The researchers witnessed powerful learning among the elementary school students who for the first time saw the coordinate plane, but "re-invented" it by adding numerical values to the coordinate axes to be able to quantitatively compare the teams' wins and losses.

The researchers concluded that "the non-prescriptive nature of the task [such as the Team Ranking Problem] opens up a space in which problem solvers decide the math they find useful in establishing a path from givens to goals" (p. 253). Such an open-modeling task allows students to "travel from givens to goals [in an] iterative and nonlinear" fashion. Moreover, the "powerful ideas" the students came up with "provide an opportunity for assessment of any number of standards, not necessarily at their [or even single] grade level" (p. 253). Although the solutions differed in sophistication between the two groups, the elementary students arrived at very similar ideas as their older counterparts (i.e., models based on the ratio of wins to losses, and in the case of a tie, using the total number of games played as the deciding factor) and developed models that were developmentally appropriate for their age group.

As would be suggested by Thomas et al. (2015), Carmona and Greenstein (2010) used modeling to teach their students mathematics. While working on the team ranking problem, students applied all four representations of algebra mentioned by Dossey (1998): structural, when trying to understand how the relative position of points related to their ranking; linguistic, when moving between different representations; functional, when concluding that the slope of a line (or the ratio between wins and losses) could be used; and modeling, when iteratively arriving at additive and multiplicative models of the teams' ranking and then comparing them.

Curriculum overemphasis on linear relations is criticized by many researchers as a likely source of the difficulties that students later have with non-linear models. The new elementary school mathematics curriculum in Ontario (2020) has improved this area compared to the earlier version by prescribing the use of both linear and non-linear regression models as part of the Data Management stream in Grades 7–8. However, in the entire curriculum, there is no mention of the term "function"; the curriculum has been developed solely around the notion of relation. This is unfortunate, as research conducted in the UK and Israel has established that the early use of a complex mathematical term (e.g., function), may help students to keep building and refining a concept image around it as they progress through the grades (Ayalon et al. 2017).

Rate of change and covariation are the founding ideas for functions. These ideas differ from those of mapping (correspondence) and input-output machines (Watson et al. 2018; Ayalon et al. 2017), through which students first learn about the relations of dependence between two quantities. According to Ayalon et al. (2015), students are capable of understanding rate and covariation in linear relationships long before they are formally exposed to them in the curriculum. However, what can prevent students from developing some more advanced concepts early are oversimplification or overuse of tasks, representations, and questions that encourage automatic and effortless responses. One such approach is a point-by-point representation of a function, usually given in a form of a table with easy integer values for one variable. To discourage students from relying on this image, teachers can use non-integer tables of values and introduce multiple representations simultaneously (e.g., graphs and equations) (Wilmot et al. 2011).

Modeling activities can help students to understand covariation while, equally, understanding covariation can help during modeling, but for this to happen, students need to see models of covariation beyond linear (Gil and Gibbs 2017). Next, we will introduce a concept of a number line, first how mathematicians see it, and then how it is presented in the school curriculum. By comparing examples in which a number line is used in different grades with examples in which it is used in research and less traditionally, we will make a case that the problem may be not in the curriculum's overemphasis of the linear model but in the limited ways in which it may be used in classrooms. We will conclude with adding a semiotics view of a number line, and mathematical models in general, which may better explain the challenges and opportunities in using this model in teaching algebra.

3.1 *Number Line: A Powerful Model*

Students are introduced to school algebra through learning the structure of a number system, which starts with arithmetic. When learning arithmetic, students use different tools (e.g., abacus, manipulatives) and models (e.g., area model, Venn diagram), including a number line. According to Thomas et al. (2015), a number line is used to teach numbers, relations between them, and operations with them.

> The number line—a conceptual tool that allows for numbers to be conceived as locations along a line mapping numerical differences onto differences in spatial extension... is mathematically simple, yet it is extraordinarily powerful. (Núñez, 2017, as cited in Rycroft-Smith and Gould 2021, p. 2)

This model is powerful but also complex, as it combines ideas from geometry and algebra. Its limitation is in its one-dimensionality, which can be overcome by simultaneously using more than one number line. In this section, we will introduce the number line by relating concepts from geometry and algebra. Then we will go over examples of using a number line as a tool for learning and doing mathematics as well as in modeling situations at different grade levels.

3.2 *Number Line in Mathematics*

In their *Foundations of Geometry*, Borsuk and Szmielew (1960) use sets to introduce primitive objects, such as points, (straight) lines, and planes. Two complementary half-lines are defined by a point on the line. Such a point is called an origin for both half-lines and does not belong to either of them. Figure 1 presents $L = A \cup a \cup A^*$, where A and A^* are the half-lines defined by point a on the line L. To create an axis, it is necessary to orient a half-line by adding an arrow, which introduces the

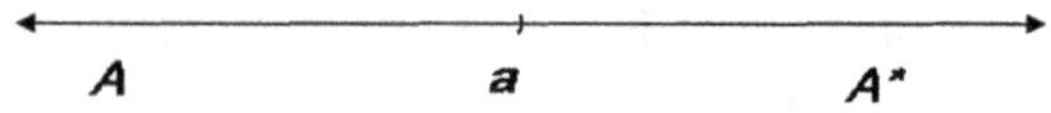

Fig. 1 Half-lines and axes (Borsuk and Szmielew 1960)

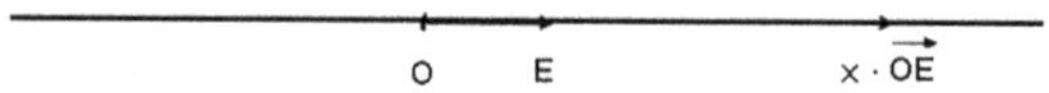

Fig. 2 Emergence of a number line from a geometric line

order between the points. As Fig. 1 shows, the ordering starts from *a* and goes in the direction of arrows.

Algebraist Djuro Kurepa (1969) takes an axis from geometry and adds numbers to it. After introducing the set of real numbers *R*, Kurepa takes a (unit) vector $\overrightarrow{OE}$ on a line *L* and for every real number, $\forall x \in R$, he defines a point *X* on *L*, such that it is an end point of the vector $x \bullet \overrightarrow{OE}$ (Fig. 2). He calls the relation between the points on *L* and the set of real numbers *R* "close or intimate" (*prisna* in Serbo-Croatian). This relation allows us to call the point *O*, 0 (zero); the point *E*, 1; and the point *X*, x. In other words, every real number *x* has its opposite number –*x*, such that$-x = -1 \bullet x$. In that way, the opposite number is found after multiplying the number by -1, while the opposite point to *X* is found by finding its symmetric point on *L* with respect to *O*. In that way, the arithmetic operation of a multiplication of *x* by -1 and the geometric operation of a reflection of *X* with respect to *O*, correspond to each other.

In a nutshell, to get a number line, one needs to select an origin (*O*) and a unit size ($\overline{OE}$) on an axis. After that, the position of any real number x will be uniquely determined. While this brief description of the mathematical origins of a number line omits the associated axioms and theorems, it connects the basic concepts from different mathematics disciplines in a consistent way. Or, as Strauss (2014) wrote: "[T]his concept of a number line combines two distinct perspectives: the numerical and the spatial" (p. 192). When developing an algebraic concept of a number line based on a geometric line, "the discourses on numbers and of geometry can be distinguished in terms of the objects they deal with and the strategies for the production of such objects" (Herbst 1997, p. 37). Herbst calls a number line a metaphor for a number system since many properties of the line as well as points on the line are used to describe properties of both the number line and numbers. Speaking from a semiotician's perspective, Danesi (2022) sees in a number line "an example of an indexical model ..., which is a linear model relating numbers to each other in a vectorial (left-right) way."

3.3 *The Number Line in School Mathematics*

In school mathematics, a number line is approached concretely and intuitively, leaving both teachers and students unaware of its mathematical complexity that we saw in the previous section. According to Rycroft-Smith and Gould (2021),

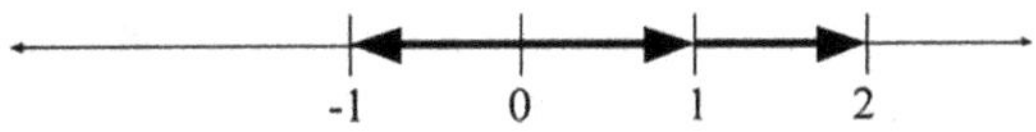

Fig. 3 Number line with vectors

the early conceptualization of a number line starts with "[y]oung children [using] bead strings or linear number tracks..., later moving to notched number lines" (p.1). Bead strings can be used for counting by 1, 2, 5, 10, or any other "easy" number to develop fluency with addition and the relative position of integers (i.e., "If the beads represent groups of 10s, show me the 8th bead. What is the number it represents?"). Having beads on elastic strings can help children develop the idea that when they stretch the string, the spaces between the beads can be filled with other beads (numbers) ad infinitum.

Following Herbst's (1997) notion of a number line as a metaphor for a number system, when children are instructed to dissect the segment between two successive integers into equal parts, it is implicit that the line is continuous and that it can always be dissected that way. A conscientious teacher may inform students that any fraction may be uniquely represented on the number line, but that if we place all the fractions on the number line, there will be holes between them, these holes being places held for other decimal numbers that are not fractions. As Herbst points out, the teacher may leave no doubt in the students that all the decimal numbers can be placed on the number line and that they can fill it without holes; in other words, there is matching (one-to-one correspondence) between the points on a line and the set R of real numbers. Or as Strauss (2014) puts it, the "infinite divisibility [of a line] embodies another way in which the meaning of space points backward to the meaning of number" (p. 203).

When introducing students to operations with integers, it is useful to use a number line with vectors, similar to how Kurepa (1969) described it. Negative numbers are at the tips of the vectors that start from the origin and point to the left (Fig. 3). In this representation, addition and subtraction are conducted by concatenating two vectors tip to end. From Fig. 3, we can see that by doing so, the following statements would follow: $1 + 1 = 2$, $2–1 = 1$ (or $2 + (-1) = 1$, $-1 + 2 = 1$, $-1 + 1 = 0$).

Thompson and Carlson (2017) problematize a notion of the number line as a line full of numbers. Alternatively, a number line may be presented without numbers, but full of positions, a conceptualization that is crucial for understanding numbers as magnitudes and for developing an understanding of the continuum of positions. Galileo Galilei (1564–1642) considered a line to be a path along which a physical object moves (Sinkevich 2015). This conceptualization is similar to how continuity in school mathematics is described: as tracing the graph of a function in one sweep, without ever having to lift the tip of the pencil from the paper.

School examples that follow the idea of a line full of positions include asking students to draw a numberless line and then to choose positions for general numbers a and b (i.e., without specifying what a and b are). From there, students discuss how the positions of $-a$, $-b$, $-(-a)$, $3a$, and $b/4$ change after changing the positions of a and b. In this task, students learn that to find a position for $-a$, they need to know the

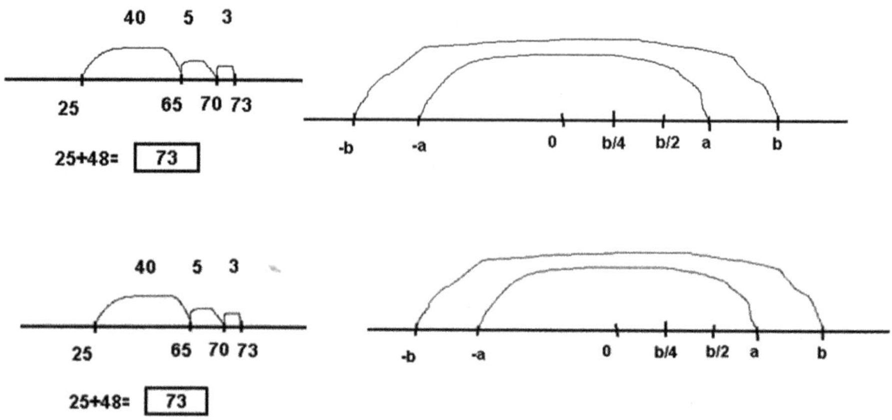

Fig. 4 Different uses of an empty number line: on the left, as an aid in adding concrete numbers, and on the right, to work with general numbers

position of 0. Then they need to find a symmetrical point for a with respect to the reference point 0. Such an approach follows Czarnocha et al.'s (2012) suggestion that for students to bridge a cognitive gap between arithmetic and algebra, they need to experience three faces of a variable: as a specific unknown (e.g., in $a + 2 = 4$), as a general number (e.g., $a + b = b + a$), and as the functional relationship (e.g., $a = 2b$).

Depending on the purpose of using an empty number line (and on the student level), teachers can ask students to be aware of and to follow certain rules of placement. For example, if both a and b are to the right of 0, and b is to the right of a, $-a$ and $-b$ should be to the left of 0, $-b$ being on the left of $-a$. Students can then estimate the position of the missing general numbers by applying geometrical transformations of reflection, dilation, and translation. However, as van den Heuvel-Panhuizen (2008) advises, when the empty number line is used to visualize operations (e.g., 25 + 48), "[i]n no way are the children asked to put the numbers on the empty number line in a way that is proportionally correct. [. . .] [T]he empty number line is not a measuring line" (p. 27).

The difference between working with general numbers and working with concrete numbers on an empty number line is that in the first case, students would need to mark the position of 0, while in the second case (i.e., 25 + 48), they do not need 0, but can place the numbers arbitrarily on the number line (Fig. 4). In neither case, however, do the students need to know the unit size.

Of course, the empty number line can also be used in tasks that combine general and concrete numbers, for example, when solving linear equations (Dickinson and Eade 2004). Figure 5 shows some approaches to visualizing relationships when solving $3x - 4 = 2$ and $4x + 1 = x + 4$. On the top part of the line, students can represent the left side of the equation, and on the bottom part, the right side. Both expressions start and end at the same place on the number line. While in the first

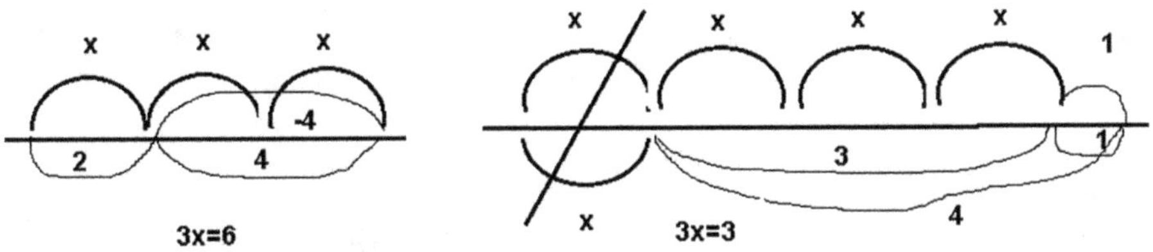

Fig. 5 Possible approaches in using an empty number line to solve linear equations

example, students cannot anticipate how the magnitudes of x and 4 relate, because 3x − 4 is positive they can conclude that there is some leftover of 3x when they subtract 4 from it. They can further infer that 3x = 2 + 4, leading to x = 2. While, following van den Heuvel-Panhuizen's (2008) advice, students are not expected to use the exact proportionality between general and concrete numbers, the same numbers should be represented as segments of similar length.

Dickinson and Eade (2004) demonstrated that using a number line in this way allows teachers to introduce solving linear equations with variables on both sides earlier than the curriculum prescribes (e.g., in Ontario, from Grade 7). In the study conducted in the UK, the eleven-year-old students who were on different achievement levels enjoyed and gained skill in solving linear equations. The authors noted how:

> We also see comfort with 'different' forms of the same equation as further evidence of students beginning to see algebraic statements as objects as well as processes. For example, the move from 2x + 1 = 5 to 1 + 2x = 5 represents an increasing complexity and sometimes a serious difficulty for students. The flexibility to be able to interpret 2x + 1 as both a process (to be evaluated) and an object (to be manipulated) is crucial for algebraic progress and seemed to be developing in our students. (p. 46)

In another example from the Shell Centre for Mathematical Education (1985, p. 70), the middle school students use a coordinate system consisting of two orthogonal axes without numbers. Only the origin is given as the point where the axes intersect. One axis stands for length and the other for width. The teacher asks the students to imagine a rectangle with the area of 36 square units that fits into the corner between the axes. The teacher marks a point for the top right vertex of the imagined rectangle and then asks a student to add another point to represent his or her own rectangle. How does the student know where to put his or her point? Can the point be anywhere? Where would the point representing a rectangle that is extremely long (wide) be? If a rectangle is very long, what does the student know about its width? Can the length be 0? And so on.

In this activity, the students are encouraged to develop an idea of the regions in the first quadrant that make sense for placing points (i.e., (Length, Width)) and whose coordinates covariate according to the rule, Length * Width = 36. The first point V that the teacher marks determines impossible areas (Fig. 6) for future points. The top right shaded area is defined by an inequality Length * Width >36 and the area within the first bottom left shaded rectangle by the inequality Length * Width

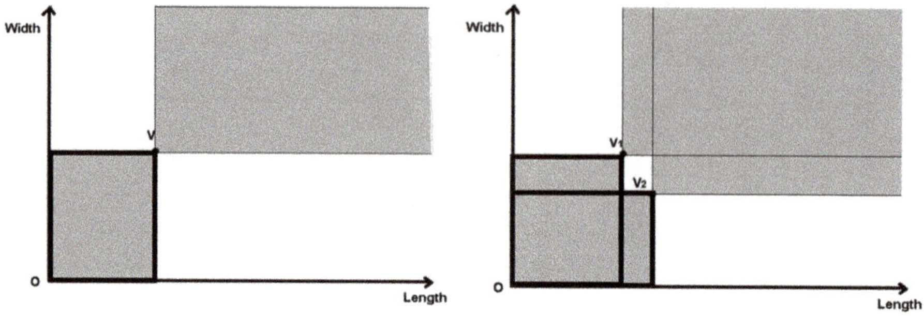

Fig. 6 Visualization of the rectangles with an area of 36: on the left, representing the teacher's rectangle, and on the right, with another rectangle added by the student

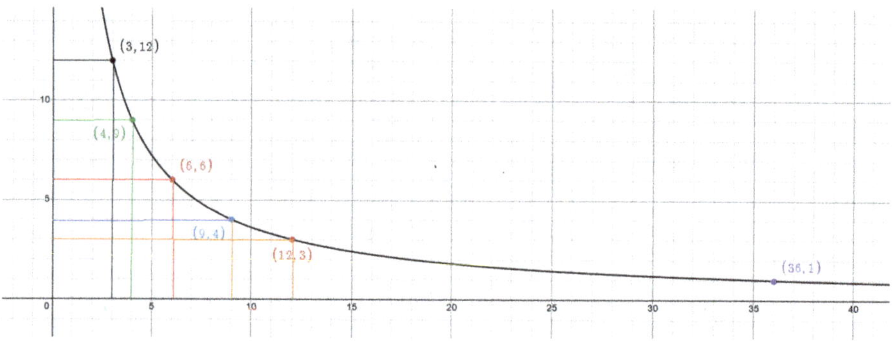

Fig. 7 The right top vertices of all rectangles with an area of 36 square units lie on the hyperbola

<36. Therefore, the future points cannot be placed in any of the two shaded areas. Following this reasoning, the student can choose to place their rectangle only in the non-shaded area.

On the left of Fig. 6, the shaded areas are forbidden for future points since any additional rectangle cannot have both sides smaller or both sides greater than those of the first rectangle. Each new point increases the restricted area for future points; ultimately, a rectangular hyperbola $y = \frac{36}{x}, x > 0$ emerges (Fig. 7).

Following Gil and Gibbs's (2017) recommendation, this modeling activity can help students to understand covariation, as the lengths and widths of these rectangles change in unison: If one side increases, the other must decrease, so that their product, the area, remains constant. The significance of this model is that it is non-linear but still approachable even to elementary school students.

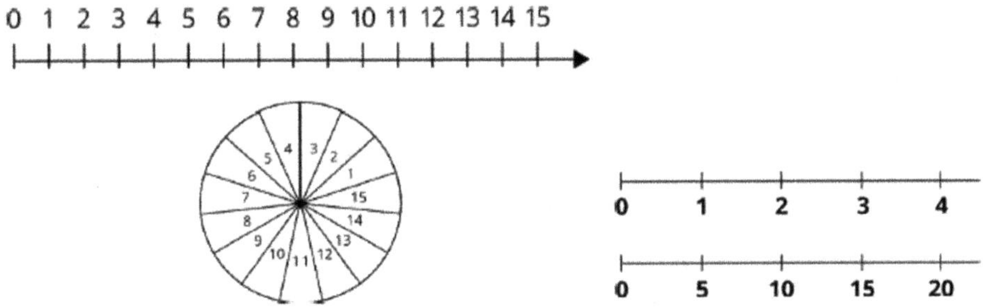

Fig. 8 Number lines featured in the Ontario Grade 1 curriculum: regular (top left), circular (bottom left), and double (right). (MoEd 2020, p. 121, p. 525)

3.4 Complexities in Using a Number Line in School Mathematics

> Before children are introduced to the empty number line, they should already be familiar with a linear representation of number – a number line with numbers. (Bobis 2007, p. 412)

The Grade 1 Ontario curriculum (MoEd 2020) introduced numbers as quantities that could be composed and decomposed, compared, and ordered. To understand numbers, children must understand both cardinality and ordinality. According to Strauss (2014), it makes sense that to develop the intuition of cardinality, one needs to first use ordinality. This aspect is visible in the curriculum, which describes cardinality of a collection in the following way:

> Each object in a collection must be touched or included in the count only once and matched to the number being said (one-to-one-correspondence).
>
> The numbers in the counting sequence must be said once, and always in the standard order (stable order).
>
> The last number said during a count describes how many there are in the whole collection (cardinality). (MoEd 2020, p. 116)

For Grade 1 students who learn counting to 50 (by 1s, 2s, 5s, and 10s) as well as addition and subtraction of the whole numbers that add up to 20, using a number line seems intuitive. However, how number lines are visualized in the curriculum document and presented in Fig. 8 differs from how a line is defined in the same curriculum, that is, as "a geometric figure that has no thickness whose length goes on infinitely in both directions" (MoEd 2020, p. 545). Therefore, one must question to what extent policymakers and teachers understand the effect of using models that refer to "a line" (as in Fig. 8), while grossly diverging from the definition of a line in the same document.

To be more precise, the top left diagram in Fig. 8 is a ray (i.e., extends indefinitely in one direction only), the circle on the bottom left does not extend indefinitely, and the diagram on the right is similar to the top left one but does not use arrow to show the direction. Similarly, the Grade 3 curriculum introduces a "probability line" as "[a] line with 0 at the left-hand end (for 'impossible') and 1 at the right-hand end (for 'certain')" (p. 565), thus equating it with a [0, 1] closed segment. Such inconsistencies in using the term "a line" within the curriculum document may instill in both teachers and students the impression that in mathematics, deviations from the established narrative are allowed. Moreover, such discrepancies move the said "number lines" further away from their geometrical origins, which may result in increased difficulties for students in conceptualizing a "true" number line as it is gradually introduced in the later grades.

The elementary school mathematics curriculum in Ontario is split into six strands, three of which are particularly relevant to our discussion; these are Number (numbers and operations), Spatial Sense (geometry and measurement), and Algebra (patterns, variables, expressions, equations, and inequalities). The curriculum's separation of numbers from geometry, and even from measurement, creates gaps that are not easy for teachers and students to bridge. A number line is defined as "[a] line that represents a set of numbers using a set of points" (MoEd 2020, p. 554), a definition that presupposes the existence of geometrical (a set of points) to numerical (a set of numbers). In the curriculum, a number line is considered a mathematical model, or "[a] structured representation that illustrates mathematical ideas" (p. 546), in this case, from arithmetic, geometry, and algebra. On the other hand, the curriculum's definition of a double number line seems too narrow, being "[a] visual model used to represent the equivalencies between two quantities" (MoEd 2020, p. 523).

The examples from Dickinson and Eade (2004) suggest that to present equivalencies of quantities, such as in equations, one line is enough. A double number line, as is the case with the orthogonal lines in the Cartesian coordinate system, presents a relationship between two quantities, which is best described as covariation, a concept that is absent from the Ontario curriculum, even though it has been strongly suggested by the research (e.g., Ayalon et al. 2017) as essential.

Differences also exist between the mathematical and curriculum definitions of a Cartesian coordinate system. While Lakoff and Núñez (2000) describe "Cartesian plane [. . . as] a conceptual blend of (a) two number lines and (b) the Euclidean plane, with two lines perpendicular to each other" (pp. 384–5), the school curriculum defines it as "[a] system that identifies the points where the lines on a grid intersect" (MoEd 2020, p. 517), further defining grid as "[a] plane that contains regularly spaced lines that cross one another at right angles to form squares or rectangles" (p. 538). The curriculum definition thus appears not open for teachers using two empty orthogonal axes when the points emerge from some relationship (e.g., as in the Shell Centre for Mathematical Education's 1985 example).

The 2020 curriculum creates other difficulties for students. Grasping that a number line has no gaps may be problematic for them because the curriculum starts by introducing natural (counting) numbers without mentioning what is happening between natural numbers on a number line. Things become even more confusing when students start working with fractions and present them on the same number line, which is presumably full of fractions. The idea that the average of two fractions is a fraction between them and that such a process of halving could be conducted forever suggests that there are no gaps between fractions. However, the gaps do exist, and their total is vast!

Finally, although mathematical concepts are first introduced through the concrete (i.e., physical; a line as a necklace of beads), a mathematical space is in fact "both continuous and infinitely divisible, while [the physical] is neither continuous nor infinitely divisible" (Strauss 2014, p. 199). In effect, while one can determine the number that precedes and the number that follows any whole number, this is not possible for fractions and decimals. Thus, students need to somehow grapple with these inconsistencies arising from limitations in models used during their schooling, inconsistencies that may not necessarily be resolved before they hopefully get a fuller picture during their post-secondary mathematics courses.

For some of the stated reasons, students and teachers are finding a number line challenging to use. However, the ways in which students respond to using multiple models for the same concept may inform us of their understanding of the concept. In two studies conducted by Reeve and Pattison (1996) with 7–8 and then 7–9 grade students in Australia, the researchers found that when asked to visualize fractions, students seldom used number lines and rather opted for discrete sets or area models (e.g., squares, rectangles, or circles). Reeve and Pattison posited that "a comprehensive understanding of number-line representations depends on a sophisticated mental model of fractions" (p. 148). Their findings demonstrate that "progressively more-complex mental models of fraction understanding underlie increases in fraction problem-solving sophistication" (p. 167). While these conclusions inform pedagogical approaches to teaching fractions, they also contribute to answering a more general question, such as "What is the relation between modeling and knowing?" (Sebeok and Danesi 2000, p. 6). A final example in this chapter is suitable for both elementary and secondary school students and is done with a graphics calculator or any corresponding emulator technology (e.g., GeoGebra, Desmos).

3.5 Example: A Train Problem

Time-distance-speed problems appear frequently in mathematics curricula. A simple formula for speed, $s = \frac{d}{t}$, allows elementary school teachers to explore concepts such as ratios, time-distance graphs and units, and model real-life situations of varied complexity. Typical problems have students comparing the movement of two objects for which some of the initial conditions differ. In our experience, students have difficulties with such problems, mostly because they cannot move beyond the first step in modeling: understanding the situation. Using a double number line, followed by a graphics calculator, may help.

A double number line is used to illustrate multiplicative/scaling relationships, such as $f(x) = k \cdot x$, where k is a constant. A double number line almost looks like a table of values put horizontally, allowing students to perceive the pattern in the change of one quantity compared to another quantity. For example, distance traveled

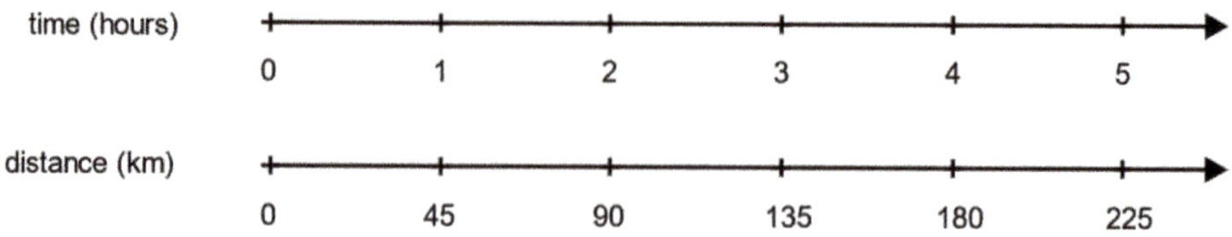

Fig. 9 The change in distance in uniform motion following $d = 45 \cdot t$ km

Fig. 10 Using an online map tool to visualize the railway route

in t hours by a train running between two places at a uniform speed of 45 km/h can be expressed as $d = 45 \cdot t$ km. This situation is presented in Fig. 9, where at the end of every hour, the train is 45 km further from its initial station (when $t = 0$).

Having two trains with different uniform speeds and starting times may require a third number line. In modeling such situations, we could use a coordinate system and apply parametric functions, which are not in the curriculum but are handy for understanding the following problem:

"Windsor and Oshawa are 420km apart. At 9:00AM, a local train leaves Windsor for Oshawa at a uniform speed of 75km/h. An hour later, an express train leaves Windsor and follows the same route, traveling at a uniform speed of 100km/h. At what time will the express train catch the local?" (Fig. 10).

In Desmos graphing calculator (https://www.desmos.com/calculator), we first create the two moving points: $(75t,\ 10)$ and $(100(t\text{–}1),\ 50)$, which represent two trains moving along the parallel tracks ($y = 10$ and $y = 50$) with the distance expressed in terms of time (i.e., $x = 75\,t$ and $x = 100(t\text{–}1)$; $0 \le t < 6$ h). The students see two horizontal parallel number segments and the points moving along with different starting times and uniform speeds. The segments represent the railway's route from Windsor to Oshawa and are parallel to the x-axis that represents the train's distance from Windsor. Animation in any graphics calculator helps students to follow the trains, visually compare their distances from the start for any value of

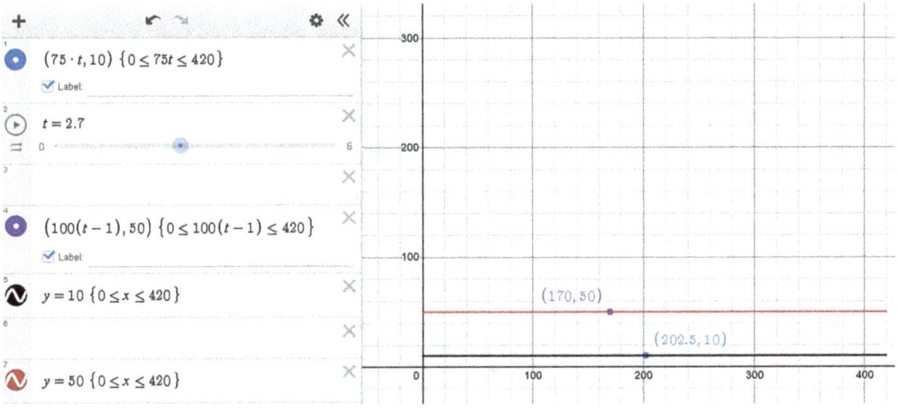

Fig. 11 Presenting the problem on a graphics calculator using parametric representation

parameter *t*, and understand that under the given circumstances, the faster train will overcome the slower one after three hours (or four hours after the local train has left the station).

In comparison to the graph in Fig. 11, a usual time-distance graph (e.g., $y = 75x$, where y is the distance and x is the time) is suitable for gaining a different insight into the problem. It is easy to see that where the two lines intersect (Fig. 12), the express train catches up with the local one and that the steeper line represents the faster train, but the intersection points on x-axis are difficult to interpret. The Grade 9–10 students will likely have trouble reading this graph, especially interpreting the points in which the lines intersect the x-axis. What is the meaning that points (0, 0) and (1, 0) have in relation to the trains' scenario? Comparing the two graphs (Figs. 11 and 12) could help students understand that each model describes the same situation in a different way, allowing for different insights. Figure 11 is much easier to interpret, as students observe that one train starts later but is faster. The trains start from the y-axis, so they could conclude that the y-axis "represents" a station in Windsor and x-axis shows distance from Windsor. The graph on Fig. 12 clearly shows when one train catches with the other, but it is harder to understand that the time on the horizontal axis is given in terms of the local train. In other words, the point of intersection (4, 300) shows that the local train traveled for 4 h before it was overcome by the express train that traveled only 3 h. At that time, however, both trains were 300 km from Windsor, their initial station. Note that in this model, Windsor is "on" x-axis. The grid that the Ontario curriculum mentions in its definition of a coordinate system allows students to see that the distance between the two trains becomes smaller, but after becoming 0, it steadily increases.

So, while the parametric representation is not in the curriculum and may be unfamiliar to teachers, it provides a very useful connection to the double number line that students used in elementary school, allowing teachers to start using parametric

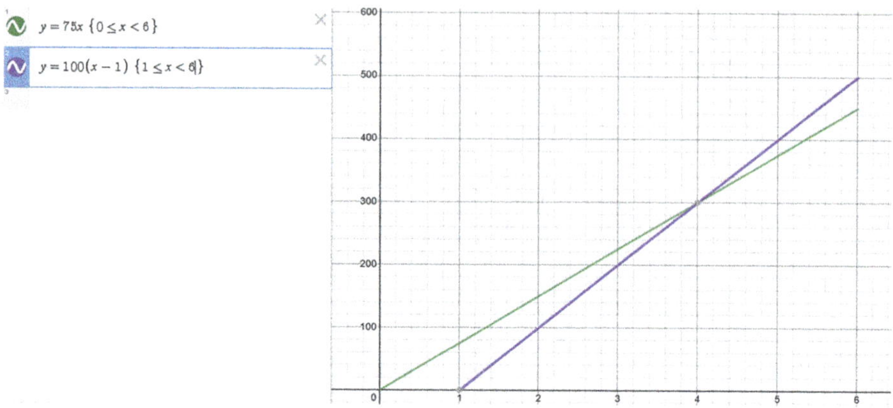

Fig. 12 The usual time/distance representation of the two trains problem

representation more frequently to bridge the multiplicative/scaling representations introduced in elementary school with covariations taught in secondary school.

4 Conclusions

In this chapter, we discussed algebra and modeling from the standpoint of a mathematics educator. The relation between algebra and modeling is multifaceted. For teachers, modeling can be a pedagogy: a way of clarifying and reinforcing mathematical concepts. For mathematicians, modeling is about applying mathematics in real-life situations. For semioticians, algebra itself is a modeling system (Danesi 2022). Sebeok and Danesi (2000) posit that "algebra is really no more than 'arithmetic with letters'" (p. 140), a view to which we add Radford's (2015), for whom algebraic thinking goes beyond the use of symbols. While "arithmetic with letters" may be how algebra is introduced to mid-school students, this creates opportunities for students to learn more advanced mathematics (Mason 2018). Danesi (2022) clarifies the position of algebra with respect to increasing modeling levels of counting, numeration, and arithmetic, by stating that "algebra constitutes a fourth-order modeling system that has the capacity to generalize the arithmetic by giving it its abstract form." He also posits that "algebra is the upper limit of modeling orders. Now, this does not mean that algebraic models are static. Indeed, there are countless such models—linear algebra, matrix algebra, abstract algebra, etc."

Models are created during modeling; taken that way, they are an outcome of a modeling process. Also, teachers of mathematics introduce various models to their students as tools for thinking. These models may appear simple, but they can have deep and sophisticated mathematical meanings. In this chapter, we have compared the curriculum conceptualization of a number line to that in the mathematics and education research literature. This comparison presents the concept of a number line as complex, bringing together the notions of space and number, of geometry and algebra.

In the education literature, a number line is sometimes called a model, a tool, or a mathematical object, but it is also a metaphor for mathematical ideas studied in school. As a model, it allows one to visualize, explain, and manipulate numbers, reflecting both their magnitude and order. Despite its name, a number line can include variables and relational statements, such as inequalities. Operations with numbers and/or variables can be presented on one or two parallel number lines, while two or three mutually orthogonal number lines create 2D or 3D coordinate systems that allow for new insights into relationships between variables and their covariance.

In semiotics, a number line is a sign (Kralemann and Lattmann 2013); it represents the set of numbers that exists in one's mind. Depending on the developmental level of a person and on the place in a curriculum, a number line can represent different (sets of) numbers and their relative positioning; it can have additional elements (e.g., jumps, arrows) that suggest how the numbers could be combined—each example that we mentioned represents numbers from a different perspective.

According to Danesi (2022), the *modeling principle* humans are guided by, implies that for numbers to be understood and remembered, a form in the shape of a number line was invented. Based on the *extensionality principle*, it arose "in the embodied world from a physical line drawn with pencil and ruler to a 'perfect' platonic construction that has length but no thickness (Tall 2008, p. 14) ... consistent with the colloquial notion of 'giving a body' to an abstract idea (Tall 2004, p. 32)" (Presmeg et al. 2016, p. 23). It is also concretely experienced by children when counting pebbles and putting them in one row. Its composition into more complex structures consisting of multiple lines, sometimes parallel and sometimes orthogonal and not always containing similar numbers or units, follows the *structuralist principle* (Danesi 2022).

The education research strongly suggests that algebraic thinking is suitable for even the youngest students. Carraher et al. (2006) recommend that early elementary school mathematics have an algebraic character. Their proposal to teach "algebrafied arithmetic" in early grades can be added to the voices of other educators mentioned in this chapter. It is our wish that some of the ideas presented here will help teachers use models and modeling in ways that are as true to mathematics as is developmentally appropriate for their students. Remembering Resnik's (1981) words cited in the introduction, this would help students experience mathematics, take classroom learning and apply it to real-world life, and offer options for their imagination to explore the yet unexperienced/unexperientable. Danesi's essay in this book suggests that such an approach would align with the "modeling instinct" that exists in all of us.

Appendix: Glossary of Notation and Definitions

Epistemology of mathematics informs us how mathematical knowledge is created, negotiated, and justified.

Mathematical modeling includes abstracting real-life relationships in the form of models, which are then assessed and compared.

A number line is the result of blending of points and numbers, resulting in number points (Lakoff and Núñez 2000). By pairing two orthogonal number lines, the concept of a coordinate system emerges. Pairing of two parallel number lines, illustrates multiplicative/scaling relationship or covariation.

Algebra is a domain of mathematics and a school subject. In this chapter, we support the idea that teaching students to think algebraically is appropriate even for early grades.

References

Ayalon, M., Watson, A., & Lerman, S. (2015). Functions represented as linear sequential data: Relationships between presentation and student responses. *Educational Studies in Mathematics*, *90*, 321–339.

Ayalon, M., Watson, A., & Lerman, S. (2017). Students' conceptualisations of function revealed through definitions and examples. Research in Mathematics Education, 19(1), 1–19.

Bobis, J. (2007). The empty number line: A useful tool or just another procedure? Teaching Children Mathematics, 13(8), 410–413.

Borsuk, K., & Szmielew, W. (1960). Foundations of Geometry. Revised English translation. Amsterdam, The Netherlands: North Holland.

Carraher, D.W., Schliemann, A.D., Brizuela, B.M., & Earnest, D. (2006). Arithmetic and algebra in early mathematics education. *Journal for Research in Mathematics Education*, *37*(2), 87–115.

Carmona, G., & Greenstein, S. (2010). Investigating the Relationship Between the Problem and the Solver: Who Decides What Math Gets Used? In R. Lesh, P. L. Galbraith, C. R. Haines, and A. Hurford (Eds.), Modeling students' mathematical modeling competencies: ICTMA 13 (pp. 245–254). New York, NY: Springer.

Czarnocha, B., Baker, W., Dias, O., & Prabhu, V. (2012). Learning trajectories from the arithmetic/algebra divide. *Mathematics Teaching-Research Journal*, *5*(3). Retrieved from https://commons.hostos.cuny.edu/mtrj/wp-content/uploads/sites/30/2018/12/v5n3-Learning-Trajectories-from-the-Arithmetic-Algebra-Divide.pdf

Danesi, M. (2022). Algebra as a semiotic modeling system. In Mahouton Norbert Hounkonnou, Dragana Martinovic, Melanija Mitrovic, and Philippa Pattison (Eds.), Mathematics for Social Sciences and Arts - Algebraic Modeling. Mathematics in Mind Series. Springer Nature Switzerland,

Dickinson, P., & Eade, F. (2004). Using the Number Line to Investigate the Solving of Linear Equations. *For the Learning of Mathematics*, *24*(2), 41–47.

Dossey, J. A. (1998). Making algebra dynamic and motivating: A national challenge. In The nature and role of algebra in the K–14 curriculum (pp. 17–22). Washington, DC: National Academy Press.

Gil, E., & Gibbs, A. L. (2017). Promoting modeling and covariational reasoning among secondary school students in the context of big data. *Statistics Education Research Journal*, *16*(2), 163–190.

Grønmo, L. S. (2018). The Role of Algebra in School Mathematics. In Kaiser G., Forgasz H., Graven M., Kuzniak A., Simmt E., and Xu B. (Eds.), Invited Lectures from the 13th International Congress on Mathematical Education. ICME-13 Monographs. Springer, Cham. https://doi.org/10.1007/978-3-319-72170-5_11

Herbst, P. (1997). The number-line metaphor in the discourse of a textbook series. *For the Learning of Mathematics*, *17*(3), 36-45.

Højgaard, T. (2010). Communication: The essential difference between mathematical modeling and problem solving. In R. Lesh, P. L. Galbraith, C. R. Haines, and A. Hurford (Eds.), Modeling students' mathematical modeling competencies: ICTMA 13 (pp. 255–264). New York, NY: Springer.

Jupri, A., Drijvers, P., & van den Heuvel-Panhuizen, M. (2014). Difficulties in initial algebra learning in Indonesia. *Mathematics Education Research Journal*, *26*(4), 683–710.

Kaput, J. J. (2000). Teaching and Learning a New Algebra with Understanding. National Center for Improving Student Learning and Achievement in Mathematics and Science, Dartmouth, MA.

Kralemann, B., & Lattmann, C. (2013). Models as icons: Modeling models in the semiotic framework of Peirce's theory of signs. *Synthese, 190*, 3397–3420.

Kurepa, Dj. (1969-in Serbo-Croatian). Visa Algebra [Higher Algebra]. Zavod za Izdavanje Udzbenika.

Lakoff, G., & Núñez, R. E. (2000). *Where mathematics comes from: How the embodied mind brings mathematics into being.* Basic Books.

Lesh, R., Galbraith, P. L., Haines, C. R., & Hurford, A. (2010, Eds.). Modeling students' mathematical modeling competencies: ICTMA 13. New York, NY: Springer.

Mason, J. (2016). Overcoming the algebra barrier: Being particular about the general, and generally looking beyond the particular, in homage to Mary Boole. In S. Stewart (Ed.), And the rest is just algebra (pp. 97–117). New York: Springer.

Mason, J. (2018). How Early Is Too Early for Thinking Algebraically? In Kieran C. (ed.) Teaching and Learning Algebraic Thinking with 5- to 12-Year-Olds. ICME-13 Monographs. Springer, Cham. https://doi.org/10.1007/978-3-319-68351-5_14

Mason, J., Graham, A., & Johnston-Wilder, S. (2005). Developing thinking in algebra. Sage, London.

MoEd. (2020). The Ontario Curriculum, Grades 1–8: Mathematics, 2020. *Queen's Printer for Ontario, 2020*. https://shorturl.at/czALS

Presmeg, N., Radford, L., Roth, W. M., & Kadunz, G. (2016). Semiotics in mathematics education. ICME-13 Topical Surveys. Hamburg: Springer.

Radford, L. (2015). Early Algebraic Thinking: Epistemological, Semiotic, and Developmental Issues. In Cho S. (ed.), The Proceedings of the 12th International Congress on Mathematical Education. Springer, Cham. https://doi.org/10.1007/978-3-319-12688-3_15

Reeve, R. A., & Pattison, P. E. (1996). The referential adequacy of students' visual analogies of fractions. *Mathematical Cognition*, *2*(2), 137–169.

Resnik, M. D. (1981). Mathematics as a Science of Patterns: Ontology and Reference. *Noûs, 15*(4), 529–550. https://doi.org/10.2307/2214851

Rycroft-Smith, L., & Gould, T. (2021). The number line: A flexible and useful model. Cambridge Mathematics: Espresso, *35*. Retrieved from https://www.cambridgemaths.org/Images/espresso_35_the_number_line.pdf

Sebeok, T. A., & Danesi, M. (2000). *The Forms of Meaning: Modeling Systems Theory and Semiotic Analysis*. Berlin: Mouton de Gruyter.

Shell Centre for Mathematical Education. (1985). The Language of Functions and Graphs. Retrieved from https://www.mathshell.com/publications/tss/lfg/lfg_teacher.pdf

Sinkevich, G. (2015). On the History of Number Line. arXiv:1503.03117v1 [math.HO] https://arxiv.org/pdf/1503.03117

Strauss, D. F. M. (2014). What is a Line? *Axiomathes*, *24*, 181–205.

Thomas, M.O.J., de Freitas Druck, I., Huillet, D., Ju, M.-K., Nardi, E., Rasmussen, C., & Xie, J. (2015). Key Mathematical Concepts in the Transition from Secondary School to University.

In S.-J. Cho (Ed.), The Proceedings of the 12th International Congress on Mathematical Education: Intellectual and Attitudinal Challenges (pp. 285–304). New York, NY: Springer.

Thompson, P. W., & Carlson, M. P. (2017). Variation, covariation, and functions: Foundational ways of thinking mathematically. In J. Cai (Ed.), Compendium for research in mathematics education (pp. 421–456). Reston, VA: National Council of Teachers of Mathematics.

Usiskin, Z. (1995). Why Is Algebra Important to Learn? *American Educator*, *19*(1), 30–37.

van den Heuvel-Panhuizen, M. (2008). Learning From "Didactikids": An Impetus for Revisiting the Empty Number Line. *Mathematics Education Research Journal*, *20*(3), 6–31.

Watson, A., Ayalon, M., & Lerman, S. (2018). Comparison of students' understanding of functions in classes following English and Israeli national curricula. Educational Studies in Mathematics. 97(1), 255–272.

Wilmot, D. B., Schoenfeld, A., Wilson, M., Champney, D., & Zahner, W. (2011). Validating a learning progression in mathematical functions for college readiness. Mathematical Thinking and Learning, 13(4), 259–291.

Gödel's Incompleteness as an Argument for Dualism

Zvonimir Šikić

1 Introduction

At the end of the first half of the nineteenth century, formal logic was being developed mainly in Great Britain, under the influence of the newly raising abstract algebra. It has just begun to appear in the works of G. Peacock (1830), W. R. Hamilton (1837), and A. de Morgan (1849). Attempts to apply mathematical analysis in studying the laws of quantity already proved very fruitful and convincing. The similar attempts to apply mathematical analysis to the laws of quality, i.e., to formal logic, became characteristic of the period. Setting a suitable symbolic apparatus and founding the laws of its manipulation, in the same way algebra does it, was the ultimate purpose of the use of the new method. The final result was a Boole-Schroeder algebra which represented, in its various interpretations, the logic of concepts and the logic of one-place propositional functions (G. Boole 1847), as well as the logic of propositions (C. S. Pierce 1880 and E. Schröder 1890). It has been founded as an abstract mathematical system, independent of its particular interpretation, and it substantially influenced the character of mathematics. Namely, it used the method of deduction from a small number of premises, and this method has become a characteristic of mathematics regardless of the quantitative or the qualitative character of the matter being researched.

In this spirit, many tried to use these new algebraic and logical methods to solve classical philosophical problems. One of the most famous such attempt is understanding of Gödel's incompleteness theorem as an argument for dualism.

Z. Šikić (✉)
University of Rijeka, Center for Logic and Decision Theory, Rijeka, Croatia
e-mail: zsikic@math.hr

© The Author(s), under exclusive license to Springer Nature Switzerland AG 2023
M. N. Hounkonnou et al. (eds.), *Mathematics for Social Sciences and Arts*,
Mathematics in Mind, https://doi.org/10.1007/978-3-031-37792-1_5

2 Naturalism vs. Dualism

For many, it is still hard to conceive how the world of subjective experiences spring out of merely physical events. This problem of qualia is the hardest and the main part of the mind-body problem. The problem is today summed up in the question "how matter, i.e. body and brain, becomes mind," but it has a long history.

Although it is not entirely clear when people began to realize that the brain is the source of thought and behavior, it is certain that the consequences of severe head injuries have for a long time indicated many brain functions. Faced with such cases, Hippocrates already in the fifth century BC concluded that the brain is the foundation of all our thoughts, feelings, and ideas (cf. Breitenfeld et al. 2014). He was a naturalist – seeking explanations of how things work in the natural world. He considers invocations to gods and other supernatural things futile.

In contrast, Plato was a super-naturalist who believed that each of us has a soul that lives before birth, is embodied in the course of our lives, and after our bodily death returns to the realm of souls. In this realm, all absolute truths reside. Building on this belief, he called the body the prison of the soul (cf. Plato 1966 Vol. 1, p. 82). Only the soul comes to true knowledge, remembering the absolute truths with which it was close while dwelling in their realm. Plato is, at least in the Western tradition, the initiator of the idea of the incorporeal soul – the initiator of the idea of the dualism of things.

Aristotle, although Plato's best student, is firmly rooted in the physical world. Like Hippocrates, he favors naturalism. His ideas about psychological states are complex and have various interpretations, but he undoubtedly believes that all emotional states (anger, fear, joy, sadness, love, hate, etc.) are bodily states. It is less clear whether he places the mind in bodily functions when, for example, it deals with mathematics. He modestly observes that to say anything credible about the soul is one of the most difficult tasks (cf. Aristotel 1931 Vol III, De Anima, 1, 402a10-11).

Two Western traditions emerge from Plato's mysticism and Aristotle's science. Dualism and naturalism.

Early Christianity did not require the idea of a Platonic soul. The body was enough, because the body is resurrected. Many questions arise in this regard. In what condition will the body be resurrected, young or old, with or without an amputated leg, will the ancestors be the same age as their descendants, etc. The idea of a resurrected body is also in conflict with the obvious fact that after death the body decays. One way to reconcile these opposites is to call on Christ's power to transform our decaying bodies into spiritual, indestructible, and eternal bodies (cf. Maas 1911). Thus Plato's soul was reconciled with the Christian belief in a bodily resurrection. Of course, the idea of a spiritual body looks like the idea of a square circle, so the details of how exactly it works have remained undeveloped. But the dualism of the whole construction is unquestionable.

In the seventeenth century, Descartes held that the use of language and the making of reasonable decisions were achievements for which no physical mechanism was capable. He concluded that mental functions – perception, thinking, decision making, dreaming, feeling etc. – are functions of the soul, which is distinct from the body and can exist without it (cf. Descartes 1984–1991 Vol. II, p. 54.) He placed the exchange of information between the brain and the soul in the pineal gland, conveniently located in the middle of the head. Today we know that its basic function is to create melatonin that regulates sleep.

Why was Descartes' imagination so limited? In the seventeenth century, the finest physical mechanisms were clocks and fountains. They were impressive, but they lacked the ability to create something new. In contrast, human minds were capable of all kinds of novelties, especially in speech. If Descartes had today's knowledge of the brain and its capabilities, perhaps his imagination would have been richer. (He was more troubled by problems with physics: when the soul causes events in the physical body, then energy is transferred from the soul to the body, and this violates the laws of conservation.)

In the nineteenth century, Helmholtz realized that souls, occult forces, and other supernatural things were a dead end when it came to explaining mental functions such as perception, thinking, and feeling. He realizes, before Freud, that most brain operations take place below the level of consciousness (cf. Helmholtz 2000 Vol. III, p. 27). He was prompted to do so by the fact that in a moment, without any conscious effort, we successfully perceive a very complex visual scene. He did not understand the nature of these operations, but he realized that the brain had to do a lot on an unconscious level and that it was not enough to consider conscious activities alone. Moreover, if conscious and unconscious operations are interdependent, the identification of the soul with conscious activities becomes questionable, as is any identification of the part with the whole.

In the middle of the twentieth century, dualism ceased to be an acceptable explanation for consciousness, thought, and decision-making. The deep rooted paradigm does not disappear overnight, but it changes unstoppably with the accumulation of new knowledge. It becomes clear that physical changes in the brain lead to changes in the supposed functions of the soul. Inhalation of ether causes loss of consciousness. Consumption of LSD causes hallucinations. A stroke in a specific place leads to a loss of the ability to recognize previously well-known faces. In some other areas, it causes a loss of the ability to speak or abolishes social inhibitions. All these phenomena point to the nervous system, not to the disembodied soul.

Of particular concern among the remaining dualists were Roger Sperry's experiments with split-brain patients (cf. Myers and Sperry 1958), who had their cerebral hemispheres surgically separated for the purpose of preventing continuous epileptic seizures. Their hemispheres became cognitively independent. If the visual stimulus is presented only to the right hemisphere, the patient cannot confirm this in words because it is not seen by the left hemisphere that generates speech. It was a stunning result. Did you divide the brain or did you divide the soul? The soul should be indivisible. It's obviously a divided brain – if the cerebral hemispheres are separated, their mental states are separated. These results were strong support for the hypothesis that mental states are states of the bodily brain rather than states of the disembodied soul.

What would a disembodied soul be anyway? How would it, for example, stop feeling the toothache when we inject procaine? The bodily mechanism is clear. Procaine temporarily inhibits sodium flow so there is no action potential necessary to transmit a nerve signal. That explanation is easy to test, and the details fit into what we already know about neurons and pain. Can dualists explain why procaine blocks pain? What does this bodily substance do to the disembodied soul? Is there any explanation for the mechanism of this action? There is no such explanation, because even deeply convinced dualists do not even try to figure it out. As if it were enough to say, "this is what the soul does." Of course it is not enough.

The science of the brain exists, and the science of the soul does not. Maybe because, unlike the brain, the soul does not exist and naturalism is right.

Nevertheless, some remaining twentieth-century dualists, like Lucas (1961) and Penrose (1994, 1999), thought that they found a new argument for their thesis. They think that Gödel's incompleteness theorems prove the dualist thesis. Their argument is that Gödel's theorems imply human-machine non-equivalence in the following sense:

There is no machine which could capture all our mathematical intuitions. Hence, we are not just machines, there is something beyond that.

3 Bare Bones of Incompleteness

We start with the bare bones of incompleteness (as presented in Šikić 2005). Let **M** be a machine which is programmed to print finite sequences of three symbols: –, P, and D. (In what follows these sequences will be simply called sequences.) At each stage one sequence is printed into a square and each square is part of the tape unending in one direction.

We say that **M** prints a sequence if **M** prints it at some stage. We say that **M** does not print a sequence if **M** does not print it at any stage. Some of the sequences are meaningful and we call them sentences. Here is the definition.

Definition of Sentences and Their Meanings
A sentence is a sequence of the form PX, –PX, PDY *or* –PDY, *where X is any sequence not starting with* D *and* Y *is any sequence.*

(i) *The meaning of a sentence of the form* PX *is "***M** *prints* X*".*
(ii) *The meaning of a sentence of the form* –PX *is "***M** *does not print* X*".*
(iii) *The meaning of a sentence of the form* PDY *is "***M** *prints* YY*".*
(iv) *The meaning of a sentence of the form* –PDY *is "***M** *does not print* YY*".*

Remark
It helps to think of –, P *and* D *as words meaning "not", "prints" and "double".*

Sentences have meanings, so they are true or false. All other sequences are neither true nor false, they are meaningless. Our main interest concerning machines are their correctness and completeness. These notions are defined as follows.

Definition of Correctness
Machine **M** *is correct if it prints only true sentences* (i.e., *if* **M** *prints a sentence then this sentence is true). Remark: A correct machine may print a lot of non-sentences. For example, a machine which prints only non-sentences is trivially correct.*

Definition of Completeness
Machine **M** *is complete if it prints all true sentences. Remark: A complete machine may print a lot of false sentences. For example, a machine which prints all sequences is trivially complete.*

The most interesting machines would be those which are correct and complete. Unfortunately, there is no such machine (i.e., it is impossible to construct such a machine).

Theorem on Correctness and Completeness Impossibility
Every machine is either incorrect or incomplete.

Proof
Take a look at sentence –PD–PD. *It means "***M** *does not print* –PD–PD*". Of course,* **M** *either prints* –PD–PD *or not. If* **M** *prints* –PD–PD *then it prints a false sentence* i.e. **M** *is not correct. If* **M** *does not print* –PD–PD *then it does not print a true sentence* i.e. **M** *is incomplete.*

What is the link between this very simple theorem, Gödel's incompleteness theorems and mechanization of our mathematical intuitions? To mechanize our mathematical intuitions means to construct a machine which would prove (print) all mathematical theorems which are normally derived using these intuitions. A formalized mathematical theory with explicitly defined language, axioms, and deductive rules is such a machine. Hence, this is the machine to which we may apply the simple theorem.

But there is a serious problem. Mathematical machines do not have the property of self-reflexivity in the sense in which machines from a simple theorem have it. These machines produce sentences which assert something about the machines themselves. Our mathematical theories assert many things about various mathematical objects, but nothing about the theories themselves. If we are interested in a theory itself, we usually construct another theory, called meta-theory, to deal with it. But here comes Gödel. In his famous Gödel (1931), he proved that a mathematical theory **M**, which includes an appropriate amount of arithmetic, may represent its own meta-theory. This is the hardest part of Gödel's proof.

Now we can apply our simple theorem and what we get is Tarski's version (cf. Boolos et al. 2002, p. 223) of the first Gödel's incompleteness theorem.

Tarski-Gödel First Incompleteness Theorem If mechanized mathematical theory **M** *includes an appropriate amount of arithmetic, it is either incorrect or incomplete,* i.e., *if it is correct, then it is incomplete. Even more, there is an explicitly definable sentence* G *(corresponding to* –PD–PD*) which is true but not provable in the theory,* i.e., $\nvdash_{\mathbf{M}}$ G.

4 Gödel's First Incompleteness Theorem

Gödel's original version of the first incompleteness theorem avoids the truth concept and reads as follows:

Gödel First Incompleteness Theorem
If mechanized mathematical theory **M** *includes an appropriate amount of arithmetic, there is an explicitly definable sentence* G *which asserts its own unprovability and is such that:*

(i) *If* **M** *is consistent then* $\nvdash_{\mathbf{M}}$ G.
(ii) *If* **M** *is* ω*-consistent then* $\nvdash_{\mathbf{M}} -$ G.

M is ω*-consistent if whenever* $\vdash_{\mathbf{M}} -$ S(0), $\vdash_{\mathbf{M}} -$ S(1), $\vdash_{\mathbf{M}} -$ S(2), ... *then* $\nvdash_{\mathbf{M}} \exists$ (x) S(x).
(In what follows $\vdash_{\mathbf{M}}$ *is abbreviated to* $\vdash$*.)*

Gödel defined arithmetical predicate Prv (which represents the meta-concept of provability in **M** within **M** itself) and proved that it has the following properties:

(a) *n is the numerical code of a provable formula* $\Longrightarrow \vdash$ Prv (n, m) *for some m.*
(b) *n is not the numerical code of a provable formula* $\Longrightarrow \vdash -$Prv (n, m) *for every m.*

Gödel then defined Pr (x) as $\exists\, y$ Prv (x, y) and proved the diagonal lemma: there is a sentence G such that

$$\text{(DG)} \qquad \vdash \mathrm{G} \iff -\mathrm{Pr}\ (\mathrm{G})\,.$$

(Instead of "Pr ('X'), *where* 'X' *is the numerical code of the formula* X" we write "Pr (X)".)

The Gödel's first incompleteness theorem now easily follows from the first Hilbert-Bernays provability condition (B1) and its converse (B):

$$\text{(B1)} \qquad \vdash \mathrm{X} \Rightarrow \vdash \mathrm{Pr}\ (\mathrm{X})$$

$$\text{(B)} \qquad \nvdash \mathrm{X} \Rightarrow \nvdash \mathrm{Pr}\ (\mathrm{X})\,.$$

(B1) and (B) almost immediately follow from (a) and (b). By (a), $\vdash$ X implies $\vdash$ Prv (X, m) for some m. Hence $\vdash \exists$y Prv (X, y), i.e. $\vdash$ Pr (X). This proves the assertion of (B1). By (b), $\nvdash$ X implies $\vdash -$Prv (X, m) for every m. Hence, by ω-consistency, $\nvdash \exists$y Prv (X, y), i.e. $\nvdash$ Pr (X). This proves the assertion of (B).

The proof of the Gödel's first incompleteness theorem is now as follows.

(i) Assume $\vdash$ G. Then, by (B1), $\vdash$ Pr (G), which implies $\vdash -$ G. This contradicts the consistency. Hence, $\nvdash$ G.

(ii) Assume $\vdash -$ G, hence $\vdash$ Pr (G). Then, by (B), $\vdash$ G. This contradicts the consistency. Hence, $\nvdash -$ G.

5 Gödel's Second Incompleteness Theorem

Gödel's sentence G associated with **M** (a mechanized mathematical theory which includes an appropriate amount of arithmetic) asserts its own unprovability in **M**. It is provably equivalent to the sentence Con (**M**), short for $-$Pr ($\bot$), where $\bot$ is a contradiction. $-$Pr ($\bot$) asserts the consistency of **M**, so

$$\vdash \text{Con } (\mathbf{M}) \leftrightarrow \text{G}$$

This easily follows from the diagonal lemma

$$(\text{DL}) \vdash \text{G} \leftrightarrow -\text{Pr } (\text{G})$$

and Hilbert-Bernays provability conditions

$$(\text{B1}) \qquad \vdash \text{X} \Rightarrow \vdash \text{Pr } (\text{X}),$$

$$(\text{B2}) \qquad \vdash \text{Pr } (\text{X} \rightarrow \text{Y}) \rightarrow (\text{Pr } (\text{X}) \rightarrow \text{Pr } (\text{Y})),$$

$$(\text{B3}) \qquad \vdash \text{Pr } (\text{X}) \rightarrow (\text{Pr } (\text{Pr } (\text{X}))).$$

The conditions are valid for our **M**, although (B3) is not so easy to prove (cf. Boolos et al. 2002, p. 234). Now, from (DL) we easily get

$$\vdash \text{G} \rightarrow (\text{Pr } (\text{G}) \rightarrow \bot).$$

Applying first (B1) and then (B2), we get

$$\vdash \text{Pr}(\text{G}) \rightarrow (\text{Pr } (\text{Pr } (\text{G})) \rightarrow \text{Pr } (\bot)).$$

By (B3) it then follows

$$\vdash \text{Pr}(\text{G}) \rightarrow \text{Pr}(\bot), \text{ i.e. } \vdash -\text{Pr } (\bot) \rightarrow -\text{Pr } (\text{G}).$$

It means that $\vdash$Con (**M**) $\rightarrow$ G.

On the other hand, from $\vdash \bot \rightarrow$ G, by applying (B1) and (B2), we get

$$\vdash \text{Pr}(\bot) \rightarrow \text{Pr } (\text{G}), \text{ i.e. } \vdash -\text{Pr } (\text{G}) \rightarrow -\text{Pr } (\bot).$$

It means that $\vdash G \to \text{Con}(\mathbf{M})$.

Now it is easy to prove the second Gödel's incompleteness theorem.

Gödel's Second Incompleteness Theorem

A mechanized mathematical theory **M** *which includes an appropriate amount of arithmetic cannot prove its own consistency,* i.e.

$$\nvdash \text{Con}(\mathbf{M}).$$

Namely, from $\vdash \text{Con}(\mathbf{M}) \leftrightarrow G$ and $\nvdash G$, it immediately follows that $\nvdash \text{Con}(\mathbf{M})$. Notice further that from $\vdash G \leftrightarrow -\text{Pr}(G)$ and $\vdash G \leftrightarrow \text{Con}(\mathbf{M})$, it follows that

$$\vdash -\text{Pr}\ (G) \leftrightarrow \text{Con}\ (\mathbf{M})\,.$$

So, unprovability of G in **M** is provably equivalent to the consistency of **M**.

What do we know about the unprovability of $-G$, which is the other part of the first incompleteness theorem? First note that $\vdash -G \leftrightarrow \text{Pr}(\bot)$, because $\vdash -G \leftrightarrow -\text{Con}(\mathbf{M})$. By (B1) and (B2), we get

$$\vdash -\text{Pr}\ (-G) \leftrightarrow -\text{Pr}\ (\text{Pr}\,(\bot))\,.$$

But it is easy to see that $-\text{Pr}(\text{Pr}(\bot))$ expresses the consistency of $\mathbf{M} + \text{Con}(\mathbf{M})$. Namely, the consistency of **M** extended with Con (**M**) is expressed by $-\text{Pr}_{\mathbf{M}+\text{Con}(\mathbf{M})}(\bot)$, where $\text{Pr}_{\mathbf{M}+\text{Con}(\mathbf{M})}$ is the provability predicate of $\mathbf{M} + \text{Con}(\mathbf{M})$. On the other hand, $-\text{Pr}_{\mathbf{M}+\text{Con}(\mathbf{M})}(\bot)$ is equivalent to $-\text{Pr}(-\text{Con}(\mathbf{M}))$, i.e., to $-\text{Pr}(\text{Pr}(\bot))$, hence

$$\vdash -\text{Pr}\ (-G) \leftrightarrow \text{Con}\ (\mathbf{M} + \text{Con}\ (\mathbf{M}))\,.$$

It means that it is provable in **M** that the unprovability of $-G$ in **M** is equivalent to the consistency of $\mathbf{M} + \text{Con}(\mathbf{M})$.

Notice that $\text{Con}(\mathbf{M} + \text{Con}(\mathbf{M}))$ is stronger than Con (**M**). $\text{Con}(\mathbf{M}) \vdash \text{Con}(\mathbf{M} + \text{Con}(\mathbf{M}))$ would contradict second Gödel's incompleteness theorem, because $\text{Con}(\mathbf{M} + \text{Con}(\mathbf{M}))$ would be provable in $\mathbf{M} + \text{Con}(\mathbf{M})$.

We also offer a formal proof of this fact to the interested reader. Namely, from the logical truth $-S \to -\text{Pr}(S) \vdash \text{Pr}(S) \to S$, we get $\text{Pr}(-S \to -\text{Pr}(S)) \vdash \text{Pr}(\text{Pr}(S) \to S)$. Applying Löb's theorem:

$$\text{(L)} \qquad \text{Pr}\ (\text{Pr}\ (S) \to S)\ \vdash \text{Pr}\ (S)\,,$$

we get $\text{Pr}(-S \to -\text{Pr}(S)) \vdash \text{Pr}(S)$ and then by contraposition $-\text{Pr}(S) \vdash -\text{Pr}(-S \to -\text{Pr}(S))$. Substituting $\text{Pr}(\bot)$ for S, we get

$$-\text{Pr}\ (\text{Pr}\,(\bot))\ \vdash -\text{Pr}\ (-\text{Pr}\,(\bot) \to -\text{Pr}\ (\text{Pr}\,(\bot)))\,,\ \text{i.e.}$$

$$\text{Con }(\mathbf{M}+\text{Con }(\mathbf{M})) \vdash -\text{Pr }(\text{Con }(\mathbf{M}) \rightarrow \text{Con }(\mathbf{M}+\text{Con }(\mathbf{M}))) .$$

This is a formal proof that Con (**M** + Con (**M**)) is stronger than Con (**M**).
By the way, substituting ⊥ for S we get

$$-\text{Pr}(\perp) \;\; \vdash -\text{Pr }(- \perp\rightarrow -\text{Pr}(\perp)) \text{, i.e.}$$

$$\text{Con }(\mathbf{M}) \vdash -\text{Pr}(\text{Con }(\mathbf{M})) .$$

It proves that the second Gödel's theorem is a consequence of Löb's theorem. On the other hand, Kripke proved that Löb's theorem is a consequence of the second Gödel's theorem:

$$\nvdash \text{S} \qquad \text{(by 2.Godel)} \Rightarrow \qquad -\text{S} \nvdash -\text{Pr }(\text{S}) \qquad \Rightarrow \qquad \nvdash \text{Pr }(\text{S}) \rightarrow \text{S}$$

6 Gödelian Dualist Arguments and Their Refutations

Lucas and Penrose's argument has been criticized by many. Here we recall the most convincing critiques.

Roger Penrose, in Penrose (1999, pp. 107–108), claims that we can always see that G is true in the following way. If G is provable in Peano arithmetic PA, then it is false, but that is impossible "because our formal system should not be so badly constructed that it actually allows false propositions to be proved." Hence, G is unprovable and therefore true. But, Boolos (1990) and many others warn that even if we concede that we can see the truth of the Gödel sentence for PA, it is not to concede that we can see the truth of Gödel sentences for more powerful theories, e.g., ZFC set theory. What we can see, says Boolos, is that Gödel sentence for ZFC is ZFC unprovable, and therefore true, **if** ZFC is consistent. And we do not know that. We could be in the same situation regarding ZFC that Frege was in with respect to his Frege (1893–1903) before receiving the letter from Russell, showing that Frege (1893–1903) is inconsistent.

Furthermore, Feferman showed that Penrose's argument in Penrose (1994) is marred by a number of errors, which are enumerated and explained in Feferman (1995). We will not repeat that. Feferman also noticed that Gödel (in the Gibbs lecture Gödel 1991/1995) countenances the possibility that "the human mind (in the realm of pure mathematics) is equivalent to a finite machine that, however, is unable to understand completely its own functioning." In footnote 13. p. 309 of Gödel (1991/1995) Gödel says:

> It is conceivable (although far outside the limits of present-day science) that brain physiology would advance so far that it would be known with empirical certainty

1. that the brain suffices for the explanation of all mental phenomena and is a machine in the sense of Turing;
2. that such and such is the precise anatomical structure and physiological functioning of the part of the brain which performs mathematical thinking.

So Lucas and Penrose would have a hard time convincing Gödel himself.

Good (in Good 1969) considered the argument that a mechanized mathematical theory **M** will never prove its Gödel sentence G and that that this shows that humans, qua mathematicians, transcend machines in at least this respect. This is easily refuted by the observation that Gödel's construction could itself be carried out by another machine. But further Gödel propositions can then be appended and a complete treatment leads inevitably to questions concerning the mechanization of transfinite counting. If the dualists still wish to make a case, they must base it on transfinite counting rather than on Gödel's theorem. Any machine for transfinite counting can be improved in the sense that there exists another machine that can go further, and it is on this basis that a dualist should build his case. What dualist must prove is that he personally can always make the improvement. But no such proof is possible since, if it were given, it could be used for the design of a machine that could always do the improving. This is impossible since it would lead to the conclusion that the smallest non-constructible ordinal is constructible. Good concludes that Gödel's theorem is a red herring, and Lucas's article should have been called "Minds, machines and transfinite counting" instead of "Minds, machines and Gödel."

Taking this into account, Lewis (1989) gave an account of Lucas' argument, which is substantially clearer than the original. He pointed out that in accordance with Gödel second incompleteness theorem, we can define an effective function Con from mechanized mathematical theories **M** (which include an appropriate amount of arithmetic) to their sentences, such that we can prove the following:

C1. Con (**M**) is true if and only if **M** is consistent.
C2. If **M** is correct then Con (**M**) is true.
C3. Con (**M**) is provable if and only if **M** is inconsistent.

Call C a *consistency sentence* for set of sentences S if and only if there is some **M** such that S is the set of provable sentences of **M** and C is Con (**M**). Lewis introduced the following rule of inference:

R. If S is a set of sentences and C is a consistency sentence for S, infer C from S.

Rule R is a sound rule of inference: if the premises S are all true, then by C2 so is the conclusion C. According to Lewis, Lucas is extending Peano arithmetic PA with the rule R. The resulting theory he calls Lucas arithmetic LA and he assumes that Lucas thinks that he has reason to believe that the theorems of his arithmetic are true for the same reason that the theorems of PA are true: the theorems of both theories come from Peano axioms by truth-preserving rules of inference.

Now, if LA is a mechanized mathematical theory **M**, it would have a consistency sentence C = Con (**M**). Since LA is closed under R, C would be a theorem of LA and, by C3, LA would be inconsistent. Lucas arithmetic would contain falsehoods, by C1, and so the falsehoods would follow from the Peano axioms themselves. Therefore, insofar as we trust the Peano axioms, we know that Lucas arithmetic is not the output of any mechanized mathematical theory **M**. So if Lucas can verify all the theorems of Lucas arithmetic, then Lucas is no machine. This is Lewis' benevolent reconstruction of Lucas' argument.

But we are given no reason to believe that the argument is valid. Lucas arithmetic is not an ordinary theory. In order to check whether Lucas's rule R has been used correctly, a checking procedure would have to decide whether a given set S of sentences is the output of a mechanized theory **M**. But a general method for deciding that could be converted into a general method for deciding whether any given Turing machine will halt on any given input—and that, we know, is impossible. So we do not know how many theorems of LA Lucas can produce as true. He can certainly go beyond PA, but he can go beyond it and still be a machine, because limitations on his ability to verify theoremhood in LA may leave him unable to recognize a lot of theorems of LA.

McCall's reasoning in McCall (1999) differs from the earlier Gödelian arguments in his admission that the recognition of truth of Gödel sentence G for a mechanized theory **M** depends essentially on the unproved assumption that the theory **M** under consideration is consistent. But McCall notes that we have two different cases:

1. If **M** is consistent then G is not provable.
2. If **M** is consistent then $-$G is not provable.

He claims that both sentences are true, the difference being that the formal version of 1. is a theorem, whereas the formal version of 2. "to the best of our knowledge" is not, i.e.

1. $\vdash \text{Con}\ (\mathbf{M}) \rightarrow -\text{Pr}\ (\text{G})$,
2. $\nvdash \text{Con}\ (\mathbf{M}) \rightarrow -\text{Pr}\ (-\text{G})$.

Moving from 1. to 2. yields the candidate for the true but unprovable sentence. That is what McCall thinks. But, to the extent that we can recognize the truth of $-$Pr ($-$G), we are assuming the consistency of $\mathbf{M} + \text{Con}\ (\mathbf{M})$; cf. 4. paragraph. Hence, for the comparison to be fair, it would have to involve a mechanized theory equipped with whatever assumptions we ourselves have employed in order to see the truth of 2. But we proved in 4. paragraph that the mechanized theory **M** can prove $-$Pr ($-$G) by assuming the consistency of $\mathbf{M} + \text{Con}\ (\mathbf{M})$, i.e.

$$\vdash \text{Con}\ \ (\mathbf{M} + \text{Con}\ \ (\mathbf{M})) \rightarrow -\text{Pr}\ \ (\text{G})\,.$$

So, what we can prove, the machine can also prove.

My own account of dualists' argument is as follows.

Dualists argue that no machine **M** can be identical to a human mathematician **H**, in the following way. Let M_p be the set of arithmetical sentences provable by **M** and, similarly, H_k is to be the set of arithmetical sentences knowable by **H**. By knowledge we mean true, justified belief (this definition has its problems but they are not relevant here). Then $\text{M}_\text{p} \subseteq \text{H}_\text{k}$ or $\text{M}_\text{p} \nsubseteq \text{H}_\text{k}$. In the second case $\text{M}_\text{p} \neq \text{H}_\text{k}$ and therefore **M** $\neq$ **H**. In the first case whatever is provable by **M** is knowable by **H** and that means that all sentences in M_p are true. Therefore **H** knows that **M** is a correct system. But then **H** knows that it is a consistent system, i.e., **H** knows that Con (**M**). So Con (**M**) $\in H_k$. On the other hand, by the second incompleteness theorem, Con (**M**) $\notin M_p$. It follows that $M_p \neq H_k$ and therefore **M** $\neq$ **H**. Hence, **M** $\neq$ **H** in every case.

But the above conclusion "Therefore **H** knows that **M** is a correct system." is not justified. From the fact that every sentence provable by **M** is knowable by **H**, it does not follow that **H** knows that, because truth does not entail knowledge. It is possible that $M_p \subseteq H_k$ and that **H** does not know that.

In some specific cases, we may know just enough to conclude that **M** is a correct system. On the other hand, it remains possible that there may exist, and even be empirically discoverable, mathematical machines which in fact are equivalent to our mathematical intuitions. For example, we could be such machines.

So, dualists like Lucas (1961), Penrose (1994, pp. 189 and 641), or Penrose (1999, pp. 107–108) confused the incorrect argument (1) with the correct argument (2):

1. *There is no machine which could capture all our mathematical intuitions.*
2. *There is no machine which could capture all our mathematical intuitions and which we could understand well enough to know that its Gödel's sentence is true.*

We may conclude. As far as Gödel's incompleteness theorem is concerned, we could well be machines. But if we are, then we are definitely not capable of the complete knowledge of the machines, i.e., of the complete knowledge of ourselves. It is very close to Gödel's understanding of the problem; cf. above.

Appendix: Glossary of Notation and Definitions

Sentence. An expression which according to its meaning is either true or false.

Theory. A formal system with axioms and deduction rules which are decidable (it means it is always decidable whether a finite sequence of sentences is a proof).

Provability symbol. $\vdash_T S$ means that sentence S is provable in theory T. $\nvdash_T S$ means that sentence S is not provable in theory T.

Correct theory. A theory is correct if it proofs only true sentences.

Complete theory. A theory is complete if it proofs all true sentences.

Tarski-Gödel first incompleteness theorem. If formal mathematical theory T includes an appropriate amount of arithmetic, it is either incorrect or incomplete, i.e., if it is correct, then it is incomplete. Even more, there is an explicitly definable sentence G which is true but not provable in the theory, i.e., $\nvdash_T G$.

ω-consistency. A theory T is ω-consistent if whenever $\vdash_T -S(0)$, $\vdash_T -S(1)$, $\vdash_T -S(2)$, ... then $\nvdash_T \exists (x)S(x)$.

Gödel numbering. Gödel numbering of the expressions of a theory T is 1–1 correspondence between these expressions and natural numbers. If $n \in \mathbb{N}$ corresponds to expression E, we say that n is the numerical code of E.

Predicate Prv_T. Predicate Prv_T is a predicate defined in theory T which within T itself represents the meta-concept of provability in T. So, the meaning of Prv_T (x, y) is "the finite sequence of sentences with code y is a T-proof of the sentence with code x."

Predicate Pr_T. Predicate Pr_T (x) is defined as $\exists$ y Prv_T (x, y). The meaning of Pr_T (x) is "the sentence with code x is provable in T."

Diagonal lemma. If formal mathematical theory T includes an appropriate amount of arithmetic, there is a sentence G in T such that $\vdash \mathrm{G} \iff -\mathrm{Pr}_\mathrm{T}\ (\mathrm{G})$.

Gödel first incompleteness theorem. If formal mathematical theory T includes an appropriate amount of arithmetic, there is an explicitly definable sentence G which asserts its own unprovability, i.e., $\vdash \mathrm{G} \iff -\mathrm{Pr}_\mathrm{T}\ (\mathrm{G})$, and is such that (i) if T is consistent, then $\nvdash_\mathrm{T} \mathrm{G}$ and (ii) if T is ω-consistent then $\nvdash_\mathrm{T} -\mathrm{G}$.

Gödel's second incompleteness theorem. A formal mathematical theory T which includes an appropriate amount of arithmetic cannot prove its own consistency, i.e., $\nvdash$ Con (T). The sentence Con (T) is defined as $-\mathrm{Pr}_\mathrm{T}\ (\bot)$, where $\bot$ is a contradiction.

References

Aristotel (1931). *The Works of Aristotle, Vol III, De Anima,* Ross, W.D. (Ed.), Smith, J. A. (Transl.), Oxford, Clarendon Press.

Boole, G. (1847). *The Mathematical Analysis of Logic, Being an Essay towards a Calculus of Deductive Reasoning*, Macmillan, Barclay, & Macmillan, Cambridge.

Boolos, G. (1990). On seeing the truth of the Gödel sentence, *Behavioral and Brain Sciences* 13, 655-656.

Boolos, G. S. Burgess, J. P. Jefrey, R. C. (2002). *Computability and Logic*, 4th Ed, Cambridge University Press.

Breitenfeld, T. Jurasic, M.J. Breitenfeld, D. (2014). Hippocrates: the forefather of neurology. *Neurol Sci.* 35(9), 1349-52.

Descartes, R. (1984–1991). *The Philosophical Writings of Descartes*, 3 vols, Cottingham, J. Stoothoff, R. Murdoch, D. (Transl.), Cambridge University Press.

Feferman, S. (1995). Penrose's Gödelian Argument – A Review of Shadows of the Mind by Roger Penrose, *PSYCHE: An Interdiscipl. Journal of Research on Consciousness* 2, 21–32.

Frege, G. (1893–1903). *Grundgesetze der Arithmetik,* Verlag von Hermann Pohle, Jena.

Gödel, K. (1931). Über formal unentscheidbäre Sätze I. *Monatshefte für Mathematik und Physik*, 38, 173-198.

Gödel, K. (1995). Gibbs Lecture (1951), in *Collected Works, Vol. 3: Unpublished Essays and Lectures*, ed. S. Feferman, Oxford University Press.

Good, I. J. (1969). *Gödel's theorem is a red herring,* The British Journal for the Philosophy of Science 18, 359–73.

Hamilton, W.R. (1837). Theory of Conjugate Functions, or Algebraic Couples; with a Preliminary and Elementary Essay on Algebra as the Science of Pure Time, *Trans. Royal Irish Academy*, vol. 17, 293–422, 1837.

Helmholtz, H. (2000). *Helmholtz's Treatise on Physiological Optics,* Wade, N. (Ed.), Southall, J. P. C. (Transl.), Thoemmes Continuum.
Lewis, D. (1989). Lucas against mechanism II. *Canadian Journal of Philosophy* 9, 373–6.
Lucas, J.R. (1961). Minds, machines and Gödel. *Philosophy*, 36, 112-137.
Maas, A. J. (1911). *Resurrection*, in Catholic Encyclopaedia, Robert Appleton Company,
McCall, S. (1999). Can a Turing machine know that the Gödel sentence is true? *Journal of Philosophy* 96, 525-532.
Morgan, de A. (1849). *Trigonometry and Double Algebra*, London.
Myers, R. E. Sperry, R. W. (1958). Interhemispheric communication through the corpus callosum: mnemonic carry-over between the hemispheres, *AMA Arch. Neurol. Psychiatry* 80(3), 298–303.
Peacock, G. (1830). *Treatise on Algebra*, Cambridge.
Penrose, R. (1999). *The Emperor's New Mind.* Oxford University Press.
Penrose, R. (1994). *Shadows of the Mind: A search for the missing science of consciousness*, Oxford University Press.
Pierce, S. C. (1880). On the algebra of logic, *American Journal of Mathematics*, 3, 15–57.
Plato (1966). *Plato in Twelve Volumes, Vol. 1, Phaedo,* Fowler, H. N. (Transl.), London, William Heinemann Ltd.
Schröder, E. (1890). *Vorlesungen über die Algebra der Logik (Exakte Logik).* Bd. 1. Leipzig, 1890.
Šikić, Z. (2005). Gödel's Incompleteness and Man-Machine Non-Equivalence, *Grazer Mathematische Berichte*, 304, 75-78.

Vetoing: Social, Logical and Mathematical Aspects

Branislav Boričić and Marija Srećković

1 Introduction

An alternative title of this chapter could be 'On Vetoer in Language, Logic and Mathematics' because this text can be considered as an example comparing definitions of a veto player in a natural language, i.e. in real world, in first-order logic and, generally, in mathematics. It is intuitively quite simple to understand how the vetoer can be defined by means of a natural language as a person with absolute negative power, i.e. a person who can impede any social initiative. Here we show how in first-order logic the existence of a vetoer is acknowledged by a first-order formula which is inconsistent with some well-known axioms of traditional Social Choice Theory. In addition, we present how the exact power of a vetoer can be mathematically calculated and transparently defined by means of a weighted voting system.

The subject of voting systems makes an important part of both Decision Theory and Social Choice Theory, and it is relevant to politics, economics, business, law and mathematics. At the same time, voting systems present the essence of functioning of democratic institutions and procedures, legislatures, corporations etc. In this chapter we deal with the position of veto-agents in decision-making systems with a pure ordinal approach like Arrow–Sen type Social Choice Theory focusing on the issue of consistency. We also consider some abstract viewpoints on the presence of vetoing in weighted voting systems and possibilities of its quantification and measuring. We consider these two aspects, vetoer theoretical treatment and its modeling, as two parts of totality presenting a triangle 'theory–model–reality', which will be discussed later.

B. Boričić (✉) · M. Srećković
Faculty of Economics and Business, University of Belgrade, Belgrade, Serbia
e-mail: boricic@ekof.bg.ac.rs; marija.sreckovic@ekof.bg.ac.rs

© The Author(s), under exclusive license to Springer Nature Switzerland AG 2023
M. N. Hounkonnou et al. (eds.), *Mathematics for Social Sciences and Arts*, Mathematics in Mind, https://doi.org/10.1007/978-3-031-37792-1_6

The central teacher's problem, during the process of transferring knowledge to his students, is how to simplify and present complex and difficult concepts and ideas and simultaneously preserve their essence. Through a very common example, the notion of vetoing in real social life, we demonstrate the logical and mathematical complexity of this phenomenon in the course of its theoretical treatment. Logical complexity of vetoing can be measured, in an objective way, by complexity of language used in analysis, while mathematical complexity is evident from the difficulties that appear in formulating and proving basic statements regarding the vetoing concept. This approach could be considered a good example of an immediate application of modern mathematical and logical techniques in social sciences.

After this introductory note, we proceed with reviewing the impossibility phenomenon through the history of mathematics, preparing the reader for the more difficult part dealing with impossibility in social choice theory, followed by a general discussion on interdependence between practice and theory, foundations and applications.

The elements of the traditional Arrow–Sen approach to social choice theory are given in the fourth part, with the aim to better understand the needs for its simplified presentation for educational purposes.

The fifth part is devoted to the methodological background for simplification of traditional Social Choice Theory axioms. Although the vetoing has quite a simple definition in a natural language, its theoretical processing causes significant complications. Logical analysis of vetoing contributes to better understanding of vetoer conditions in the context of other known social choice theory axioms and their influence on the consistency and deductive interdependence between axioms. We justify the process of simplifying and substituting a traditional axiom A of a higher order language by a new one, say A' (A-prime), expressed in the pure first-order language. As 'prime'-axioms A' will be constructed so that A logically implies A', we obtain a simpler context in which impossibility results can be considered in an easy but appropriate way.

In the sixth part of this chapter, we analyze the relationships between dictatorship, Pareto rule and vetoer condition, as axioms of the traditional Arrow–Sen theory, and give some new approachable and simple examples of impossibility theorems. We express dictatorship, vetoer and Pareto rule, and their negations, explicitly as first-order axioms. Their descriptive presentations, as usually given in papers and textbooks (see Arrow 1963; Chichilnisky 1982; Fishburn 1973; Kelly 1978; Mas–Colell and Sonnenschein 1972 etc.) as a combination of a natural language with elements of mathematical formalism, give possibilities for various interpretations (and even misunderstanding), but their pure logical formulations (see Murakami 1968; Routhley 1979; Boričić 2009, 2014a,b, 2023) enable us to work with them more formally.

The seventh part of this chapter deals with the status of a veto player in a cardinal context enabling to measure the influences of particular players, as suggested in Taylor (1995). Bearing in mind that there is a strong intuition regarding deep similarities between 'one agent—one vote' systems, weighted voting systems and systems that include agents with the veto power, we prove that these systems are mutually equivalent and, as a particular result, we propose a method to calculate exactly the weights of agents with the veto power as a function of the number of ordinary players, veto players, their weights and quotas. This mathematical analysis of vetoing gives possibilities for pure quantification of the veto power.

Finally, we discus some general questions regarding relationships between reality, model and theory, referring to examples given in this chapter, as well as the influences of social sciences and humanities on development of contemporary mathematics.

Both the role of the vetoer condition in the context of Arrow–Sen social choice theory, as an ordinal concept, and an exact measuring of the veto power in a weighted voting system, corresponding to the cardinal approach, are discussed. We present some impossibility theorems, such as impossibility of a dictatorial non-vetoer and impossibility of a Paretian non-vetoer as new determinants of a veto player in the presence of (non-)dictatorship and (non-)Pareto rule in the ordinal theory. On the other side, a certain kind of representation theorem, in the context of cardinal theory, is proposed: for each weighted voting system, containing some agents with the veto power, there exists an equivalent weighted voting system with no agents with the transparent formal veto power.

We believe that these explanations enable one to better understand the position of veto-agents in decision-making and voting systems.

It is supposed that the reader is familiar with the basic notions of social choice theory (see Arrow 1963; Fishburn 1973; Kelly 1978; Murakami 1968; Schofield 1985; Sen 1969, 1970b; or Taylor 1995), related to the first part of this chapter, and with concepts given in Blau and Deb (1977), Fishburn (1973), Kang (2010), Moulin (1981), Moulin (1982), Taylor (1995) or Winter (1996), regarding the second part dealing with the weighted voting systems, although we include all basic definitions and explanations of notation in the appropriate places of this chapter.

2 Impossibility through History

Impossibility phenomenon essentially matches with inconsistency. Namely, there is no model or fragment of reality corresponding to an inconsistent theory, so a non-existing situation cannot be described by a consistent set of statements.

We give a sketch of the inconsistency notion by means of deduction relation $\vdash$, expressing the fact that a statement Y can be derived from the statement X, denoted by $X \vdash Y$. In traditional logic, if we are able to infer an empty conclusion from two hypotheses X and Y, denoted by $X, Y \vdash$, then we can state that the set $\{X, Y\}$ of statements X and Y, i.e. that the theory based on axioms X and Y, is inconsistent; otherwise, it will be consistent.

In particular, the fact $X \vdash Y$ is logically equivalent to $X, \neg Y \vdash$, meaning that the set $\{X, \neg Y\}$ presents an inconsistent entity, where $\neg Y$ denotes logical negation of Y. By focusing on $X \vdash Y$, we have a way to take into consideration explicitly just positive forms of statements.

This notation with deduction relation will be used in the sequel of this chapter.

Therefore, when we establish that some set of statements is inconsistent, i.e. that the coexistence of statements under consideration is impossible, then we can conclude that there is no model or part of reality which can be described by those statements. Here we will bring to mind some of the famous impossibility results through the history of mathematics (see Artemidiadis 2004; Boyer 1968; Boričić 2007, 2018).

Incommensurability of the diagonal and a side of a square was the first deep and non-trivial example of impossibility in the history of science. Pythagoreans believed that X: 'everything can be expressed by fractions of two integers', but they discovered that Y: 'the diagonal of a square with a side 1 is not a fraction of two integers'. Consequently, $X, Y \vdash$, i.e. a theory based on X and Y is inconsistent, and it is impossible to express the length of the diagonal of a square as a rational function of the length of its side. This was the first known use of an indirect method of argumentation, *reductio ad absurdum* method. This antinomy was resolved by introducing irrational numbers.

The next three famous unsolvable problems also arrive from Ancient Greece: Squaring the Circle, Trisecting the Angle and Doubling the Cube. Impossibility of both trisecting the angle and the doubling the cube was proved in 1837 by Wantzel (P. L. Wantzel (1814–1848)). An immediate consequence of Lindemann's (C. L. F. Lindemann (1852–1939)) proof that π is transcendental, in 1882, is the impossibility of squaring the circle.

Abel–Galois–Ruffini (N. H. Abel (1802–1829), E. Galois (1811–1832) and P. Ruffini (1765–1822)) Impossibility Theorem, stating that there is no algebraic solution to the general polynomial equations of degree five or higher, preceded the finding of the final status of the three ancient problems mentioned above.

Integral calculus enriched our experience with impossibilities by nice new examples. For instance, Liouville (J. Liouville (1809–1882)) proved that the indefinite integral of any real function of one variable cannot always be expressed as an elementary function. Some of his examples were the famous Fresnel (A.–J. Fresnel (1788–1827)) integral $\int \sin x^2 dx$ and $\int \cos x^2 dx$ and Gauss (C. F. Gauss (1777–1855)) integral $\int e^{x^2} dx$.

The Gödel incompleteness theorems (1931) (K. Gödel (1906–1978)) were among the most important and influential discoveries in the fields of logic, methodology, physics, art and philosophy of science of modern epoch during the first half of the twentieth century (see Boričić 2011; de Swart 2018; van Dalen 1980; Hofstadter 1979, or Takeuti 1975). The first one states that it is not possible to prove all true informal number theory statements in any reasonable axiomatization of number theory, while the second theorem can be roughly formulated as follows: if a theory is consistent, then the sentence asserting its consistence is not provable in this theory.

Finally, the Arrow–Sen theory, with impossibilities of social choice, will be the central point of this chapter. Let us note here that the first known valuable impossibility example in social choice theory was the so-called Condorcet paradox dealing with a group, consisting of three individuals a, b and c, with equal rights, having to make a choice between the three alternatives: X, Y and Z. Suppose that individual a prefers alternative X to Y, and alternative Y to Z, meaning that, by transitivity, a also prefers X to Z, denoted as follows: $X >_a Y >_a Z$. If we suppose, similarly, for individuals b and c that $Y >_b Z >_b X$ and $Z >_c X >_c Y$, then, it is clear that it is impossible to define any decision on behalf of the group $\{a, b, c\}$ over the alternatives X, Y and Z because of the cyclic structure of social preference generated by given individual preferences. A. N. C. de Condorcet was a French sociologist and mathematician of eighteenth century.

3 Social Decision: Practice and Theory

Each practice needs grounding on some theoretic basis, and each theory has to satisfy some elementary methodological expectations, such as a specific (formal) language, axiomatics, consistency and relationship with reality and practice. The foundation of any scientific theory requires explicit answers regarding the existence (or non-existence) of the basic notions closely connected with consistency as well. Brilliant examples of such type of results in the field of social decision topics are non-existence (and existence) of social welfare functions in Social Choice Theory (see Arrow 1963; Arrow and Sen 2002; Arrow et al. 2010, and Sen 1969) and existence (and non-existence) of utility functions in Utility Theory (see Debreu 1959, and Skala 1975).

The basic problem of Social Choice Theory is how to transform a finite number of diverse individual 'preference lists' of members of a society into a unique integrated social 'preference list', respecting some ethical and democratic principles. But the previous question is always the following one: is (and when is) this transformation possible? Or, as Sen (1970b) posed the question, do procedures for social decision-making exist that reasonably respect individual values and preferences? He also pointed out that 'there are several deep–seated difficulties' regarding the problem of rationality in social decisions.

In order to give a rough picturesque abstract description of social welfare or social decision function, we introduce a notion of a quasi-matrix as a scheme having a usual matrix form but with some empty places for missing elements fulfilled by a symbol '–'. A scheme like the following one

$$\begin{pmatrix} a & y & z & - \\ x & z & - & \{y, z\} \\ - & b & x & - \end{pmatrix}$$

presents an example of 4×3 quasi-matrix. Obviously, each matrix is a quasi-matrix, but not vice versa. The intended meaning of each column of a quasi-matrix A, in this context, will be the order of alternatives defined as preferences from above to below, where the elements of the form $\{x, y\}$ express the case of indifference between the alternatives x and y. Let A be an $m \times n$ quasi-matrix containing preferences of n individuals, through n columns, over m alternatives. The central point is to define a transformation of A into an $m \times 1$ quasi-matrix B, presenting the social preference, so that B depends on A, satisfying some additional ethical and rational conditions. This definition shows how difficult this problem is from the combinatorial point of view. On the other side, if we have, for instance, a 3×3 matrix A, expressing 3 individual preferences, each consisting of just 3 alternatives, given through their cyclic permutations

$$A = \begin{pmatrix} x & y & z \\ y & z & x \\ z & x & y \end{pmatrix}$$

the problem is to generate a column matrix B consisting of these 3 alternatives expressing the resulting social preference. A possible result is

$$B = \begin{pmatrix} \{x, y, z\} \\ - \\ - \end{pmatrix}$$

but is this satisfiable? This is Condocet paradox considered above.

A possible explanation of differences between ordinal and cardinal concepts of preferences anyhow has to characterize the ordinal preferences as pure abstract orderings (or, even, pre-orderings) without any explicit quantification, while the cardinal one must be connected with some numeric (real) values. The ranking of alternatives in a quasi-matrix column, from top to bottom, enabling just to conclude if 'x is at least as good as y', would be typical for an ordinal treatment of alternatives, while any ranking including some numerical (or even descriptive) evaluation, giving the possibility to calculate, e.g. the difference $|u(x) - u(y)|$ between the utilities $u(x)$ and $u(y)$ of alternatives x and y, is specific for a cardinal interpretation. The ordinal concept is typical of Arrow–Sen approach, but the cardinal one appears in classical works related to utility theory.

We think it is natural that individual preferences have ordinal (but not cardinal) character, due to the fact that individuals usually have vague opinions about most alternatives. For instance, in political life individuals are commonly able to rank just a few parties, and most of them leave unranked, so that the individual preferences are even incomplete (non-linear), i.e. there are alternatives which some individuals cannot rank. On the other side, in practice, social preference resulting from individual one's is quite natural to be complete (linear) and even numerically ranked. For instance, parliamentary elections generate a linear social numerical ranking from preferences presenting non-linear non-numerical descriptive individual rankings. Namely, voters participating at political elections usually do not have any attitude about all parties or political groups taking parts in the campaign, but only about a few of them and those attitudes are just on a rough descriptive level, as 'good', 'better than' or 'bad'. In contrast to the vague and incomplete individual preferences, the results of elections are commonly expressed as complete numerical ranking of all participants of elections through the percentage of votes or final number of seats in parliament.

A possible conclusion that can be drawn is that, in practice, individuals do not satisfy Arrow's rational choice axioms (e.g. linearity), but groups and societies do.

Another widespread field of application of formally founded fragments of Social Choice Theory is artificial intelligence, which includes as its essential part the automated reasoning. Namely, the formalism of axiomatic method, particularly this one based on simple axioms over the propositional language, opens a perspective for the development of mechanical reasoning deductive systems. In more detail, multiple von Wright preference logic enriched by varying axioms and initial views of individuals gives a framework to form a knowledge database which can be used as supporting tool for effective decision-making, simulation and prediction of groups' or individuals' behavior. This logic can be considered a dynamic system of algorithmic steps, defined by formal logical principles and rules, which is not only sensitive to changes in the knowledge base caused by changes in individual opinions but is also taking into account of possible and impossible combinations of axioms.

4 Social Choice: Traditional Theory

Here we follow the spirit of our presentations given in Boričić (2007, 2009). Traditional Arrow–Sen Social Choice Theory can be considered an extension of classical first-order logic. The extension is reflected in enriching the language and adding some new special axioms related to a social choice. The language contains two finite lists of symbols: for binary relations, $P, P_1, \ldots, P_n$, and for alternatives, $x, y, z, \ldots$. It is supposed that each binary relation P or P_i is asymmetric, $\forall x, y(xPy \to \neg yPx)$, i.e. $\forall x, y(xP_iy \to \neg yP_ix)$, and transitive, $\forall x, y, z \in X(xPy \wedge yPz \to xPz)$, i.e. $\forall x, y, z \in X(xP_iy \wedge yP_iz \to xP_iz)$, over all alternatives $x, y, z, \ldots$. These are so-called rational choice axioms. In this case, binary relations under consideration will be called the strict preference relations. $P_1, \ldots, P_n$ are individual, and P is a social preference relation. Therefore, social choice theory presets a formal description of relationships between individual and social preference relations over a given set of alternatives. The central problem is how to define a procedure assigning an adequate social preference P to any profile $\mathcal{P} = (P_1, \ldots, P_n)$ of individual preferences. More accurately, this procedure deals with social choices over a finite set of alternatives, respecting preferences of a finite set of individuals.

Each strict preference relation P, social or individual, with or without subscripts, generates an indifference relation I, as follows: xIy iff $\neg xPy \wedge \neg yPx$, and a weak preference relation R, by definition: xRy iff $xPy \vee xIy$. Here we use symbols for universal, $\forall$, and existential, $\exists$, quantifiers, as well as, propositional connectives for negation, $\neg$, conjunction, $\wedge$, disjunction, $\vee$, implication, $\rightarrow$, and equivalence, $\leftrightarrow$, with the usual meaning they have in classical logic.

As we can see, a wider context for the Arrow–Sen theory is the classical set theory, more descriptive than the formal one.

Weak preference relation R enables us to define a choice set $C(Y, R) = \{x | x \in Y \wedge (\forall y \in Y) x R y\}$, presenting the set of the best alternatives, with respect to R and $Y \subseteq X$, while $C(Y, R)$ will present a choice function, over the set of all alternatives, if $C(Y, R)$ is non-empty for every non-empty $Y \subseteq X$. According to Sen (1969) a rule is defined as a functional relation specifying one and only one social binary relation R for each profile of individual ordering $(R_1, \ldots, R_n)$, with one R_i for each i $(1 \leq i \leq n)$. A social welfare function is defined as a rule the range of which is restricted to the set of orderings. A social decision function is defined as a rule ranged over relations R generating a choice function $C(Y, R)$ over entire X.

Arrow introduced the following four conditions—axioms—as the fundament of social choice theory.

Axiom **U** of 'the unrestricted domain' requires that the choice function can be applied to **any profile**, $\forall \mathcal{P}$, of logically possible profiles of individual preferences.

Axiom **IIA** of 'the independence of irrelevant alternatives' ensures that for any binary relations R and R' generated, respectively, by **any two profiles**, n–tuples of individual preferences $(R_1, \ldots, R_n)$ and $(R'_1, \ldots, R'_n)$, and for all pairs of alternatives $(x, y) \in Y^2$, where Y is any subset of the set X of all possible alternatives, if $(\forall i \in V)(x R_i y \leftrightarrow x R'_i y)$, then $C(Y, R) = C(Y, R')$, where V denotes the set of all individuals.

Non-dictatorship axiom **TND**, **T** for 'traditional' and **ND** for 'non- dictatorship ', states that there is no person $i \in V$, dictator, having such power that, for **all profiles** and each of the two alternatives x and y, if i prefers x to y, the society must prefer x to y as well.

The Pareto property **TP** claims that, for **all profiles**, if every individual $i \in V$ prefers x to y, then the society must prefer x to y. This is, in fact, a weak version of the Pareto principle, as introduced by Arrow (see Arrow 1963; Sen 1969, 1970b).

In denotation of some axioms, we use the prefix **T** pointing out that this is a 'traditional' form of axiom, e.g. **TND** and **TP**. It is necessary because we will exploit the differences between 'traditional' and 'simplified' form of axioms.

Here we get an appropriate context in which Arrow's Impossibility Theorem, or as it is originally called 'General Possibility Theorem', can be formulated correctly, as given in Arrow (1963) or Sen (1969, 1970b).

Arrow's Impossibility Theorem *There is no social welfare function satisfying axioms* **U**, **IIA**, **TND** *and* **TP**.

We will also present here Sen's famous result (see Sen 1970a,b) known as the 'impossibility of Paretian liberal' or the 'liberal paradox'. Namely, Sen has taken into consideration, following the spirit of J. S. Mill's liberalism comprehension, 'the liberalism axiom' **TL** (see Sen 1970a,b; Boričić 2009, 2014b) by which, for **all profiles** and each individual $i \in V$, there is at least one pair of alternatives $(x, y) \in X^2$ such that $x \neq y \wedge (x P_i y \rightarrow x P y) \wedge (y P_i x \rightarrow y P x)$, which means that every member of society has absolute power to decide on at least one pair of alternatives.

Sen's theorem can be formulated in the following way:

Sen's Impossibility Theorem *There is no social decision function satisfying axioms* **U**, **TL** *and* **TP**.

Let us emphasize that Sen (1969) makes a subtle difference between a social welfare function and a social decision function. Namely, a social welfare function is ranged over the set of orderings, while a social decision function is ranged only over those binary relations R each of which generates a choice function $C(X, R)$ over entire Y. But such details will not be essential for our approach given in the sequel of this chapter.

We also note that Sen's theorem does not refer to **IIA**, 'the independence of irrelevant alternatives', and this fact presents an essential simplification of this fragment of Social Choice Theory. On the other side, many authors (see Routhley 1979), including Sen (1970b), deal with the controversial role of **IIA** in the proof of Arrow's theorem, since it is never even mentioned in the original proof, although 'the independence of irrelevant alternatives' is a principle of a high logical complexity.

Sen's Impossibility Theorem deals with the fragment of theory in which non-dictatorship axiom **TND** is replaced by the liberalism axiom **TL**. Non-dictatorship is a logical negation of a positive sentence, but the liberalism axiom is by itself a positive sentence. Moreover, non-dictatorship is problematic in cases of a majority decision when there is an individual who always votes with the majority, and this requires additional philosophical arguments to accept non-dictatorship in this form. All of these are the additional reasons why Sen's treatment can be considered more approachable than Arrow's one.

A logical analysis of Arrow–Sen axioms (see Routhley 1979) shows that they can be divided naturally, on the basis of their complexity, into two groups: {**U**, **IIA**} and {**TND**, **TP**, **TL**}. The first two axioms, **U** and **IIA**, have a deeply metalogical character and we will use them as general properties of our system. On the other side, the statements, such as **TND**, **TP**, **TL** and their negations, present formally simpler and mutually similar structures enabling an easier logical analysis of their deductive interdependencies.

In the sequel of this chapter, we will present some new simpler examples of impossibilities of social choice theory more approachable to a wider circle of readers.

5 A Method of Axioms Simplification

In many biographical notes about A. K. Sen, one can find that "his work simultaneously embraces social choice theory and economic development, thus breaking the barrier between mathematized 'high theory' and 'real world' economics" (see

Boričić 2004b). In the same spirit we consider Sen's approach to formalism of Social Choice Theory. Namely, in his motivation for 'impossibility of a Paretian liberal', a twin theorem for Arrow's 'general (im)possibility theorem', we see an easier approach for a wider circle of readers to the essence of impossibilities in axiomatic theory of social choice. With Sen's methodological simplifications and necessary philosophical explanations, the theory of social choice, together with a very prominent corpus of impossibility results, became an attractive and propulsive scientific discipline, as opposed to the heightened pessimism that followed the publication of the first impossibility result of social choice. Here we have an almost identical development of events as in case of Goedel's incompleteness theorems when the power of the axiomatic method was 'questioned'.

Our paper on preference logic on rough and fuzzy sets (see Boričić and Konjikušić 2004a) was one of our attempts to simplify some fragments of Social Choice Theory with the aim to give more approachable facts to students.

The need for a deeper examination and discussion of the basic mathematical and logical assumptions of Social Choice Theory is visible particularly in Fishburn's conditions classification as well (see Fishburn 1973). In this classification, conditions for a social choice function were divided into structural, existential and universal.

We treat 'the unrestricted domain' and 'the independence of irrelevant alternatives' as conditions, belonging to a higher order formal language, necessary defined and moved in metalanguage.

The aim of this text is to demonstrate how to simplify some logical axioms or mathematical conditions, expressed in a pure formal language, without losing the core of the spirit of a theory. Roughly, we can speak about two approaches to impossibility in social choice: one, say set-theoretic, based on a social choice function and assaying its existence, and another one, say pure logical, concerning with the consistency of basic axioms. Here we nurture the use of logical analysis believing that this is easier and more natural for students who are not mathematicians. Besides, we see Sen's approach more as logical than as set-theoretic one.

Subtle differences between various forms of social (welfare, choice, decision etc.) functions, technically necessary, but causing difficulties for students in understanding the general concept, may be minimized by an emphasized logical approach to the subject. From a logical view point, Social Choice Theory can be considered an analysis of deductive interdependence between various (groups of) axioms appearing in a social choice context. This deductive analysis covers at least the two following phenomena: impossibility and complexity. An interesting aspect of each theory is connected with its (in)consistency ((im)possibility), and we know for many such examples in social choice (see Boričić 2009; Fishburn 1973; Kelly 1978; Sen 1969, 1970a). On the other side, the complexity of a theory defines the limits in understanding and applicability of this theory. Our experience says that teaching elements of Social Choice Theory to students who are not mathematicians is a great challenge because of their complexity. Namely, the complexity of Social Choice Theory presents a difficult barrier for reasonable teaching of this subject. We try to simplify some things, but every procedure of simplification brings the danger of banalization.

In one of the first notable logical analyses of Arrow's original proof of impossibility theorem, Routhley (1979) notes that 'what has to be shown is of the form $\forall pA(p) \rightarrow \forall pB(p)$, not—as the standard "proofs" assume—of the stronger erroneous form $A(p) \rightarrow B(p)$. It is enough to show, however, how to repair standard proofs.' He also emphasizes that 'the textbooks have failed to produce what it is essential to have, especially in the case of a theorem with such far–reaching consequences (even if it is after all only an exercise in second-order quantificational logic), namely a correct and rigorous proof.' While Routhley's 'repairing proofs' focuses on the slight but essential differences between two logical forms

$$\forall x A(x) \rightarrow \forall x B(x) \text{ and } A(x) \rightarrow B(x)$$

and their roles in presenting the proof of Arrow's general impossibility theorem, we deal with the problem of defining some fragments of Social Choice Theory which are not too complex and which can be obtained by substituting some traditional axioms of the form

$$\exists x \forall y A$$

by simpler ones

$$\exists x A$$

relying on the general logical fact that

$$\exists x \forall y A \rightarrow \forall y \exists x A$$

and then moving the universal quantification $\forall y$ to some kind of metatheoretical level. This operation can be of great importance when the object 'y' belongs essentially to the higher order language. By this procedure we can obtain a similar but simpler fragment of the theory which could be more approachable than the original one.

The treatment of this higher order universal quantification over sentences is quite similar to the usual defining propositional calculi, where instead of $\forall A \forall B(A \wedge B \rightarrow A)$ we write just $A \wedge B \rightarrow A$ pointing out that this axiom has a schematic character.

In case of simple axioms of Social Choice Theory, such as, for instance, Pareto rule, dictatorship, vetoer and liberalism, there are two typical quantifier prefixes in traditional approach:

1. There exists an individual i, such that **for all profiles** $\mathcal{P}$ and all alternatives x and y, A.
2. **For all profiles** $\mathcal{P}$ and all alternatives x and y, A,

which, respectively, can be expressed symbolically as

$$\exists i \forall \mathcal{P} \forall x \forall y A \text{ and } \forall \mathcal{P} \forall x \forall y A$$

and we propose to substitute the first form by its consequence

$$\forall \mathcal{P} \exists i \forall x \forall y A$$

in order to get the universal quantification over profiles as a prefix, which does not belong to the first-order language. In this way, we obtain a uniform form for all these axioms, each of them prefixed by the universal quantification over profiles. In the next iteration, we move this quantification over profiles onto the level of metatheory, and then we work with fragments or, say, sub-theories based on simple first-order axioms.

In short, roughly, from '$\exists i \forall \mathcal{P} \forall x \forall y A$' we proceed to '$\forall \mathcal{P} \exists i \forall x \forall y A$' and finally to 'we suppose that, for all profiles $\mathcal{P}$, we have an axiom: $\exists i \forall x \forall y A$'.

This is a way to define an essential simplification of some parts of Social Choice Theory, but we still hope that this simplification has preserved the basic spirit of the traditional theory. Moreover, we believe that this approach is logically quite justified and that the corpus of impossibility results can be represented correctly in this way, due to the obvious fact that if a subtheory is inconsistent, then, *a fortiori*, each its extension will be inconsistent as well.

It might seem that in this way we have desecrated the authentic Arrow–Sen tradition, but we believe that (1) we preserved the spirit of a formal logical treatment of descriptive social choice conditions, (2) we reduced the complexity degree of conditions under consideration, and (3) we enabled simplified argumentations for some social choice impossibilities.

More accurately, instead of a traditional vetoer condition

$$\mathbf{TV}: \quad (\exists i \in V)(\forall \mathcal{P})(\forall x, y \in X)(x \neq y \wedge (x P_i y \rightarrow \neg y P x))$$

and traditional Pareto rule

$$\mathbf{TP}: \quad (\forall \mathcal{P})(\forall x, y \in X)((\forall i \in V) x P_i y \rightarrow x P y)$$

we accept their simplified variations

$$\mathbf{SV}: \quad (\exists i \in V)(\forall x, y \in X)(x \neq y \wedge (x P_i y \rightarrow \neg y P x))$$

and

$$\mathbf{SP}: \quad (\forall x, y \in X)((\forall i \in V) x P_i y \rightarrow x P y)$$

S for 'simplified' and **V** and **P** for 'vetoer' and 'Pareto', respectively, supposing that these variations hold for all profiles $\mathcal{P}$, which is in line with the general assumption about the schematic character of axioms. We emphasize that in both cases we have

$$\textbf{TV} \vdash \textbf{SV} \text{ and } \textbf{TP} \vdash \textbf{SP}.$$

A brief explanation of our idea, in general, is that, for a theory $\mathcal{T}$, we define its subtheory, i.e. its fragment, in order to understand it better. Namely, if a part of $\mathcal{T}$ is based on axioms A and B and if we are familiar with some simpler statements A' and B', then, in case when $A \vdash A'$ and $B \vdash B'$, we can consider a subtheory $\mathcal{T}'$ based on axioms A' and B', instead of A and B, respectively. Obviously, $\mathcal{T}' \subseteq \mathcal{T}$ and if $\mathcal{T}'$ is inconsistent, then $\mathcal{T}$ will be inconsistent, but not conversely.

6 Impossibilities with Vetoing, Dictatorship and Pareto Rule

Here we consider the role and the relationships between simplified Arrow's Pareto rule **SP**, dictatorship condition **SD** and Fishburn's vetoer condition **SV** in Arrow–Sen social choice theory (see Fishburn 1973; Kelly 1978; Sen 1970b) and then, in the next section, their counterpart in the context of a weighted voting system.

As was mentioned above, the central problem is how to define a procedure generating a social preference relation, denoted by R, from a finite list, profile, say profile $(R_1, \ldots, R_n)$, of individual preference relations. This procedure deals with social choices over a finite set of alternatives, respecting preferences of finitely many individuals. It is supposed that individuals and society satisfy rational choice axioms, i.e. that each individual preference relation R_i characterizing the behavior of an individual $i \in V$ is linear,

$$(\forall x, y \in X)(x R_i y \vee y R_i x)$$

and transitive,

$$(\forall x, y, z \in X)(x R_i y \wedge y R_i z \rightarrow x R_i z)$$

and that the corresponding preference relation R characterizing the behavior of society is also linear and transitive. Such relations R_i and R are called the weak preference relations, and xRy stands for 'x being regarded as at least as good as y'. Each weak preference relation R defines corresponding strict preference P and indifference I, as follows: xPy iff $xRy \wedge \neg yRx$, and xIy iff $xRy \wedge yRx$. A strict preference P defined in this way will be transitive and asymmetric $(\forall x, y \in X)(xPy \rightarrow \neg yPx)$. In this chapter we will focus dominantly on impossibility statements expressed in terms of strict preferences. Here we use symbols for universal, $\forall$, and existential, $\exists$, quantifiers, as well as, propositional connectives for negation, $\neg$, conjunction, $\wedge$, disjunction, $\vee$, implication, $\rightarrow$, and equivalence, $\leftrightarrow$, with the usual meaning they have in classical logic. Also, we use the turnstile symbol, $\vdash$, for deduction relation in an informal way, $A, B \vdash C$, in order to express that 'C can be derived from A and B', as described earlier.

The Arrow–Sen theory is based on some axioms establishing connections between individual and social preferences.

Let V and X be finite sets of individuals and alternatives, respectively, and P_i and P—the strict individual and social preference relations on X. After moving 'for all profiles' onto the metalevel, as described above, we can define modified Arrow's dictatorship condition **SD**, Pareto rule **SP** and vetoer condition **SV**, similarly as given by Sen and Fishburn (see Sen 1969, 1970b; Fishburn 1973), all in style of Arrow–Sen social choice theory, but purely in first-order language:

SD $(\exists i \in V)(\forall x, y \in X)(x P_i y \rightarrow x P y)$
SP $(\forall x, y \in X)((\forall i \in V) x P_i y \rightarrow x P y)$
SV $(\exists i \in V)(\forall x, y \in X)(x \neq y \wedge (x P_i y \rightarrow \neg y P x))$

Let us note that, for instance, the dictatorship axiom, as given above, apparently has two sorts of variables and quantifiers, over alternatives and over individuals (i.e. individual preferences), but this axiom can be presented alternatively as first-order formula $\bigvee_{i \in V}(\forall x, y \in X)(x P_i y \rightarrow x P y)$, bearing in mind that the set of individuals V is a finite set and similarly for vetoer and Pareto rule.

We also emphasize that the complete presentation of Arrow–Sen theory supposes the presence of two more axioms: 'the unrestricted domain' and 'the independence of irrelevant alternatives'. 'The unrestricted (or universal) domain' allows to work with each finite number of individuals $n \geq 2$ and every possible combination of individual preferences, i.e. for all profiles, while 'the independence of irrelevant alternatives' provides the way how a society decides $x P y$, for a pair of alternatives x and y, should depend on the individual preferences, i.e. all profiles, only over that pair, but not on how the other, 'irrelevant'. Our opinion is that these two conditions have a deeply meta-theoretical character and we will treat them as general (meta-)properties of our system, while, on the other side, the statements, such as **SD**, **SP** and **SV**, present formally simpler and mutually similar structures enabling easier logical analysis of their deductive interdependencies. This impression may be due to the fact that the above-mentioned axioms are evidently expressible in the first-order language, and for these two conditions, 'the unrestricted domain' and 'the independence of irrelevant alternatives', this is not clear at all.

Here we give some new facts completing the puzzle of logical relationships between basic axioms of social choice theory (see Boričić 2009, 2014a,b, 2023), enriching the list of simple examples of impossibility results in social choice theory reformulated in terms of inconsistency, as follows:

Lemma 1 (Fishburn's Note on Impossibility of a Dictatorial Non-vetoer) **SD** $\vdash$ **SV**

Proof Obviously, from **SD**: $x P_i y \rightarrow x P y$, bearing in mind that P is asymmetric $(\forall x, y \in X)(x P y \rightarrow \neg y P x)$, we conclude **SV**: $x P_i y \rightarrow \neg y P x$, i.e. that **SD** $\vdash$ **SV**, but not conversely. □

This Lemma can state that the dictatorship and a non-vetoer make an inconsistent, i.e. impossible, system.

On the other side, if we denote by **SD**(i) and **SV**(i) the variations $(\forall x, y \in X)(xP_iy \rightarrow xPy)$ and $(\forall x, y \in X)(xP_iy \rightarrow \neg yPx)$ of dictatorship axiom **SD** and vetoer axiom **SV**, respectively, although, for each i, **SD**(i) $\vdash$ **SV**(i), the set of axioms {**SD**(i), **SV**(j)}, for $i \neq j$, is inconsistent, due to the fact that from **SD**(i): $(\forall x, y \in X)(xP_iy \rightarrow xPy)$ we can infer xPy, and from **SV**(j): $(\forall x, y \in X)(yP_jx \rightarrow \neg xPy)$ we infer its negation $\neg xPy$.

A similar conclusion will be derived in the sequel of this chapter regarding dictatorship and the presence of the veto power in weighted voting systems in general. This is because the dictatorship condition and the vetoer condition in weighted voting systems are defined in a slightly different way, and their mutual conditioning has to be reconsidered. Namely, we will see that our definition of dictatorship in weighted voting systems is acceptable only in case when the system under consideration does not contain agents with the veto power.

Lemma 2 (Impossibility of a Non-Paretian Dictator) **SD** $\vdash$ **SP**

Proof By following the proof presented in Boričić (2009), from $xP_iy \rightarrow xPy$, with the help of weakening antecedent rule n-times, by which from $p \rightarrow q$ one can always derive $p \wedge r \rightarrow q$, we infer $\bigwedge_{1 \leq j \leq n} xP_jy \rightarrow xPy$, i.e. $(\forall j \in V)xP_jy \rightarrow xPy$. It means that **SD** $\vdash$ **SP**. The same conclusion can be justified by simple formal logical argumentation, by means of the following general well-known equivalence $(\forall xA \rightarrow B) \leftrightarrow \exists x(A \rightarrow B)$, if x is not free in B. □

This Lemma can state that the dictatorship and a non-Pareto rule make an inconsistent, i.e. impossible, system.

Bearing in mind that Arrow's impossibility of a Paretian non-dictator can also be expressed roughly as **SP** $\vdash$ **SD**, we glimpse a logical equivalence between the Pareto rule and the dictatorship. This result was obtained by Chichilnisky (1982) in a topological context.

Let us note here that Mas–Colell and Sonnenschein (1972) introduced the weak dictatorship condition, whose simplified version looks as follows: **SWD**: $(\exists i \in V)(\forall x, y \in X)(xP_iy \rightarrow xRy)$. It is slightly inconvenient because it mixes strict and weak preferences.

Lemma 3 *Conditions* **SV** *and* **SWD** *are logically equivalent.*

Proof If we suppose **SV**, then from $\neg yPx$, bearing in mind that $\neg yPx \leftrightarrow \neg yRx \vee xRy$ and that R is linear, we infer xRy, i.e. that **SV** $\vdash$ **SWD**. Conversely, if we suppose **SWD**, then, bearing in mind that $xRy \leftrightarrow xPy \vee xIy$, from the fact that P is asymmetric and that $xIy \rightarrow \neg yPx$, we conclude **SWD** $\vdash$ **SV**. □

We are also able to give a simple argumentation for Mas–Colell—Sonnenschein's result:

Lemma 4 (Impossibility of a Paretian Non-vetoer) **SP** $\vdash$ **SV**.

Proof This conclusion, **SP** $\vdash$ **SV**, can be inferred directly from Arrow's 'impossibility of a Paretian non-dictator', **SP** $\vdash$ **SD**, and Fishburn's 'impossibility of a dictatorial non-vetoer', **SD** $\vdash$ **SV**, i.e. that each dictator is a vetoer. □

This Lemma can state that the Pareto rule and a non-vetoer axiom make an inconsistent, i.e. impossible, system.

It is interesting that a wider context enriched with Sen's liberalism axiom (see Sen 1970a) defines a more precise position of a vetoer in social choice theory.

7 Measuring the Veto Power in Voting Systems

In this section we deal with the status of a veto player in the context of pure weighted voting systems.

A voting system can be defined as an ordered triple (A, f, q), where A is a finite set of n agents, f—mapping from A into the set of natural numbers assigning to each agent $a \in A$ its weight $w_a \geq 1$, i.e. the number of votes held by an agent a, and a natural number $q \geq 1$—a quota, i.e. the number of votes needed to pass a motion. Let m be the total number of votes of the system (A, f, q), $m = \sum_{a \in A} w_a$. Then we can define some usual properties of the voting system (A, f, q). If $(\exists a \in A) w_a > 1$, then the system (A, f, q) can be characterized as *a (essentially) weighted voting system*; otherwise, when $(\forall a \in A) w_a = 1$, (A, f, q) will be a 'one agent—one vote' system. In case when $(\exists a \in A) w_a > q$, for $2q > m$, the system is called *dictatorial*; otherwise, it is *non-dictatorial*. *A consensus system* can be characterized by the claim that $q = m$. Let us note that in a consensus system each agent can block the decision-making process. Inequalities $2q > m$ and $3q \geq 2m$ present some of the usual general conditions for a quota, known as a simple majority and a two-third majority, respectively. If there is an agent $a \in A$ with an additional power, the veto power, having the possibility to stop any action and block any decision-making process, then such a system is called *a system with the veto*. This is the so-called an absolute veto. In practice, there are various modalities of the veto power. In contrast to the absolute veto, there are the systems with a limited veto. For instance, it can be defined that the two-third votes override the veto power in a legislative process. Obviously, each consensus system can be considered a system with a veto, where each agent has the veto power. Systems with a veto will be of particular interest here. Namely, we will show that each system with a veto can be represented as a pure weighted voting system (without a veto).

We say that two voting systems (A, f, q) and (A, g, r) are *equivalent* if, for each group of agents $B \subseteq A$, holds: $\sum_{b \in B} f(b) \geq q$ *iff* $\sum_{b \in B} g(b) \geq r$. This means that each decision made by votes of agents belonging to the group B in (A, f, q) can be made by the same group B in (A, g, r) and vice versa.

First, we will present, after adapting our approach used by Srećković (2017) to this context, how each 'one agent—one vote' system containing agents with the veto power is equivalent to a 'one agent—many votes' system. We give an accurate connection between these two kinds of systems.

Lemma 5 *Let* (A, f, q) *be an 'one agent—one vote' system, i.e.* $(\forall a \in A) f(a) = 1$, *consisting of* n *agents, with* $k (1 \leq k \leq n)$ *agents having the veto power. Then this system is equivalent to the weighted system* (A, g, r), *where* $g(a) = n - q + 1$, *if* $a \in A$ *is an agent with the veto power, and* $g(a) = f(a) = 1$, *otherwise, and* $r = k(n - q) + q$.

Proof Our aim is to define weights x of agents with the veto power. We suppose that $1 \leq k \leq q \leq n$. It is clear that $(k - 1)x + n - k$ votes are not sufficient for decision-making, i.e. we have that (1) $(k - 1)x + n - k < r$. On the other side, the minimal number of votes necessary to reach a decision is $kx + q - k$, meaning that we have (2) $r \leq kx + q - k$. From (1) and (2) we infer immediately: $(k - 1)x + n - k < kx + q - k$, wherefrom we have $n - q < x$, so that we can conclude that the minimal possible weight of an agent possessing the veto power is (3) $x = n - q + 1$. By substituting (3) in (1) and (2), respectively, we obtain (4) $(k - 1)(n - q + 1) + n - k < r$ and (5) $r \leq k(n - q + 1) + q - k$, wherefrom we derive (6) $k(n - q) + q - 1 < r$ and (7) $r \leq k(n - q) + q$ giving finally $r = k(n - q) + q$. In other words, formula (3), together with the last formula, presents, respectively, the weight of agent with the veto power and a new quota in the new system. □

We have just demonstrated how the weights of agents with the veto power in a 'one agent—one vote' voting system can be measured and transformed into an equivalent weighted voting system. Let us note here that the weight of agents with the veto power does not depend on their number.

The next step is to formulate a similar statement for weighted voting systems containing agents with the veto power. This can be obtained and justified as a consequence of the previous Lemma.

Theorem *Let* (A, f, q) *be a weighted system consisting of* n *agents with* $m \geq n$ *votes and* k $(1 \leq k \leq n)$ *agents having the veto power. Then this system is equivalent to the weighted system* (A, g, r), *where* $g(a) = (m - q + 1) f(a)$, *if* $a \in A$ *is an agent with the veto power, and* $g(a) = f(a)$, *otherwise, and* $r = k(m - q) + q$.

Proof By using the previous Lemma and considering our system (A, f, q) as a 'one agent—one vote' system consisting of m agents, we can justify this Theorem immediately. □

Our Theorem gives a suitable framework to obtain a necessary and sufficient condition for the existence of a vetoer in a weighted voting system:

Corollary *A weighted voting system* (A, f, q) *contains a vetoer iff* (A, f, q) *is equivalent to a weighted system* (A, g, r), *with* $g(a) = (m - q + 1) f(a)$, *for some agent* $a \in A$.

We also have the following immediate consequences:

Corollary *Each dictatorial voting system is a voting system with a veto, but not conversely.*

Corollary *Each consensus voting system is a voting system with a veto.*

Note that in a joint stock company the so-called golden share, which is in the possession of a member of the company, appears throughout the history of joint stock companies. Possessing the golden share is an equivalent to the veto power, as well.

We conclude our note with the following statement:

Lemma 6 *Each weighted voting system with a vetoer excludes the existence of a dictator, except in case when the system contains exactly one vetoer who is a dictator simultaneously.*

Proof Let (A, f, q) be a weighted voting system, with k $(1 \leq k \leq n)$ agents having the veto power, dictator $d \in A$ and vetoer $v \in A$, for $d \neq v$. Suppose that $q \leq f(d) < m$. In the worst case, for $k = 1$ and $f(v) = 1$, we have that $r = m$, and, as $f(d) < r$, d cannot be a dictator in (A, g, r). □

Example Let us consider the following system (A, f, q), where $A = \{a, d, v\}$, $a \neq d \wedge a \neq v \wedge d \neq v$, $f(a) = f(v) = 1$, $f(d) = 3$, $m = 5$ and $q = 3$, and individual v has the veto power. Obviously, individual d, with $f(d) = 3 \geq q$, looks like he has the power of a dictator, but, in the presence of a vetoer v, in the new equivalent system (A, g, r), we have $g(a) = f(a) = 1$, $g(d) = f(d) = 3$, $g(v) = 3$ and $r = m = 5$, so that no one has a majority.

In other words, there is no weighted voting system containing a dictator and a vetoer simultaneously. This means that the presence of dictatorship and vetoing in the same voting system is impossible, i.e. that such a requirement is contradictory.

8 A Note on Possible Further Research

Combining traditional mathematical theories with various logical concepts has always been challenging. Traditional Social Choice Theory is based on classical two-valued logic, but there have also been attempts to investigate formalism of social choice founded on some non-classical logics. Usually this is a quest to find some new and simpler context for proving (or refuting) some well-known theorems. One significant direction in Social Choice Theory was to avoid inconsistencies between the axioms, i.e. to obtain possibility instead of impossibility results. Although many fruitful ideas have appeared in a fuzzy logic approach to Social Choice Theory (see De Baets et al. 1998a,b; Bass and Kwakernaak 1977; Boričić and Konjikušić 2004a; Bufardi 1999; Li and Yen 1995), other non-classical logic alternatives have not been considered yet. It seems that the social choice based on (super)intuitionistic (see Gabbay 1981; Hosoi 1967, 1969) and relevant logics (see

Anderson and Belnap 1975) would be of great interest, particularly with a pure proof-theoretic treatment (see Boričić 1988a,b; Ilić 2016, 2017, 2021,b; Ilić and Boričić 2014, 2021; Routhley 1979). The advantage of proof–theoretic syntactic analysis lies in the fact that such an approach will be comprehensible to a wider circle of readers, because of its simplicity, and it would be applicable in expert systems for decision support as well.

Inconsistency of any mathematical theories based on (super)intuitionistic or relevant logic, including Social Choice Theory, implies inconsistency of a corresponding theory based on classical logic, but not vice versa. On the other hand, subtle and sensitive concepts of inconsistency, such as the one developed in paraconsistent logics (see da Costa et al. 2007), may shed an additional light on inconsistencies in Social Choice Theory. In particular, we expect that some traditional impossibility theorems of Social Choice Theory cannot be proved in a non-classically based Social Choice Theory. For example, our proof of 'impossibility of a non-Paretian dictator', as given in Boričić (2014a) or indicated in Boričić (2009), although intuitionistically acceptable, is clearly incorrect from a relevant logic point of view. Within this approach we can anticipate many such new conclusions.

9 Concluding Remarks

The influence of physics on the development of mathematics was evident over the centuries. During the twentieth century and later, we notice an increasing role of chemistry, biology, computer science, but most of all the needs and expectations of the social sciences to use mathematics in analysis, examination, description, modeling and prediction of social phenomena. Although the Nobel Prize is not awarded for achievements in mathematics, it is known that many mathematicians are Nobel Prize winners, but in other disciplines, mainly in economics and physics. This testifies to how much mathematics is in the service of not only natural but also social sciences today.

We see the main role of mathematics in connecting real phenomena, natural or social, which are studied by a scientific discipline, with the very theoretical concepts of these phenomena. In Boričić (2018) we discuss the perception of modern science as 'the Holy Trinity' of the following three entities: the Reality, the Model, and the Theory (Rl, Md, Th). We explained how this 'Holy Trinity' has been an appropriate skeleton of scientific research through the centuries and described how the natural path leads from the real environment, through the model, to its theoretical epilogue. Either natural or social reality obtains its description in the framework of some more or less formal theory based on a specific language and a clear logical structure. The connection between the Reality and the Theory is realized by means of some measuring processes and contemplations, usually denoted as modeling. In that sense, the Model can be seen as a bridge between the Reality and the Theory and the triangle (Rl, Md, Th) as the fundament of each scientific research.

As we have pointed out in Boričić (2018), the Unique physical reality, natural or social environment, inspires Many of its models which, then, lead to One theory. Finding every new fact regarding the Reality implies changes of Models and, consequently, causes the reconsideration of the Theory. This is a continuous dynamic process of transforming the relationships between the vertices of our triangle (Rl, Md, Th). The Model can be understood as the first iteration in a partial description of the Reality, and the Theory is its final description. Also, the communications and interactions between vertices Rl and Md, and between Md and Th, are very intensive, while the immediate connections between Rl and Th are quite rare. The interactions between Rl and Th occur almost exclusively through the intermediation of Md. This is the reason why Md occupies a central place in 'the Holy Trinity' (Rl, Md, Th) of science.

Here we see mathematics as the essence of modeling Md and logic as the soul of theory Th. The basic requirement that each theory must satisfy is consistency. This requirement can be met only by the adequate application of mathematical modeling. Namely, the twentieth century introduced consistency as the main quality of a scientific theory. Consistency becomes the fundamental criteria of existence. The explanation of existence was possible by means of models, and the role of a model with respect to a theory has become clear through the unity of syntax and semantics, i.e. the soundness and completeness of proof systems with respect to the corresponding classes of models expressed through their validity and satisfaction relations. The only stable base of a deductive method was established by axiomatic approach.

Hence, parts of this text entitled 'Social Choice—Traditional Theory' and 'Impossibilities with Vetoing, Dictatorship and Pareto rule' can be treated as a theoretical description of some social phenomena, logically founded on the part under the title 'A Method of Axioms Simplification' and mathematically analyzed through the model given in the part entitled 'Measuring the Veto Power in Voting Systems'.

A model and a theory always share elements of a specific language, but the status of each modeling can be understood as one (thought) experiment, while each theoretical conclusion represents a proposition of general nature. As we already stated, a theory is founded on a pure logical structure, and modeling uses mathematical tools and methods. A new understanding of reality often requires the development of new mathematical methods and an adjusted context for defining an adequate model. This is a reversible influence of reality on mathematics. For example, mathematics for finance has developed a stochastic differential and integral calculus to meet the demand for adequate modeling of modern financial markets. And this is not an isolated example.

On the other hand, there are many cases when some abstract (and consistent) mathematical theories, theories that emerged as *l'art pour l'art* later, found their application. Lobachevskian geometry (N. I. Lobachevsky (1792–1856)) is one of the most famous such examples, and now here we have a new example of applying semigroup theory in social sciences and humanities in combination with intuitionistic or other non-classical logics.

Let us conclude with quite a personal opinion. We believe that each abstract consistent theory built up and founded consequently on some mathematical ideas will find its specific application in the real world.

Appendix: Glossary of Notation and Definitions

Axiomatic System The context that enables accurate argumentation based on given statements (axioms) and rules. After Euclid, the axiomatic approach is used in many sciences to avoid doubts and ambiguities. Basic formal properties of a deduction relation are induced by nature of logic on which *axiomatic system* is build up.

Classical Logic An *axiomatic system* defining logical rules of reasoning based on the Aristotle's two-valued principle. The law of excluded middle, that each statement or its negation is provable, is characteristic for classical logic.

Deduction Relation Relationship between a set of hypotheses Γ and its conclusion A, denoted by $\Gamma \vdash A$, with meaning that A can be derived from Γ in a corresponding *axiomatic system*.

Dictatorship A form of social organization in which one person possesses absolute power without effective limits. Particularly, in *social choice theory*, *dictator* is defined as an individual who can impose her/his personal preferences to the whole society.

Impossibility In this book, this is a synonym for *inconsistency*.

Inconsistent Statements Two or more statements are inconsistent if they always enable inferring untruth. In this book, the fact that X and Y are inconsistent is denoted as $X, Y \vdash$, by means of a *deduction relation*.

Logical Connectives The usual starting point of a logical formalism are *propositional connectives*, such as: *and, or, not, if ... then, if and only if*, and *quantifiers: for all* and *for some* (or *there exists*). The list of basic logical symbols is the following one: $\wedge$—*conjunction* (for *and*), $\vee$—*disjunction* (for *or*), $\neg$—*negation* (for *not*), $\rightarrow$—*implication* (for *if ... then*), $\leftrightarrow$—*equivalence* (for *if and only if*), $\forall$—*universal quantifier* (for *for all*), and $\exists$—*existential quantifier* (for *there exists*).

Non-classical Logic An *axiomatic system* defining a logic denying the Aristotle's two-valued principle. The most popular *non-classical logics* are classified as relevant, intuitionistic, linear, substructural or paraconsistent.

Pareto Rule In *social choice theory, Pareto rule* means that if each member of a society prefers the same state, then the society prefers this state.

Preference Logic An *axiomatic system*, based on *rationality choice axioms*, which enables formal reasoning about individual or social preferences.

Preference Relation A binary relation expressing relationship between two alternatives. For instance, 'xPy' expresses that 'the alternative x is preferred to the alternative y'.

Rationality Choice Axioms Asymmetric, i.e., $\forall x, y(xPy \to \neg yPx)$, and transitive, i.e., $\forall x, y, z \in X(xPy \land yPz \to xPz)$, binary relation P is called the strict preference relation, and we say that it satisfies rationality choice axioms. Each strict preference relation P generates an indifference relation I, as follows: xIy iff $\neg xPy \land \neg yPx$, and a weak preference relation R, by definition: xRy iff $xPy \lor xIy$.

Social Choice Theory Also known as *theory of social decision-making* or *theory of public choice*, is treated in this book as a mathematically founded study of collective decision-making procedures. It is focused on the problem of how to define a social state derived from the personal states, and when it is possible.

Vetoer A person participating in the process of decision-making with veto power, i.e., a voter having the possibility to stop any action or block any decision-making process or its result.

Voting System Set of rules and procedures defining process of decision-making by means of voting. This is a part of *social choice theory*.

Weighted Voting System A more general form of a *voting system* which enables expressing various powers of voters including the possibility that some voters possess more than one vote.

References

Anderson, A. R., Belnap, Jr., N. D. (1975). *Entailment: The Logic of Relevance and Necessity.* Vol. I, Princeton University Press, Princeton.

Arrow, K. (1963). *Social Choice and Individual Values*. John Wiley, New York.

Arrow, K. J., Sen, A. (2002). *Handbook of Social Choice and Welfare.* Vol 1, Gulf Professional Publishing, Houston.

Arrow, K. J., Sen, A., Suzumura, K. (2010). *Handbook of Social Choice and Welfare*. Vol 2, Elsevier, Amsterdam.

Artemidiadis, N. (2004). *History of Mathematics*. American Mathematical Society, Providence.

Bass, S. M., Kwakernaak, H. (1977). Rating and Ranking of Multiple–Aspect Alternatives Using Fuzzy Sets. *Automatica,* 18, 47–58.

Blau, J. H., Deb, R. (1977). Social Decision Functions and the Veto. *Econometrica,* 45 (4), 871–879.

Boričić, B. (1988a). On certain normalizable natural deduction formulations of some propositional intermediate logics. *Notre Dame Journal of Formal Logic* 29, 563–568.

Boričić, B. (1988b). A note on sequent calculi intermediate between LJ and LK. *Studia Logica,* 47, 151–157.

Boričić, B., Konjikušić, S. (2004a). Preference logic on rough and fuzzy sets. *Economic Annals,* XLIV 160, 131–146.

Boričić, B. (2004b). Nobel prize in economics—Amartya K. Sen. *Economic Annals,* XLIV 160, 215–220.

Boričić, B. (2007). Logical and historical determination of impossibility theorems by Arrow and Sen. *Economic Annals,* LI 172, 7–20.

Boričić, B. (2009). Dictatorship, liberalism and the Pareto rule: possible and impossible. *Economic Annals,* LIV 181, 45–54.

Boričić, B. (2011). *Logic and Proof.* Faculty of Economics, University of Belgrade, Belgrade, 2011. (Zbl 1214.03043; MR 2011k:03001)

Boričić, B. (2014a). Multiple von Wright's preference logic and social choice theory. *The Bulletin of Symbolic Logic,* 20, 224–225.

Boričić, B. (2014b). Impossibility theorems in multiple von Wright's preference logic. *Economic Annals,* LIX 201, 69–84.

Boričić, B. (2018). Model, proving and refuting, in *Quantitative Models in Economics* (Ed. J. Kočović et all), Faculty of Economics, University of Belgrade, Belgrade, 3–19.

Boričić, B. (2023). A note on dictatorship, liberalism and the Pareto rule, *Economic Annals* (to appear)

Boyer, C. B. (1968). *A History of Mathematics.* J. Wiley and Sons, New York.

Bufardi, A. (1999). On the fuzzification of the classical definition of preference structure. *Fuzzy Sets and Systems,* 104, 323–332.

Chichilnisky, G. (1982). The topological equivalence of the Pareto condition and the existence of a dictator. *Journal of Mathematical Economics*, 9, 223–233.

da Costa, N. C. A., Krause, D., Bueno, O. (2007). Paraconsistent logics and paraconsistency. *Handbook of the Philosophy of Science*, Elsevier, 791–911.

De Baets, B., Van de Walle, B., Kerre, E. E. (1998a). A plea for the use of Lukasiewicz triplets in the definition of fuzzy preference structures, Part I: General argumentation. *Fuzzy Sets and Systems,* 97, 1998, 349–359.

De Baets, B., Van de Walle, B., Kerre, E. E. (1998b). Part II: The identity case. *Fuzzy Sets and Systems,* 99, 303–310.

Debreu, G. (1959). *The Theory of Value: an Axiomatic Analysis of Economic Equilibrium.* Wiley, New York.

de Swart, H. (2018). *Philosophical and Mathematical Logic.* Springer Undergraduate Texts in Philosophy, Springer, Berlin.

Gabbay, D. M. (1981). *Semantical Investigations in Heyting's Intuitionistic Logic.* D. Reidel Publ. Comp., Dordrecht.

Fishburn, P. C. (1973). *The Theory of Social Choice.* Princeton University Press, Princeton.

Hofstadter, D. R. (1979). *Gödel, Escher, Bach: An Eternal Golden Braid.* Basic Books, New York.

Hosoi, T. (1967). On intermediate logics I. *J. Fac. Sci. Univ. Tokyo Sect. I,* 14, 293–312.

Hosoi, T. (1969). On intermediate logics II. *J. Fac. Sci. Univ. Tokyo Sect. I,* 16, 1–12.

Ilić, M. (2016). An alternative Gentzenization of RW^o_+. *Mathematical Logic Quarterly,* 62(6), 465–480.

Ilić, M. (2017). A natural deduction and its corresponding sequent calculus for positive contractionless relevant logics. *Reports on Mathematical logic,* 52, 93–124.

Ilić, M. (2021a). A note on an alternative Gentzenization of R^o_+. *Mathematical Logic Quarterly,* 67(2), 186–192.

Ilić, M. (2021b). A cut–elimination proof in positive relevant logic with necessity, **Studia Logica** 109(3), 2021, pp. 607–638.

Ilić, M., Boričić, B. (2014). A cut–free sequent calculus for relevant logic $RW^\star$. *Logic Journal of IGPL,* 22(4), 673–695.

Ilić, M., Boričić, B. (2021). A note on the system GRW with the intensional contraction rule. *Logic Journal of IGPL,* 29(3), 333–339.

Kang, M. S. (2010). Voting as veto. *Michigan Law Review,* 108, 1221–1281.

Kelly, J. S. (1978). *Arrow Impossibility Theorems.* Academic Press, London.

Li, H. X., Yen, V. C. (1995). *Fuzzy Sets and Fuzzy Decision-Making.* CRC Press, New York.

Mas–Colell, A., Sonnenschein, H. (1972). General possibility theorems for group decisions. *The Review of Economic Studies,* 39, 185–192.

Moulin, H. (1981). The Proportional Veto Principle. *The Review of Economic Studies,* 48, 407–416.

Moulin, H. (1982). Voting with Proportional Veto Power. *Econometrica,* 50, 145–162.
Murakami, Y. (1968). *Logic and Social Choice.* Routledge, London.
Routhley, R. (1979). Repairing proofs of Arrow's general impossibility theorem and enlarging the scope of the theorem. *Notre Dame Journal of Formal Logic*, 20, 879–890.
Schofield, N. J. (1985). *Social Choice and Democracy*. Springer, Berlin.
Sen, A. K. (1969). Quasi-Transitivity, Rational Choice and Collective Decisions. *The Review of Economic Studies*, 36 (3), 381–393.
Sen, A. K. (1970a). The impossibility of a Paretian liberal. *Journal of Political Economy*, 78 (1), 152–157.
Sen, A. K. (1970b). *Collective Choice and Social Welfare.* Holden–Day, San Francisco (The fourth edition: Elsevier, Amsterdam, 1995).
Skala, H. J. (1975). *Non-Archimedean Utility Theory*. Springer, Berlin.
Srećković, M. (2017). Measuring the veto power. *Ekonomske ideje i praksa*, 25, 89–96.
Takeuti, G. (1975). *Proof Theory.* North–Holland, Amsterdam (Second edition: Dover Publications, 2013).
Taylor, A. D. (1995). *Mathematics and Politics—Strategy, Voting, Power and Proof.* Springer, Berlin.
van Dalen, D. (1980). *Logic and Structure.* Springer, Berlin (Fifth edition 2013).
Winter, E. (1996). Voting and Vetoing. *The American Political Science Review*, 90, 813–823.

Part II
Semigroups and Other Algebraic Structures in Social Sciences

Constructive Semigroups with Apartness: A State of the Art

Melanija Mitrović, Mahouton Norbert Hounkonnou, and Paula Catarino

1 Introduction

The development of powerful computer systems draws attention to the intuitive notion of an effective procedure and of computation in general. This, in turn, stimulates the development of constructive algebra and its possible applications. *Apartness*, the second most important fundamental notion developed in constructive mathematics, shows up in computer science. Inspired by results obtained in interactive theorem proving the approach of formal verifications (more in Mitrović et al., 2021), a new constructive algebraic theory known as the **theory of semigroups with apartness** was developed by Mitrović and co-authors: Hounkonnou, Baroni, Crvenković, Romano, and Silvestrov—see Crvenković et al. (2013), Crvenković et al. (2016), Mitrović et al. (2021), Mitrović and Silvestrov (2020); Mitrović et al. (2019). We did it, we developed the theory of constructive semigroups with apartness, and we are still working on its development. But, is it Art?

A descriptive definition of a semigroup with apartness includes two main parts:

- The notion of a certain classical algebraic structure is straightforwardly adopted.
- A structure is equipped with an apartness with a standard operation respecting that apartness.

M. Mitrović (✉)
Faculty of Mechanical Engineering, CAM-FMEN, University of Niš, Niš, Serbia
e-mail: melanija.mitrovic@masfak.ni.ac.rs

M. N. Hounkonnou
International Chair in Mathematical Physics and Applications (ICMPA-UNESCO Chair), University of Abomey-Calavi, Benin Republic
e-mail: norbert.hounkonnou@cipma.uac.bj

P. Catarino
Department of Mathematics, University of Trás-os-Montes e Alto Douro, Vila Real, Portugal
e-mail: pcatarin@utad.pt

© The Author(s), under exclusive license to Springer Nature Switzerland AG 2023
M. N. Hounkonnou et al. (eds.), *Mathematics for Social Sciences and Arts*, Mathematics in Mind, https://doi.org/10.1007/978-3-031-37792-1_7

This chapter aims to provide a clear and understandable picture of constructive semigroups with apartness in Bishop's style of constructive mathematics, **BISH**, both to (classical) algebraists and to those who apply algebraic knowledge.

Our theory is partly inspired by the classical case, but it is distinguished from it in two significant aspects:

1. We use ***intuitionistic logic*** rather than classical thought.
2. Our work is based on the notion of **apartness** (between elements of the set and, later, between elements and its subsets).

"*Intuitionistic logic* can be understood as the logic of scientific research (rather positivistically conceived) while on the other hand the classical logic is the logic of ontological thought", Grzegorczyk (1964). Intuitionistic logic serves as a foundation to constructive mathematics. This means that mathematics is done without use of the so-called the Law of the Excluded Middle (**LEM**) and without the use of nonconstructive existence proofs.

It was A. Heyting (1956) who wrote in his famous book *Intuitionism, an Introduction* that "Intuitionism can only flourish, if mathematicians, working in different fields, become actively interested in it and make contributions to it. [...] In order to build up a definite branch of intuitionistic mathematics, it is necessary in the first place to have a thorough knowledge of the corresponding branch of classical mathematics, and in the second place to know by experience where the intuitionistic pitfalls lie". So, we present two points of view on constructive semigroups with apartness:

- The classical (**CLASS**), which plays a useful role as intuition guides and to at least link with the presentations given in constructive one.
- The constructive (**BISH**), comprising a binary system with apartness which satisfies a number of extra conditions: well-known axioms of apartness and the operations have to be strongly extensional.

"In R. B. Wells" *Introduction to Biological Signal Processing and Computational Neuroscience* (2010) - 10.3." it can be found that "The Bourbaki set out to discover the 'roots' of mathematics—something that was true of mathematics in general. They found it –or so they tell us– subsisting in three basic 'mother structures' upon which all of mathematics depends. *These structures are not reducible one to another. This just means no one of them can be derived from the other two. But by making specifications within one structure, or by combining two or more of these structures, everything else in mathematics can be generated.* These three basic structures are called **algebraic structure**, **order structure** and topological structure".

In what follows some results concerning special ordered and algebraic structures—the *binary* ones, e.g., the ordered structures with a single binary relation, the algebraic structures with a single binary operation, and the ordered algebraic structures with a single binary operation and a single binary relation –such as, certain types of ordered sets, semigroups, and ordered semigroups– within classical and constructive (Bishop's) settings will be presented.

"To have a structure, we need a set, a relation, and rules establishing how we will put them together", R. B. Wells (2010). That is, working within classical or intuitionistic logic, in order to analyze algebraic structures it is necessary to start with study on sets and ordered sets, relational systems, etc. Selected results on classical ones are given in Sect. 2.1, while constructive ones are given in Sect. 3.1.

Following Heyting, at least initially, classical semigroup theory is seen as a guide that helps us to develop the constructive theory of semigroups with apartness. The classical background is given in Sect. 2. All in all, material presented in Sect. 2 is broad rather than deep, and it is not intended to be comprehensive. It is heavily based on the treatments in other standard on set and semigroup theory textbooks. The main novelty is in the selection and arrangement of material. Development of an appropriate constructive order theory for sets and semigroups with apartness has been one of the main objectives of our work. Most of the results we have published to the date are given in Sect. 3. Applications and possible applications are given in Sect. 2.3 for the classical and in Sect. 3.3 for the constructive case. Sections 2.4 and 3.4 contain very short historical overviews and bibliographic notes, which give sources and suggestions for further reading for both cases. A comparative analysis between presented classical and constructive results is part of Sect. 4. All proofs can be found in the Appendix.

Results presented in Sect. 3 are based on the ones published in Crvenković et al. (2013), Crvenković et al. (2016), Mitrović and Silvestrov (2021), Mitrović et al. (2020), Mitrović et al. (2019), Darpö and Mitrović, arXiv:2103.07105, Romano (2005, 2007) and, shortly, on the work in progress Mitrović and Hounkonnou.

2 The CLASS Case

Can the whole of MODERN ALGEBRA be described in a couple of sentences? Paraphrasing J. Woźny (2018) yes it can; it has been designed to be elegantly simple: the story starts with **sets** (collections of objects) and **relations** on them and proceeds to the concept of a **semigroup**; each new concept is based on the previous ones, and, ultimately, the whole multistory edifice *rests on the sparse foundation of sets and relations*. **Semigroups** serve as the building blocks for the structures comprising the subject which is today called **modern algebra**. In fact, as it is written by Uday S. Reddy (2014) (https://www.cs.bham.ac.uk/udr/notes/semigroups.pdf), "Semigroups are everywhere. Groups are semigroups with a unit and inverses. Rings are 'double semigroups:' an inner semigroup that is required to be a commutative group and an outer semigroup that does not have any additional requirements. But the outer semigroup must distribute over the inner one. We can weaken the requirement that the inner semigroup be a group, i.e., no need for inverses, and we have semirings. Semilattices are semigroups whose multiplication is idempotent. Lattices are 'double semigroups' again and we may or may not require distributivity".

2.1 Sets and Relations

It is well known that all mathematical theories deal with sets in one way or another, i.e., nearly every mathematical object of interest is a set of some kind. In the words of Halmos (1960), "General set theory is pretty trivial stuff really, but, if you want to be a mathematician, you need some, and here it is; read it, absorb it, and forget it".

Many of the fundamental concepts of mathematics can be described in terms of relations. The notion of an order plays an important role throughout mathematics. A pure order theory is concerned with a single undefined binary relation ρ. This relation is assumed to have certain properties (such as, for example, reflexivity, transitivity, symmetry, and antisymmetry), the most basic of which leads to the concept of *quasiorder*, a reflexive and transitive relation. A quasiorder plays a central role throughout this short exposition. Combining quasiorder with the various properties of symmetry yields the notions of an equivalence, a symmetric quasiorder, and a (partial) order, an antisymmetric quasiorder.

Equivalences permeate all of mathematics. An equivalence relation allows one to connect those elements of a set that have a particular property in common. In universal algebra, the formulation of mapping images is one of the principal tools used to manipulate sets. In the study of mapping images of a set, a lot of help comes from the notion of a quotient set, which captures all mapping images, at least up to bijection. On the other hand, mapping is the concept which goes hand in hand with equivalences. Thus concepts of equivalence, quotient set, and mappings are closely related. Knowing that the equivalence ε on a set S is the kernel of the quotient map from S onto S/ε, we can treat equivalence relations on S as kernels of mappings with S as the domain. The relationship between quotients, mappings, and equivalences is described by the celebrated isomorphism theorems, which are a general and important foundational part of universal algebra.

On the other hand, as Davey and Priestley (2002) wrote "order, order, order—it permeates mathematics, and everyday life", as well "to such an extent that we take it as granted".

If a set carries an order relation as well as an equivalence relation, the important question arises: does the given order induce an order on the equivalence classes? It is a well-known fact that this will not be possible in general. Of course, there are conditions guaranteeing that the quotient set inherits the ordering.

Applications of mathematics in other sciences often take the form of a network of relations between certain objects, which justifies H. Poincarè "Mathematicians do not study objects, but relations between objects" (quoted in Newman, 1956).

2.1.1 Basic Concepts and Important Examples

Mappings are written on the left, that is, if we have a function f that takes an input x, its output is written $f(x)$. If two maps are composed, they are written right to left. The composition of two mappings f and g is written $f \circ g$ or simply fg, and we have $fg(x) = f(g(x))$. In connection with denoting subsets, the symbol $\subseteq$ is used to denote containment with possible equality, while $\subset$ is used to denote strict containment.

Let S and T be sets. A *mapping* f from S to T, denoted by $f : S \to T$, is a subset of $S \times T$ such that for any element $x \in S$ there is precisely one element $y \in Y$ for which $(x, y) \in f$, i.e.,

$$(\forall x, y \in S)\, x = y \Rightarrow f(x) = f(y).$$

Instead of $(x, y) \in f$, we usually write $y = f(x)$. Two mappings $f, g : S \to T$ are *equal* if they are equal as subsets of $S \times T$, that is, $f = g \Leftrightarrow (\forall x \in S)\, f(x) = g(x)$.

For the set of all mappings from S to T, the notation $M(S, T)$ or T^S is used and $\mathcal{T}_S$ for $M(S, S)$. The elements of $\mathcal{T}_S$ are called *transformations* of S.

A mapping f is

- *surjective* or *onto* $(\forall y \in T)\,(\exists x \in S)\,(y = f(x))$
- *injective* or *one–one*:

$$f(x) = f(y) \;\Rightarrow\; x = y$$

or, equivalently,

$$x \neq y \;\Rightarrow\; f(x) \neq f(y)$$

- *bijective:* surjective and injective.

The subset of all bijective transformations in $\mathcal{T}_S$ is denoted by $\mathcal{G}_S$, and its elements are called *permutations* or *bijections* of S. In particular, the *identity function* $i_S : S \to S$, $i_S(x) = x$, is always a bijection, the *trivial bijection*, of S. The term "function" is fundamental to almost all areas of mathematics, but it has become traditional to use terms "mapping" and "transformation" in algebra.

Let S be a set. A subset ρ of $S \times S$, or, equivalently, a property applicable to elements of $S \times S$, is called a *binary relation on* S. The empty subset $\emptyset$ of $S \times S$ is included among the binary relations on S. Other special binary relations worth to be mentioned are the universal relation $S \times S$ and the *equality* relation or *diagonal* $\Delta_S = \{(x, x) : x \in S\}$. Binary relations on a set S are often called *homogenous*. The set of all binary relations on S is denoted by $\mathcal{B}_S$.

2.1.2 Quotient Sets: Homomorphism and Isomorphism Theorems

In general, there are many properties that a binary relation may satisfy on a given set. For example, the relation $\rho \in \mathcal{B}_S$ might be:

(R) reflexive: $(x, x) \in \rho \quad (\Leftrightarrow \Delta_S \subseteq \rho)$
(S) symmetric : $(x, y) \in \rho \Rightarrow (y, x) \in \rho \quad (\Leftrightarrow \rho^{-1} \subseteq \rho \Leftrightarrow \rho^{-1} = \rho)$

(AS) antisymmetric: $(x, y) \in \rho \wedge (y, x) \in \rho \Rightarrow x = y \quad (\Leftrightarrow \rho \cap \rho^{-1} \subseteq \Delta_S)$
(T) transitive: $(x, y) \in \rho \wedge (y, z) \in \rho \Rightarrow (x, z) \in \rho$.

The most basic concept leads to the ***quasiorders***, reflexive and transitive relations with the fundamental concepts being introduced whenever possible in their natural properties:

- equivalence—symmetric quasiorder
- partial order—antisymmetric quasiorder.

The concept of an *equivalence* is an extremely important one and plays a central role in mathematics. Classifying objects according to some property is a frequent procedure in many fields. Grouping elements in classes so that elements in each class are of the same prescribed property as performed by equivalence relations, and the classification gives the corresponding quotient sets. Thus, abstract algebra can show us how to identify objects with the same properties properly—we have to switch to a quotient structure (a technique applicable, for example, to abstract data type theory).

Let us remember that we can define a surjective mapping starting from equivalences, i.e., we have the following lemma.

Lemma 1 *Let ε be an equivalence on S. The mapping $\pi : S \to S/\varepsilon$ defined by $\pi(x) = x\varepsilon$, $x \in S$, is a surjective mapping.*

The mapping π from the previous lemma is called the *natural mapping associated with an equivalence relation* ε and is sometimes written as π_ε or as π_{nat}. As usual $x\varepsilon$ denotes the ε-class of an element $x \in S$.

On the other hand, any mapping f between sets S and T, $f : S \to T$, gives rise to the equivalence on S

$$ker\ f = \{(x, y) \in S \times S : f(x) = f(y)\},$$

called the *kernel of mapping* f or *the equivalence relation induced by* f.

The first isomorphism theorem for sets follows.

Theorem 1 *Let $f : S \to T$ be a mapping between two sets S and T. Then the following statements are true:*

(i) *The mapping $\varphi : S/ker\ f \to T$ defined by $\varphi(x(ker\ f)) = f(x)$ is a unique injective mapping such that $f = \varphi \circ \pi_{nat}$.*
(ii) *If f is surjective, then φ is a bijection, i.e., we have $T \cong S/ker\ f$.*
(iii) *φ is surjective if and only if f is surjective.*

The next theorem concerns a more general situation and will be frequently used.

Theorem 2 *Let ε be an equivalence on S. Let $f : S \to T$ be a mapping between two sets S and T such that $\varepsilon \subseteq ker\ f$. Then there is a unique mapping $\varphi : S/\varepsilon \to T$ defined by $\varphi(x\varepsilon) = f(x)$ such that $f = \varphi \circ \pi_\varepsilon$.*

2.1.3 Ordered Sets

In many disciplines, sets that are equipped with a structure are investigated. The ordered pair (S, ρ), where S is a given set and ρ is a particular (binary) relation defined on it, is called a (binary) *relational structure*. A relational system (S, ρ) is called *quasiordered* or *ordered* if ρ is a quasiorder or an order on S. In what follows, we will automatically assume that a property of a (quasi)order is defined in the same way as the property of the (quasi)ordered set and vice versa.

Relational structures play an important role both in mathematics and in applications since a formal description of a real system often involves relations. For these considerations we often ask about a certain factorization of a relational system (S, ρ) because of its importance in enabling us to introduce the method of abstraction on S. More precisely, if ε is an equivalence relation on S, we often ask about a factor relation $\Theta_\rho = \rho/\varepsilon$ on the factor set S/ε such that the relational factor set $(S/\varepsilon, \rho/\varepsilon)$ shares some "good" properties of (S, ρ).

Let (S, ρ) and (T, σ) be two relational systems. A mapping $f : S \to T$ is called

- *isotone* if $(x, y) \in \rho \Rightarrow (f(x), f(y)) \in \sigma$
- *reverse isotone* if $(f(x), f(y)) \in \sigma \Rightarrow (x, y) \in \rho$
- *relation preserving* if $(f(x), f(y)) \in \sigma \Leftrightarrow ((x, y) \in \rho$,

for any $x, y \in S$. Clearly f is relation preserving if and only if it is isotone and reverse isotone.

A mapping $f : S \to T$ between two relational structures (S, ρ) and (T, σ) is called an *order bijection* if it is an isotone and reverse isotone bijection. We will write $S \cong_o T$ if there is an order bijection $f : S \to T$ between two relational structures (S, ρ) and (T, σ).

Any mapping $f : S \to T$ between a set S and a relational system (T, σ) gives rise to the following important relation on its domain:

$$\eta_f \overset{\text{def}}{=} \{(x, y) \in S \times S : (f(x), f(y)) \in \sigma\}.$$

Certain connections between η_f and the basic relation ρ of the domain set S, given in the lemma below, are from a practical point of view useful criteria in recognizing types of monotonicity.

Lemma 2 *Let (S, ρ) and (T, σ) be relational systems, and let $f : S \to T$. Then*

(i) *f is isotone iff $\rho \subseteq \eta_f$.*
(ii) *f is reverse isotone iff $\eta_f \subseteq \rho$.*
(iii) *f is relation preserving iff $\eta_f = \rho$.*

Let (S, ρ) be a relational system and let ε be an equivalence on S. Let us define a binary relation $\Theta_\rho = \rho/\varepsilon$ on the set S/ε as

$$(x\varepsilon, y\varepsilon) \in \rho/\varepsilon \overset{\text{def}}{\Leftrightarrow} (x, y) \in \rho.$$

The relational system $(S/\varepsilon; \rho/\varepsilon)$ is called a *relational factor system* of S by ε or a *quotient relational system* of (S, ρ) by ε.

Lemma 3 *Let (S, ρ) be a reflexive, antisymmetric, or transitive ordered system and let ε be an equivalence on S. Then $(S/\varepsilon, \Theta_\rho)$ is also a reflexive, antisymmetric, or transitive ordered system.*

In what follows the following elementary result will be handy and will be used without reference.

Lemma 4 *Let ρ be a relation on a set S. If ρ is reflexive (symmetric, antisymmetric, and transitive), then ρ^{-1} is also reflexive (symmetric, antisymmetric, and transitive).*

Due to Lemma 3, for a given quasiordered set (S, ρ) and an equivalence ε defined on it, $(S/\varepsilon; \Theta_\rho)$ is a quasiordered set too. The next lemma, which gives a way to construct an ordered set from any given quasiorder, is a version of the celebrated *Birkhoff lemma*, see (Birkhoff, 1967, Lemma 1, p. 21).

Lemma 5 *Let (S, ρ) be a quasiordered set. Then,*

(i) $\varepsilon_\rho = \rho \cap \rho^{-1}$ *is the greatest equivalence on S contained in a quasiorder* ρ.
(ii) *The relation* $\hat{\Theta}_\rho = \rho/\varepsilon_\rho$ *defined on* S/ε_ρ *by*

$$(x\varepsilon_\rho, y\varepsilon_\rho) \in \rho/\varepsilon_\rho = \hat{\Theta}_\rho \overset{\text{def}}{\Leftrightarrow} (x, y) \in \rho$$

is an ordering relation, i.e., $(S/\varepsilon_\rho; \hat{\Theta}_\rho)$ *is an ordered set.*
(iii) $\pi_{nat} : S \to S/\varepsilon_\rho$ *is a surjective isotone and reverse isotone mapping between the quasiordered set* (S, ρ) *and the ordered set* $(S/\varepsilon_\rho, \hat{\Theta}_\rho)$.

The order $\hat{\Theta}_\rho$ $(= \rho/\varepsilon_\rho)$ is called the *order defined by the quasiorder* ρ. Sometimes $\hat{\Theta}_\rho$ is referred to as the *kernel of the quasiorder* ρ. In addition, because of the Lemma 5, a quasiorder is often called a *preorder*. Let us remember that a quasiorder α on S is an order if and only if $\alpha \cap \alpha^{-1} = \Delta_S$.

So far we saw that the relation Θ_ρ, defined on the quotient set S/ε by an equivalence ε of a set S, is defined in terms of representatives of the equivalent classes. In the lemma which follows, it will be shown that the definition of Θ_ρ to a certain extent does not depend on the representatives.

Lemma 6 *Let (S, ρ) be a quasiordered set. Then the following conditions are equivalent:*

(i) $(x\varepsilon, y\varepsilon) \in \Theta_\rho \Leftrightarrow (\exists a \in x\varepsilon_\rho)(\exists b \in y\varepsilon_\rho)\, (a, b) \in \rho$.
(ii) $(x\varepsilon_\rho, y\varepsilon_\rho) \in \Theta_\rho \Leftrightarrow (x, y) \in \rho$.
(iii) $(x\varepsilon_\rho, y\varepsilon_\rho) \in \Theta_\rho \Leftrightarrow (\forall a \in x\varepsilon_\rho)(\forall b \in y\varepsilon_\rho)\, (a, b) \in \rho$.

Let (S, α) be an ordered set and let ε be an equivalence defined on it. Then, in general, the relation Θ_α need not be an order on the factor set S/ε. The following example (Kehayopulu & Tsingelis, 1995a) provides evidence for that.

Example 1 Let (S, α) be an ordered set with $S = \{a, b, c, d, e\}$ and

$$\alpha = \{(a, a), (a, d), (b, b), (c, c), (c, e), (d, d), (e, e)\}.$$

Let ε be an equivalence relation defined on S as

$$\varepsilon = \{(a, a), (a, e), (b, b), (c, c), (c, d), (d, c), (d, d), (e, a)(e, e)\}.$$

So $S/\varepsilon = \{\{a, e\}, \{c, d\}, \{b\}\}$, and let Θ_α be a relation on S/ε defined, as usual,

$$(x\varepsilon, y\varepsilon) \in \Theta_\alpha \Leftrightarrow (\exists a \in x\varepsilon)(\exists b \in y\varepsilon)\,(a, b) \in \alpha.$$

Then we have $(a\varepsilon, c\varepsilon) \in \Theta_\alpha$ since $(a, d) \in \alpha$ as well as $(c\varepsilon, a\varepsilon) \in \Theta_\alpha$ since $(c, e) \in \alpha$ and $a\varepsilon \neq c\varepsilon$, i.e., Θ_α is not antisymmetric. Thus $(S/\varepsilon, \Theta_\alpha)$ is not an ordered set. ◇

The *first isomorphism theorem for ordered sets* follows.

Theorem 3 *Let $f : S \rightarrow T$ be a mapping between ordered sets (S, ρ) and (T, σ). Then the following statements are true:*

(i) *Relation η_f is a quasiorder on S.*
(ii) $ker\ f = \varepsilon_f = \eta_f \cap \eta_f^{-1}$.
(iii) *The relation $\Theta_f = \eta_f/\varepsilon_f$ is an order on S/ε_f. The mapping $\varphi : S/\varepsilon_f \rightarrow T$ defined by $\varphi(x(ker\ f)) = f(x)$ is the unique isotone injective mapping between ordered sets $(S/\varepsilon_f, \Theta_f)$ and (T, σ), such that $f = \varphi \circ \pi_{nat}$ and $S/\varepsilon_f \cong_o f(S)$.*
(iv) *If f is surjective, then φ is an order bijection, i.e., we have $S/\varepsilon_f \cong_o T$.*

As a consequence of Theorem 2 and Theorem 3, we have next result.

Theorem 4 *Let $f : S \rightarrow T$ be a mapping between ordered sets (S, ρ) and (T, σ). Let ρ be a quasiorder on S such that $\rho \subseteq \eta_f$. Then the following statements are true:*

(i) $\varepsilon_\rho \subseteq ker\ f$.
(ii) *The mapping $\varphi : S/\varepsilon_\rho \rightarrow T$ defined by $\varphi(x\varepsilon_\rho) = f(x)$ is the unique isotone injective mapping between ordered sets $(S/\varepsilon_\rho, \hat{\Theta}_\rho)$ and (T, σ), such that $f = \varphi \circ \pi_\rho$.*

To conclude:

- Two isomorphism theorems for sets (Theorem 1 and Theorem 2) are based on equivalences.
- In the case of ordered sets, quasiorders play the role of equivalences (Theorem 3 and Theorem 4).

2.2 *Semigroups Within CLASS*

Although frequently fragmentary, studies of semigroups carried out at the end of 1920s are often marked as the beginning of their history. In the era without fast communication devices, semigroups took their place in development of other mathematical disciplines pretty fast. "The analytical theory of semi-groups is a recent addition to the ever- growing list of mathematical disciplines. It was my good fortune to take an early interest in this discipline and see its reach maturity. It has been a pleasant association: I hail a semigroup when I see one and I seem to see them anywhere! Friends have observed however that there are mathematical objects which are not semi-groups". See Foreword in Hille's book (1948).

2.2.1 Basic Concepts: Definitions, Examples, and Embedding Theorems

A *semigroup* $(S, \cdot)$ is a nonempty set S with a binary operation $\cdot$ called multiplication such that

$$(x \cdot y) \cdot z = x \cdot (y \cdot z),$$

for any $x, y, z \in S$. Frequently, xy is written rather than $x \cdot y$.

A mapping $f : S \to T$ between two semigroups S and T, which preserves the operations or is compatible with the operations of the semigroups, i.e.,

$$f(xy) = f(x)f(y),$$

is called a *homomorphism*. Several types of homomorphisms have a specific name. A homomorphism f is

- *embedding* or *monomorphism:* f is one–one
- *onto* or *epimorphism*: $f(S) = T$
- *isomorphism*: f is onto embedding.

If there is an isomorphism $f : S \to T$, then we say S and T are *isomorphic* and denote this by $S \cong T$.

A homomorphism $f : S \to S$ is called an *endomorphism*, and if it is, in addition, bijective, then it is called an *automorphism*.

Some of the fundamental examples of semigroups will be presented. Semigroups whose elements are the binary relations or the transformations on a given set play a central role in applications.

Let X be a set. Let $\rho, \sigma \in \mathcal{B}_X$ (Sect. 2.1). Define the composition of ρ and σ to be

$$\rho \circ \sigma = \{(x, y) : (\exists z \in X)(x, z) \in \rho \text{ and } (z, y) \in \sigma\}.$$

Proposition 1 $(\mathcal{B}_X, \circ)$ *is a monoid.*

In what follows we shall write, as usual, $\rho\,\sigma$ instead of $\rho \circ \sigma$. In addition, the ordered pair $(\mathcal{B}_X, \circ)$ will be denoted just as $\mathcal{B}_X$. Recall, a monoid is a semigroup with the identity.

The semigroup $\mathcal{T}_X$ is called the *full transformation semigroup* on X. A subsemigroup of $\mathcal{T}_X$ is called a *transformation semigroup* on X. The group $\mathcal{G}(X)$ is called the *symmetric group* on X. It is stated that permutation groups can serve as models for all groups. The fundamental importance of permutations group in algebra follows from the well-known Cayley Theorem for groups.

Theorem 5 *If G is a group, then there exists an embedding $\varphi : G \to \mathcal{G}_X$ for some X.*

Now, the Cayley Theorem for semigroups follows.

Theorem 6 *If S is a semigroup, then there exists an embedding $\varphi : G \to \mathcal{T}_X$ for some X.*

A homomorphism φ from S into some $\mathcal{T}_X$ is called a *representation of S by mappings*. A representation φ is called *faithful* if it is an embedding.

2.2.2 General Structure Results: Quotient Semigroups, Homomorphism, and Isomorphism Theorems

A relation ρ defined on a semigroup S is called:

- *left compatible* (with multiplication): $(x, y) \in \rho \Rightarrow (zx, zy) \in \rho$
- *right compatible* (with multiplication): $(x, y) \in \rho \Rightarrow (xz, yz) \in \rho$
- *compatible* (with multiplication): $(x, y), (s, t) \in \rho \Rightarrow (xs, yt) \in \rho$,

for any $x, y, z, s, t \in S$.

Lemma 7 *Let S be a semigroup and ρ a quasiorder defined on it. Then, ρ is compatible if and only if it is left and right compatible.*

A (left and right) compatible equivalence relation ε on a semigroup S is called a (*left* and *right*) *congruence*. The quotient set S/ε is then provided with a semigroup structure.

Theorem 7 *Let S be a semigroup and ε a congruence on it. Then S/ε is a semigroup with respect to the operation defined by $(x\varepsilon)(y\varepsilon) = (xy)\varepsilon$, and the mapping $\pi : S \to S/\varepsilon$, $\pi(x) = x\varepsilon$, $x \in S$, is an epimorphism.*

The semigroup S/ε from Theorem 7 is called a *quotient* or *factor semigroup* of S. The epimorphism π is called the *natural* or *canonical epimorphism* associated with the congruence ε.

The theorem that follows can be considered as a consequence of Theorems 1 and 2.

Theorem 8 *Let $f : S \to T$ be a homomorphism between semigroups S and T. Then*

(i) *$ker\ f = f \circ f^{-1}$ is a congruence on S, and the mapping $\varphi : S/ker\ f \to T$ defined by $\varphi(x(ker\ f)) = f(x)$ is an embedding such that $f = \varphi \circ \pi$. If f an epimorphism, then φ is an isomorphism.*
(ii) *If ε is a congruence on a semigroup S such that $\varepsilon \subseteq ker\ f$, then there exists a homomorphism of semigroups $\varphi : S/\rho \to T$, such that $f = \varphi \circ \pi$.*

2.2.3 Ordered Semigroups: Definitions, Ordered Homomorphism and Ordered Isomorphism Theorems

Let $(S, \cdot)$ be a semigroup and ρ relation on it. A triple $(S, \cdot; \rho)$ is called:

- a *quasiordered semigroup* if ρ is a compatible quasiorder
- an *ordered semigroup* if ρ is a compatible order.

Of course, a quasiorder or an order defined on a semigroup S need not, in general, be compatible.

Now, Birkhoff's lemma for semigroups follows next.

Lemma 8 *Let (S, ρ) be a quasiordered semigroup. Then,*

(i) *$\varepsilon_\rho = \rho \cap \rho^{-1}$ is the greatest congruence on S contained in a quasiorder ρ.*
(ii) *$(S/\varepsilon_\rho; \hat{\Theta}_\rho)$ is an ordered semigroup where*

$$(x\varepsilon_\rho, y\varepsilon_\rho) \in \rho/\varepsilon_\rho = \hat{\Theta}_\rho \overset{\text{def}}{\Leftrightarrow} (x, y) \in \rho.$$

(iii) *$\pi : S \to S/\varepsilon_\rho$ is an isotone and reverse isotone epimorphism between the quasiordered semigroup (S, ρ) and the ordered semigroup $(S/\varepsilon_\rho, \hat{\Theta}_\rho)$.*

In addition, because of Lemma 8 for quasiordered semigroup, a compatible quasiorder is sometimes called a *half-congruence*.

The results which follow are consequences of the ones given in Sect. 2.1.3 as well as in Sect. 2.2.2. The *first isomorphism theorem for ordered semigroups* follows first.

Theorem 9 *Let $f : S \to T$ be a homomorphism between ordered semigroups (S, ρ) and (T, σ). Then the following statements are true:*

(i) *$ker\ f = \varepsilon_f \subseteq \eta_f$.*
(ii) *The mapping $\varphi : S/ker\ f \to T$ defined by $\varphi(x(ker\ f)) = f(x)$ is the unique isotone embedding of the ordered semigroup $(S/ker\ f, \Theta_f)$ into (T, σ), such that $f = \varphi \circ \pi$ and $S/ker\ f \cong_o f(S)$.*
(iii) *If f is epimorphism, then φ is an order Isomorphism, i.e., we have $S/ker\ f \cong_o T$.*

Theorem 10 *Let $f : S \to T$ be a homomorphism between ordered sets (S, ρ) and (T, σ). Let ρ be a compatible quasiorder on S such that $\rho \subseteq \eta_f$. Then the following statements are true:*

(i) $\varepsilon_\rho \subseteq ker\ f$.
(ii) *The mapping $\varphi : S/\varepsilon_\rho \to T$ defined by $\varphi(x\varepsilon_\rho) = f(x)$ is the unique isotone embedding between ordered sets $(S/\varepsilon_\rho, \Theta_\rho)$ and (T, σ), such that $f = \varphi \circ \pi_\rho$.*

2.3 Applications and Possible Applications

Kline (1967) stated that "One can look at mathematics as a language, as a particular kind of logical structure, as a body of knowledge about number and space, as a series of methods for deriving conclusions, as the essence of our knowledge of the physical world, or merely as an amusing intellectual activity [...] Practical, scientific, philosophical, and artistic problems have caused men to investigate mathematics".

Within the previous subsection we presented some results concerning ordered sets, semigroups, and ordered semigroups within classical settings. The capability and flexibility of the just mentioned binary structures from the point of view of modeling and problem solving in extremely diverse situations have been already pointed out, and interesting new algebraic ideas arise with binary applications and connections to other areas of mathematics and sciences. In what follows we are going to turn our attention to their application within social sciences and humanities. However, we do not pretend to mention all of the existing applications.

Order and ordered structures enter (among other fields) into computer science and into humanities and social sciences in many ways and on many different levels. Order enters into the classifications of objects in two rather different levels:

- classifications of certain ordered sets according to various criteria
- the discipline of *concept analysis* provides, on a deeper level, a powerful technique for classifying and for analyzing complex sets of data.

Recall, formal concept analysis (**FCA**), invented by Rudolf Wille in the early 80s, built on the mathematical theories of ordered sets and lattices, is based on the mathematization of concept and concept hierarchy. "Hierarchies occur often both within mathematics and the 'real' world and the theory of ordered sets (and lattices) provides a natural setting in which to discuss and analyze them", as written in (Davey and Priestley, 2002). **FCA** has been proven successful in a wide range of applications: artificial intelligence, software engineering, chemistry, biology, psychology, linguistics, and sociology. More on applications of order theory within social sciences can be found in Davey and Priestley's book, 2002.

For application within linguistics, see, for example, Partee et al.'s book (1990 Preface – p. xvii). More precisely, as "Part C (p. 249–316) leads from the notions of order and operation to algebraic structures such as groups, semigroups, and monoids, and on to lattices and Boolean and Heyting algebras, which have played a pretty central role in certain work in the semantics of events, mass terms, collective vs. distributive actions, etc."

On the other hand, G. Birkhoff (1971) wrote "I do not wish to exaggerate the importance for computer science of lattices (including Boolean algebras), or of binary groups and fields. All of these have a quite special structure. A much more general class of algebraic systems is provided by semigroups, which are indeed basic for a great part of algebra". Almost at the same time, B. M. Schein (1970) stated that "One meets ... semigroups much more often than groups (and much more often than one thinks he does), and only the existing polarization of mind on the topic of groups and other classic algebraic structures prevents one from seeing semigroups in various processes and phenomena of mathematics and the universe".

In 1991 John Paul Boyd's book *Social semigroups a unified theory of scaling and blockmodeling as applied to social networks* (1991) appeared. One of the main goals of that book was, as Boyd wrote, "to equip the reader with powerful conceptual and analytical tools that can be used to solve other problems in the social sciences". Standard concepts of semigroup theory acquire a concrete meaning in terms of human kinship.

"Social semigroups? Are you serious?" shouted B. M. Schein (1997) in his review of Boyd's book and proceeded "semigroups do appear very naturally in various studies known under the common umbrella name of 'social sciences.' Yet, to see them, one needs eyes".

Recall, social networks are collections of social or interpersonal relationships linking individuals in a social group. The partially ordered semigroup of a network was introduced as an algebraic construction fulfilling the following two important requirements:

- it should be multirelational and so encompass different types of network relations in the description of an individual's social environment
- it should be concerned with the description of different kinds of network paths.

More about applications of partially ordered semigroups within the area can be found in Pattison (1993).

Abstract algebra can provide a great framework for analyzing music and abstracting the relationships found in modern (Western) music theory to uncover other possible music. More about the applications of semigroups in music can be found, for example, in the work of Bras-Amorós (2019, 2020).

At the very end of our short journey through the applications of semigroups within the areas of social sciences, humanities, and music, let us point out again that the list of applications given above is far from listing all of the existing applications of semigroup theory. This type of research can be a topic on its own for certain types of scientific writings.

2.4 Notes

Results presented in this section to a great extent can be seen as a kind of generalization of these given in Chajda (2004), Chajda and Snášel (1998), and Kórtesi et al. (2005) for relational structures of a certain type, Bloom (1976), Szédli and Lenkehegyi (1983a,b) for the ordered algebras, and Kehayopulu and Tsingelis (1995a,b) for the ordered semigroup case.

More about ordered sets and their history can be found, for example, in Duffus and Rival (1981) and Schröder (2003). The standard reference for semigroup theory is Howie (1995). For ordered semigroup theory and on some other algebraic structures, see Blyth (2005).

3 Within BISH

Throughout this chapter *constructive mathematics* is understood as mathematics performed in the context of intuitionistic logic, that is, without the law of the excluded middle (**LEM**). **LEM** can be regarded as the main source of nonconstructivity. It was Brouwer (1975), who first observed that **LEM** was extended without justification to statements about infinite sets. By constructive mathematics we mean Bishop-style mathematics, **BISH**. Several consequences of **LEM** are not accepted in Bishop's constructivism. We will mention two such nonconstructive principles—the ones that will be used latter.

- **The limited principle of omniscience (LPO)**: For each binary sequence $(a_n)_{n\geq 1}$, either $a_n = 0$ for all $n \in \mathbb{N}$, or else there exists n with $a_n = 1$.

- **Markov's principle (MP)**: For each binary sequence $(a_n)_{n\geq 1}$, if it is impossible that $a_n = 0$ for all $n \in \mathbb{N}$, then there exists n with $a_n = 1$.

LPO is equivalent to the decidability of equality on the real number line $\mathbb{R}$.

$$\forall_{x\in\mathbb{R}} \, (x = 0 \vee x \neq 0).$$

Within constructive mathematics, a statement P, as in classical mathematics, can be disproved by giving a counterexample. However, it is also possible to give a *Brouwerian counterexample* to show that the statement is nonconstructive. A Brouwerian counterexample to a statement P is a constructive proof that P implies some nonconstructive principle, such as, for example, **LEM** and its weaker versions **LPO** and **MP**. It is not a counterexample in the true sense of the word—it is just an indication that P does not admit a constructive proof.

Following Troelstra and van Dalen (1988), *constructive algebra* is more complicated than classical algebra in various ways: algebraic structures as a rule do not carry a decidable equality relation—this difficulty is partly met by the introduction of a strong inequality relation, the so-called *apartness* relation; there is (sometime) the awkward abundance of all kinds of substructures and hence of quotient structures.

As highlighted by M. Mitrović et al. (2019), *the theory of constructive semigroup with apartness* is a new approach to semigroup theory and not a new class of semigroups. It presents a semigroup facet of some relatively well-established direction of constructive mathematics which, to the best of our knowledge, has not yet been considered within the semigroup community.

3.1 Set with Apartness

The cornerstones for **BISH** include the notion of positive integers, sets and functions. The set $\mathbb{N}$ of positive numbers is regarded as a basic set, and it is assumed that the positive numbers have the usual algebraic and order properties, including mathematical induction. Contrary to the classical case, a set exists only when it is defined.

3.1.1 Basic Concepts and Important Examples

To define a set S, we have to give a property that enables us to construct members of S and to describe the equality $=$ between elements of S, which is a matter of convention, except that it must be an equivalence. A set $(S, =)$ is an *inhabited* set if we can construct an element of S. The distinction between the notions of a nonempty set and an inhabited set is a key in constructive set theories. The notion of equality of different sets is not defined. The only way in which elements of two different sets can be regarded as equal is by requiring them to be subsets of a third set.

A property P, which is applicable to the elements of a set S, determines a subset of S denoted by $\{x \in S : P(x)\}$. Furthermore, we will be interested only in properties $P(x)$ which are *extensional* in the sense that for all $x_1, x_2 \in S$ with $x_1 = x_2$, $P(x_1)$ and $P(x_2)$ are equivalent. Informally, it means that "it does not depend on the particular description by which x is given to us".

Let $(S, =)$ be an *inhabited* set. By an ***apartness*** on S, we mean a binary relation # on S which satisfies the axioms of irreflexivity, symmetry, and co-transitivity:

(Ap1) $\neg(x\#x)$

(Ap2) $x\#y \Rightarrow y\#x$

(Ap3) $x\#z \Rightarrow \forall_y\, (x\#y \vee y\#z)$

If $x\#y$, then x and y are different or distinct. Roughly speaking, $x = y$ means that we have a proof that x equals y, while $x\#y$ means that we have a proof that x and y are different. Therefore, the negation of $x = y$ does not necessarily imply that $x\#y$ and vice versa: given x and y, we may have neither a proof that $x = y$ nor a proof that $x\#y$.

The apartness on a set S is *tight* if

(Ap4) $\neg(x\#y) \Rightarrow x = y.$

If apartness is tight, then $\neg(x\#y) \Leftrightarrow x = y$.

By extensionality, we have

(Ap5) $x\#y \wedge y = z \Rightarrow x\#z,$

the equivalent form of which is

(Ap5') $x\#y \wedge x = x' \wedge y = y' \Rightarrow x'\#y'.$

A *set with apartness* $(S, =, \#)$ is the starting point for our considerations and will be simply denoted by S. The existence of an apartness relation on a structure often gives rise to an apartness relation on another structure. For example, given two sets with apartness $(S, =_S, \#_S)$ and $(T, =_T, \#_T)$, it is permissible to construct the set of mappings between them. A *mapping* $f : S \to T$ is an algorithm which produces an element $f(x)$ of T when applied to an element x of S, which is extensional, that is,

$$\forall_{x,y\in S}\, (x =_S y \Rightarrow f(x) =_T f(y)).$$

An important property applicable to mapping f is that of strong extensionality. Namely, a mapping $f : S \to T$ is a *strongly extensional* mapping, or, for short, an *se-mapping*, if

$$\forall_{x,y\in S}\, (f(x)\#_T f(y) \Rightarrow x\#_S y).$$

Let us remember that strong extensionality of all mappings from $\mathbb{R}$ to $\mathbb{R}$ implies the Markov principle (**MP**).

An se-mapping f is:

- an *se-surjection* if it is surjective
- an *se-injection* if it is injective
- an *se-bijection* if it is bijective.

Furthermore, f is

- *apartness injective*, shortly *a-injective*: $\forall_{x,y\in S}\, (x\#_S y \Rightarrow f(x)\#_T f(y))$
- *apartness bijective*: a-injective, se-bijective.

Given a set X with apartness, it is permissible to construct the set of all se-mappings of X into itself which inherits apartness from X.

Theorem 11 *Let $(X, =, \#)$ be a set with apartness. If $\mathcal{T}_S^{se} = M(X, X)$ is the set of all se-mappings from X to X, then $(\mathcal{T}_S^{se}, =, \#)$ with*

$$\begin{aligned} f = g &\Leftrightarrow \forall_{x\in X}\, (f(x) = g(x)) \\ f\#g &\Leftrightarrow \exists_{x\in X}\, (f(x)\#g(x)) \end{aligned}$$

is a set with apartness too.

Given two sets with apartness S and T, it is permissible to construct the set of ordered pairs $(S \times T, =, \#)$ of these sets defining apartness by

$$(s, t) \# (u, v) \overset{\text{def}}{\Leftrightarrow} s \#_S u \vee t \#_T v.$$

3.1.2 Distinguishing Subsets

The presence of apartness implies the appearance of different types of substructures connected to it. Following Bridges and Vîţă (2011), we define the relation $\bowtie$ between an element $x \in S$ and a subset Y of S by

$$x \bowtie Y \overset{\text{def}}{\Leftrightarrow} \forall_{y \in Y} (x \# y).$$

A subset Y of S has two natural complementary subsets: *the logical complement* of Y

$$\neg Y \overset{\text{def}}{=} \{x \in S : x \notin Y\}$$

and *the apartness complement* or, shortly, the *a-complement* of Y

$$\sim Y \overset{\text{def}}{=} \{x \in S : x \bowtie Y\}.$$

Denote by $\widetilde{x}$ the a-complement of the singleton $\{x\}$. Then it can be easily shown that $x \in \sim Y$ if and only if $Y \subseteq \widetilde{x}$.

If the apartness is not tight, we can find subsets Y with $\sim Y \subset \neg Y$ as in the following example.

Example 2 Let $S = \{a, b, c\}$ be a set with apartness defined by $\{(a, c), (c, a), (b, c), (c, b)\}$ and let $Y = \{a\}$. Then the a-complement $\sim Y = \{c\}$ is a proper subset of its logical complement $\neg Y = \{b, c\}$. $\diamond$

The complements are used for the classification of subsets of a given set. A subset Y of S is

- *a detachable* subset in S or, in short, a *d-subset* in S if

$$\forall_{x \in S} (x \in Y \vee x \in \neg Y)$$

- *a strongly detachable* subset of S, shortly *an sd-subset of* S, if

$$\forall_{x \in S} (x \in Y \vee x \in \sim Y)$$

- *a quasi-detachable* subset of S, shortly *a qd-subset of* S, if

$$\forall_{x\in S}\,\forall_{y\in Y}\,(x \in Y \vee x\#y).$$

A description of the relationships between those subsets of set with apartness, which, in turn, justifies the *constructive order theory* for sets and semigroups with apartness we develop, is given in the next theorem.

Theorem 12 *Let Y be a subset of S. Then,*

(i) *Any sd-subset is a qd-subset of S. The converse implication entails* **LPO**.
(ii) *Any qd-subset Y of S satisfies $\sim Y = \neg Y$.*
(iii) *If any qd-subset is a d-subset, then* **LPO** *holds.*
(iv) *If any d-subset is a qd-subset, then* **MP** *holds.*
(v) *Any sd-subset is a d-subset of S. The converse implication entails* **MP**.
(vi) *If any subset of a set with apartness S is a qd-subset, then* **LPO** *holds.*

For all subsets Y of the set with apartness S for which two distinguished complements coincide, we will adopt the following notation:

$$\boldsymbol{Y^c} = \sim Y = \neg Y.$$

3.1.3 Binary Relations

Let $(S \times S, =, \#)$ be a set with apartness. An inhabited subset of $S \times S$, or, equivalently, a property applicable to the elements of $S \times S$, is called a *binary relation* on S. Let α be a relation on S. Then

$$(a, b) \bowtie \alpha \;\Leftrightarrow\; \forall_{(x,y)\in\alpha}\,((a, b)\,\#\,(x, y)),$$

for any $(a, b) \in S \times S$. The apartness complement of α is the relation

$$\sim \boldsymbol{\alpha} = \{(x, y) \in S \times S : (x, y) \bowtie \alpha\}.$$

In general, we have $\sim \alpha \subseteq \neg\alpha$, which is shown by the following example.

Example 3 Let $S = \{a, b, c\}$ be a set with apartness defined by $\{(a, c), (c, a), (b, c), (c, b)\}$. Let $\alpha = \{(a, c), (c, a)\}$ be a relation on S. Its a-complement

$$\sim \alpha = \{(a, a), (b, b), (c, c), (a, b), (b, a)\}$$

is a proper subset of its logical complement $\neg\alpha$. ◇

The subset $\nabla_S = \{(x, y) \in S \times S : \neg(x = y)\}$ of $S \times S$ is called the *co-diagonal* of S.

Let α and β be relations on S. Then, the following operations can be defined:

- the *composition* or *product* of α and β:

$$\alpha \circ \beta = \{(x, z) \in S \times S : \exists_{y \in S} ((x, y) \in \alpha \wedge (y, z) \in \beta)\}$$

- the *co-composition* or *co-product* of α and β:

$$\alpha * \beta = \{(x, z) \in S \times S : \forall_{y \in S} ((x, y) \in \alpha \vee (y, z) \in \beta)\}$$

- α is *associated* with β:

$$\alpha \looparrowleft \beta \overset{\text{def}}{\Leftrightarrow} \forall_{x,y,z \in S} ((x, y) \in \alpha \wedge (y, z) \in \beta \Rightarrow (x, z) \in \alpha).$$

In what follows, the next lemma will be of practical use.

Lemma 9 *Let α and β be relations on S such that $\alpha \subseteq \gamma$ and $\beta \subseteq \delta$. Then $\alpha * \beta \subseteq \gamma * \delta$.*

The relation α defined on a set with apartness S is

- strongly irreflexive if $(x, y) \in \alpha \Rightarrow x \# y$
- co-transitive if $(x, y) \in \alpha \Rightarrow \forall_{z \in S} ((x, z) \in \alpha \vee (z, y) \in \alpha)$
- co-antisymmetric if $\forall_{x,y \in S} (x \# y \Longrightarrow (x, y) \in \alpha \vee (x, y) \in \alpha^{-1})$.

An alternatively way of defining properties given above is as follows:

- *irreflexive* if $\alpha \subseteq \nabla$
- *strongly irreflexive* if $\alpha \subseteq \#$
- *co-transitive* if $\alpha \subseteq \alpha * \alpha$
- *co-antisymmetric* if $\# \subseteq (\alpha \cup \alpha^{-1})$.

It is easy to check that a strongly irreflexive relation is also irreflexive. For a tight apartness, the two notions of irreflexivity are classically equivalent but not so constructively. More precisely, if each irreflexive relation were strongly irreflexive then **MP** would hold.

Lemma 10 *Let α be a relation on S. Then,*

(i) *If α (respectively, β) is strongly irreflexive, then $\alpha * \beta \subseteq \beta$ (respectively, $\alpha * \beta \subseteq \alpha$).*
(ii) *α is strongly irreflexive, then $\alpha * \alpha$ is strongly irreflexive.*

In what follows the following result will be in handy and will be used without further reference.

Lemma 11 *If α is strongly irreflexive or co-transitive, then α^{-1} is strongly irreflexive or co-transitive too.*

Now, we turn our attention to certain properties of the a-complement of given types of relations.

Lemma 12 *Let α be a relation on S. Then,*

(i) *α is strongly irreflexive if and only if $\sim \alpha$ is reflexive.*
(ii) *If α is symmetric, then $\sim \alpha$ is symmetric.*
(iii) *If α is co-transitive, then $\sim \alpha$ is transitive.*
(iv) *If α is co-antisymmetric and the apartness on S is tight, then $\sim \alpha$ is antisymmetric.*

The relation α defined on a set S is as follows:

- *co-quasiorder* if it is strongly irreflexive and co-transitive.

The co-quasiorder is one of the main building blocks for the co-order theory of set with (non-tight) apartness that we develop. We can use them to define the following important types of relations:

- *co-equivalence*: a symmetric co-quasiorder
- *co-order*: a co-antisymmetric co-quasiorder.

As in Example 3 the a-complement of a relation can be a proper subset of its logical complement. If the relation in question is a co-quasiorder, then we have the following important properties.

Proposition 2 *Let τ be a co-quasiorder on S. Then,*

(i) *τ is a qd-subset of $S \times S$.*
(ii) $\sim \tau = \neg \tau (= \tau^c)$.
(iii) *τ^c is a quasiorder on S.*

Lemma 13 *If τ and σ are co-quasiorders on a set S, then $\tau \cup \sigma$ is a co-quasiorder too.*

3.1.4 Apartness Isomorphism Theorems for Sets with Apartness

A quotient structure does not have, in general, a natural apartness relation. For most purposes, we overcome this problem using a ***co-equivalence***–symmetric co-quasiorder—instead of an equivalence. Existing properties of a co-equivalence guarantee that its a-complement is an equivalence and that the quotient set of that equivalence will inherit an apartness. The following notion will be necessary. For any two relations α and β on S, we can say that $\boldsymbol{\alpha}$ *defines an apartness on* $\mathbf{S}/\boldsymbol{\beta}$ if we have

$$\text{(Ap 6)} \qquad x\beta \# y\beta \overset{\text{def}}{\Leftrightarrow} (x, y) \in \alpha.$$

Lemma 14 *If α is a co-quasiorder and β an equivalence on a set S, then* (Ap 6) *implies*

$$\text{(Ap 6}'\text{)} \qquad ((x, a) \in \beta \wedge (y, b) \in \beta) \Rightarrow ((x, y) \in \alpha \Leftrightarrow (a, b) \in \alpha).$$

Theorem 13 *Let κ be a co-equivalence on S. Then*

(i) *The relation κ^c is an equivalence on S such that $\kappa \not\leftarrow\!\!\!\!\!\!\!\!\!\!\!\!\!\! \quad \kappa^c$.*
(ii) *$(S/\kappa^c, =, \#)$ is a set with apartness where*

$$a\kappa^c = b\kappa^c \Leftrightarrow (a, b) \bowtie \kappa$$
$$a\kappa^c \# b\kappa^c \Leftrightarrow (a, b) \in \kappa.$$

(iii) *The quotient mapping $\pi : S \to S/\kappa^c$, defined by $\pi(x) = x\kappa^c$, is an se-surjection.*

Let $f : S \to T$ be an se-mapping between sets with apartness. Then the relation

$$\text{coker } f \stackrel{\text{def}}{=} \{(x, y) \in S \times S : f(x) \# f(y)\}$$

defined on S is called the *co-kernel* of f.

Now, *the first apartness isomorphism theorem* for sets with apartness follows.

Theorem 14 *Let $f : S \to T$ be an se-mapping between sets with apartness. Then*

(i) coker f *is a co-equivalence on S.*
(ii) coker $f \not\leftarrow$ ker f *and* ker $f \subseteq (\text{coker } f)^c$.
(iii) *$(S/\ker f, =, \#)$ is a set with apartness, where*

$$a(\ker f) = b(\ker f) \Leftrightarrow (a, b) \in \ker f$$
$$a(\ker f) \# b(\ker f) \Leftrightarrow (a, b) \in \text{coker } f.$$

(iv) *The mapping $\varphi : S/\ker f \to T$, defined by $\varphi(x(\ker f)) = f(x)$, is an a-injective se-injection such that $f = \varphi\pi$.*
(v) *If f maps S onto T, then φ is an apartness bijection.*

The next theorem concerns a more general situation.

Theorem 15 *Let S be a set with apartness. Then,*

(i) *Let ε be an equivalence and κ a co-equivalence on S. Then, κ defines an apartness on the factor set S/ε if and only if $\varepsilon \cap \kappa = \emptyset$.*
(ii) *The quotient mapping $\pi : S \to S/\varepsilon$, defined by $\pi(x) = x\varepsilon$, is an se-surjection.*

Now, *the second apartness isomorphism theorem*, a generalized version of Theorem 14, for sets with apartness follows.

Theorem 16 *Let $f : S \to T$ be a mapping between sets with apartness, and let κ be a co-equivalence on S such that $\kappa \cap \ker f = \emptyset$. Then,*

(i) *κ defines an apartness on the factor set $S/\ker f$.*
(ii) *The projection $\pi : S \to S/\ker f$ defined by $\pi(x) = x(\ker f)$ is an se-surjection.*
(iii) *The mapping $\varphi : S/\ker f \to T$, given by $\varphi(x(\ker f)) = f(x)$, is an a-injective se-injection such that $f = \varphi\pi$.*
(iv) *φ is an se-mapping if and only if $\operatorname{coker} f \subseteq \kappa$.*
(v) *φ is a-injective if and only if $\kappa \subseteq \operatorname{coker} f$.*
(vi) *If φ is an se-mapping, then f is an se-mapping too.*

3.1.5 Co-ordered Sets with Apartness

A *constructive version of Birkhoff's lemma* (Lemma 5) follows.

Lemma 15 *Let $(S, =, \#; \tau)$ be a co-quasiordered set with apartness. Then*

(i) *$\kappa_\tau = \tau \cup \tau^{-1}$ is a co-equivalence on S.*
(ii) *$(S/\kappa_\tau^c, =, \#)$ is a set with apartness where*

$$x\kappa_\tau^c = y\kappa_\tau^c \Leftrightarrow (x, y) \bowtie \kappa_\tau$$
$$x\kappa_\tau^c \# y\kappa_\tau^c \Leftrightarrow (x, y) \in \kappa_\tau.$$

(iii) *$(S/\kappa_\tau^c, =, \#; \Upsilon_\tau)$ is a co-ordered set with apartness where*

$$(x\kappa_\tau^c, y\kappa_\tau^c) \in \Upsilon_\tau \overset{\text{def}}{\Leftrightarrow} (x, y) \in \tau.$$

(iv) *The quotient mapping $\pi_\tau : S \to S/\kappa_\tau^c$, defined by $\pi_\tau(x) = x\kappa_\tau^c$, is an isotone and reverse isotone se-surjection.*

Any mapping between a set with apartness $(S, =, \#)$ and a co-quasiordered set with apartness $(T, =, \#\ \sigma)$ defines the relation μ_f on S by

$$(x, y) \in \mu_f \overset{\text{def}}{\Leftrightarrow} (f(x), f(y)) \in \sigma.$$

Theorem 17 *Let $(S, =, \#)$ be a set with apartness, and let $(T, =, \#; \sigma)$ be a co-quasiordered set with apartness. If $f : S \to T$ is an se-mapping, then*

(i) *μ_f is a co-quasiorder on S.*
(ii) *$\kappa_f = \mu_f \cup \mu_f^{-1}$ is a co-equivalence on S such that $\kappa_f \subseteq \operatorname{coker} f$.*
(iii) *If σ is a co-order on T, then $\kappa_f = \operatorname{coker} f$.*

(iv) $(S/\kappa_f^c, =, \#; \Upsilon_f)$ *where*

$$(x\kappa_f^c, y\kappa_f^c) \in \Upsilon_f \overset{\text{def}}{\Leftrightarrow} (x, y) \in \mu_f$$

is a co-ordered set with apartness.

(v) *The mapping* $\psi : S/\kappa_f^c \to S/\kappa_\sigma^c$ *between co-ordered sets with apartness* $(S/\kappa_f^c, =, \#; \Upsilon_f)$ *and* $(S/\kappa_\sigma^c, =, \#; \Upsilon_\sigma)$, *defined by* $\psi(x\kappa_f) = (f(x))\kappa_\sigma$, *is an isotone and reverse isotone a-injective se-mapping such that* $\psi\pi_f = \pi_\sigma f$.

Theorem 18 *Let* $(S, =, \#, \tau)$ *be a co-quasiordered set with apartness, and let* $(T, =, \#; \sigma)$ *be a co-quasiordered set with tight apartness. If* $f : S \to T$ *is an se-mapping such that* $\mu_f \subseteq \tau$, *then the mapping* $\varphi : S/\kappa_\tau^c \to T$, *defined by* $\varphi(x\kappa_\tau^c) = f(x)$, *is an se-mapping such that* $\varphi\pi_\tau = f$.

3.2 Semigroups with Apartness

During the implementation of the **FTA** Project, Geuvers et al. (2002a,b), the notion of commutative constructive semigroups with tight apartness appeared. We put noncommutative constructive semigroups with "ordinary" apartness in the center of our study, proving first, of course, that such semigroups do exist. The initial step toward grounding the theory is done by our contributing papers, Mitrović and co-authors: Crvenković et al. (2013), Crvenković et al. (2016), Mitrović et al. (2021), Mitrović and Silvestrov (2020), and Mitrović et al. (2019). The theory of semigroup with apartness is a new approach to semigroup theory and not a new class of semigroups. It presents a semigroup facet of some relatively well-established direction of constructive mathematics which, to the best of our knowledge, has not yet been considered within the semigroup community.

3.2.1 Basic Concepts: Definitions, Examples, and Se-embeddings

A semigroup with apartness satisfies a number of extra conditions, firstly the well-known axioms of apartness, and secondly the semigroup operation has to be strongly extensional.

Given a set with apartness $(S, =, \#)$, the tuple $(S, =, \#, \cdot)$ is a *semigroup with apartness* if the binary operation $\cdot$ is associative

(A) $\forall_{a,b,c\in S}\ [(a \cdot b) \cdot c = a \cdot (b \cdot c)]$,

and strongly extensional

(S) $\forall_{a,b,x,y\in S}\ (a \cdot x \# b \cdot y \Rightarrow (a \# b \vee x \# y))$.

As usual, we are going to write ab instead of $a \cdot b$. Example 4 provides a concrete instance of a semigroup with apartness.

Example 4 Let $S = \{a, b, c, d, e\}$ be a set with diagonal Δ_S as the equality relation. If we denote $K = \Delta_S \cup \{(a, b), (b, a)\}$, then we can define an apartness $\#$ on S to be $(S \times S) \setminus K$. Thus, $(S, =, \#)$ is a set with apartness. If we define multiplication on the set S as

·	a	b	c	d	e
g	a	b	c	d	e
a	b	b	d	d	d
b	b	b	d	d	d
c	d	d	c	d	c
d	d	d	d	d	d
e	d	d	c	d	c

then $(S, =, \#; \cdot)$ is a semigroup with apartness. ◇

An important example of a class of semigroup with apartness arises in the following way. For a given set with apartness X, we can construct a semigroup with apartness $\mathcal{T}_X^{se}$ as given in the next theorem.

Theorem 19 *Let X be a set with apartness. Let $\mathcal{T}_X^{se}$ be the set of all se-functions from X to X with*

$$f = g \Leftrightarrow \forall_{x \in S} (f(x) = g(x))$$

and apartness

$$f \# g \Leftrightarrow \exists_{x \in A} (f(x) \# g(x)).$$

Then $(\mathcal{T}_X^{se}, =, \#, \cdot)$ is a semigroup with apartness with respect to the binary operation of composition of functions.

Until the end of this chapter, we adopt the convention that *semigroup* means *semigroup with apartness*. Apartness from Theorem 19 does not have to be tight. The following example shows that we cannot even prove constructively that the apartness on every *finite* semigroup is tight.

Example 5 Let $X = \{0, 1, 2\}$ with the usual equality relation, that is, the diagonal Δ_X of $X \times X$. Let

$$K = \Delta_X \cup \{(1, 2), (2, 1)\},$$

and define an apartness $\#$ on X to be $(X \times X) \setminus K$. Then, by Theorem 19, $\mathcal{T}_X^{se}$ becomes a semigroup with apartness. Define mappings $f, g : X \to X$ by

$$f(0) = 1,\ f(1) = 1,\ f(2) = 2,$$
$$g(0) = 2,\ g(1) = 1,\ g(2) = 2.$$

In view of our definition of the apartness on X, there is no element $x \in X$ with $f(x)\#g(x)$; so, in particular, f and g are se-functions. However, if $f = g$, then $1 = 2$, which, by our definition of the equality on S, is not the case. Hence the apartness on S is not tight. ◇

Let S and T be semigroups with apartness. An homomorphism $f : S \to T$ is

- an *se-embedding* if it is an se-injection
- an *apartness embedding* if it is an a-injective se-embedding
- an *apartness isomorphism* if it is an apartness bijection.

As a consequence of Theorem 19, we can formulate *the constructive Cayley Theorem for semigroups with apartness* as follows.

Theorem 20 *Every semigroup with apartness se-embeds into the semigroup of all strongly extensional self-maps on a set.*

3.2.2 Co-quasiorders Defined on a Semigroup

A relation τ defined on a semigroup S with apartness is called:

- *left co-compatible*: $(zx, zy) \in \tau \Rightarrow (x, y) \in \tau$
- *right co-compatible*: $(xz, yz) \in \tau \Rightarrow (x, y) \in \tau$
- *co-compatible*: $(xz, yt) \in \tau \Rightarrow (x, y) \in \tau \vee (z, t) \in \tau$

for any $x, y, z, t \in S$.

Let us proceed with an example of a co-quasiorder defined on a semigroup with apartness S.

Example 6 Let S be a semigroup with apartness as given in Example 4. The relation τ, defined by

$$\tau = \{(c, a), (c, b), (d, a), (d, b), (d, c), (e, a), (e, b), (e, c), (e, d)\},$$

is a co-quasiorder on S.◇

The lemma which follows will be used without reference.

Lemma 16 *Let τ be a co-quasiorder on a semigroup with apartness S. Then, τ is co-compatible if and only if τ is a left and a right co-compatible.*

3.2.3 Apartness Isomorphism Theorems for Semigroups with Apartness

Let us remember that in **CLASS** the compatibility property is an important condition for providing the semigroup structure on quotient sets. Now we are looking for the tools for introducing an apartness relation on a factor semigroup. Our starting point is the results from Sect. 3.1.4, as well as the next definition.

A co-equivalence κ is a *co-congruence* if it is *co-compatible*.

Theorem 21 *Let S be a semigroup with apartness, and let κ be a co-congruence on S. Define*

$$\begin{aligned} a\kappa^c &= b\kappa^c \Leftrightarrow (a,b) \bowtie \kappa, \\ a\kappa^c \# b\kappa^c &\Leftrightarrow (a,b) \in \kappa, \\ a\kappa^c\, b\kappa^c &= (ab)\kappa^c. \end{aligned}$$

Then $(S/\kappa^c, =, \#, \cdot)$ is a semigroup with apartness. Moreover, the quotient mapping $\pi : S \to S/\kappa^c$, defined by $\pi(x) = x\kappa^c$, is an se-epimorphism.

The *first apartness isomorphism theorem for semigroups with apartness* follows.

Theorem 22 *Let $f : S \to T$ be an se-homomorphism between semigroups with apartness. Then*

(i) coker f *is a co-congruence on S.*
(ii) coker $f \not\leftarrow$ ker f *and* ker $f \subseteq (\text{coker } f)^c$.
(iii) $(S/\ker f, =, \#; \cdot)$ *is a semigroup with apartness, where*

$$\begin{aligned} a(\ker f) = b(\ker f) &\Leftrightarrow (a,b) \in \ker f \\ a(\ker f) \# b(\ker f) &\Leftrightarrow (a,b) \in \text{coker } f. \end{aligned}$$

(iv) *The mapping $\varphi : S/\ker f \to T$, defined by $\varphi(x(\ker f)) = f(x)$, is an apartness embedding such that $f = \varphi\pi$.*
(v) *If f maps S onto T, then φ is an apartness isomorphism.*

The next theorem deals with a more general situation.

Theorem 23 *Let S be a semigroup with apartness. Then*

(i) *Let μ be a congruence and κ a co-congruence on S. Then, κ defines an apartness on the factor set S/μ if and only if $\mu \cap \kappa = \emptyset$.*
(ii) *The quotient mapping $\pi : S \to S/\mu$, defined by $\pi(x) = x\mu$, is an se-epimorphism.*

As a consequence of Theorems 16 and 23, we have the following generalization of Theorem 22.

Theorem 24 *Let $f : S \to T$ be a mapping between sets with apartness, and let κ be a co-equivalence on S such that $\kappa \cap \ker f = \emptyset$. Then,*

(i) *If S is semigroup with apartness and κ a co-congruence, then $S/\ker f$ is a semigroup with apartness and $\pi : S \to S/\ker f$ an se-epimorphism.*
(ii) *If, in addition, T is a semigroup with apartness and f an se-homomorphism, then $\varphi : S/\ker f \to T$ is also an se-homomorphism.*

3.2.4 Co-ordered Semigroup with Apartness

The next results to be presented are consequences of the ones given in Sects. 3.1.5 and 3.2.2.

A tuple $(S, =, \#; \cdot; \alpha)$ is called:

- *co-quasiordered semigroup* if α is a co-compatible co-quasiorder on S
- *co-ordered semigroups* if α is a co-compatible co-order on S.

A constructive version for Birkhoff's lemma for semigroups follows.

Lemma 17 *Let $(S, =, \#; \cdot; \tau)$ be a co-quasiordered semigroup with apartness. Then*

(i) *$\kappa_\tau = \tau \cup \tau^{-1}$ is a co-congruence on S.*
(ii) *$(S/\kappa_\tau^c, =, \#)$ is a semigroup with apartness where*

$$x\kappa_\tau^c = y\kappa_\tau^c \Leftrightarrow (x, y) \bowtie \kappa_\tau$$

$$x\kappa_\tau^c \# y\kappa_\tau^c \Leftrightarrow (x, y) \in \kappa_\tau.$$

(iii) *$(S/\kappa_\tau^c, =, \#;\ \Upsilon_\tau)$ is a co-ordered semigroup with apartness where*

$$(x\kappa_\tau^c, y\kappa_\tau^c) \in \Upsilon_\tau \overset{\text{def}}{\Leftrightarrow} (x, y) \in \tau.$$

(iv) *The quotient mapping $\pi_\tau : S \to S/\kappa_\tau^c$, defined by $\pi_\tau(x) = x\kappa_\tau^c$, is an isotone and reverse isotone se-epimorphism.*

Theorem 25 *Let $(S, =, \#)$ be a semigroup with apartness, and let $(T, =, \#;\ \sigma)$ be a co-quasiordered semigroup with apartness. If $f : S \to T$ is an se-mapping, then*

(i) *μ_f is a co-compatible co-quasiorder on S.*
(ii) *$\kappa_f = \mu_f \cup \mu_f^{-1}$ is a co-congruence on S such that $\kappa_f \subseteq \text{coker } f$.*
(iii) *If σ is a co-order on T, then $\kappa_f = \text{coker } f$.*
(iv) *$(S/\kappa_f^c, =, \#; \cdot;\ \Upsilon_f)$ where*

$$(x\kappa_f^c, y\kappa_f^c) \in \Upsilon_f \overset{\text{def}}{\Leftrightarrow} (x, y) \in \mu_f$$

is a co-ordered semigroup with apartness.
(v) *The mapping $\psi : S/\kappa_f^c \to S/\kappa_\sigma^c$ between co-ordered semigroups with apartness $(S/\kappa_f^c, =, \#; \cdot;\ \Upsilon_f)$ and $(S/\kappa_\sigma^c, =, \#; \cdot \Upsilon_\sigma)$, defined by $\psi(x\kappa_f) = (f(x))\kappa_\sigma$, is an isotone and reverse isotone a-injective se-homomorphism such that $\psi\pi_f = \pi_\sigma f$.*

Theorem 26 *Let $(S, =, \#; \cdot\ \tau)$ be a co-quasiordered semigroup with apartness, and let $(T, =, \#; \cdot;\ \sigma)$ be a co-quasiordered semigroup with tight apartness. If $f : S \to T$ is an se-homomorphism such that $\mu_f \subseteq \tau$, then the mapping*

$\varphi : S/\kappa_\tau^c \to T$, defined by $\varphi(x\kappa_\tau^c) = f(x)$, is an se-homomorphism such that $\varphi\pi_\tau = f$.

3.3 Applications and Possible Applications

There is no doubt about the deep connections between constructive mathematics and computer science. Moreover, "if programming is understood not as the writing of instructions for this or that computing machine but as the design of methods of computation that is the computer's duty to execute, then it no longer seems possible to distinguish the discipline of programming from constructive mathematics", Martin-Löf (1982)

Let us give some examples of applications of ideas presented in the previous section. We will start with constructive analysis. The proof of one of the directions of the constructive version of the Spectral Mapping Theorem is based on some elementary constructive semigroups with inequality techniques, Bridges and Havea (2001). It is also worth mentioning the applications of commutative basic algebraic structures with tight apartness within the automated reasoning area, Calderón (2017). For possible applications within computational linguistics, see Moshier (1995). Some topics from mathematical economics can be approached constructively too (using some order theory for sets with apartness), Baroni and Bridges (2008).

The theory of semigroups with apartness is, of course, in its infancy, but, as we have already pointed out, it promises the prospect of applications in other (constructive) mathematics disciplines, certain areas of computer science, social sciences, and economics. Contrary to the classical case, the applications of constructive semigroups with apartness, due to their novelty, constitute an unexplored area.

To conclude, although one of the main motivators for initiating and developing the theory of semigroups with apartness comes from the computer science area, in order to have profound applications, a certain amount of the theory, which can be applied, is necessary first. Among priorities, besides growing the general theory, are further developments of constructive relational structures—(co-)quotient structures, constructive order theory, theory of ordered semigroups with apartness, etc.

The study of basic constructive algebraic structures with apartness as well as constructive algebra as a whole can impact the development of other areas of constructive mathematics. On the other hand, it can make both proof engineering and programming more flexible.

3.4 Notes

Constructive mathematics is not a unique notion. Various forms of constructivism have been developed over time. The principal trends include the following varieties:

INT—Brouwer's intuitionistic mathematics, **RUSS**—the constructive recursive mathematics of the Russian school of Markov, and **BISH**—Bishop's constructive mathematics. Every form has intuitionistic logic at its core. Different schools have different additional principles or axioms given by the particular approach to constructivism. For example, the notion of an *algorithm* or a *finite routine* is taken as primitive in **INT** and **BISH**, while **RUSS** operates with a fixed programming language and an algorithm is a sequence of symbols in that language. The Bishop-style of constructive mathematics enables one to interpret the results in both classical mathematics, **CLASS**, and other varieties of constructivism. We regard classical mathematics as Bishop-style mathematics plus the law of excluded middle, **LEM**.

We have already emphasized that the Errett Bishop-style constructive mathematics, **BISH**, forms the framework for our work. **BISH** originated in 1967 with the publication of the book *Foundations of Constructive Mathematics* and with its second, much revised edition (Bishop and Bridges, 1985). There has been a steady stream of publications contributing to Bishop's programme since 1967. A ten-year long systematic research of computable topology, using apartness as the fundamental notion, resulted in the first book, Bridges and Vîţă (2011) on topology within **BISH** framework. `Modern algebra`, as is noticed, "contrary to Bishop's expectations, also proved amenable to a natural, thoroughgoing, constructive treatment".

Constructive algebra is a relatively old discipline developed among others by L. Kronecker, van der Waerden, and A. Heyting. For more information on the history, see Mine et al.'s book (1988). Troelstra and van Dalen's book (1988). One of the main topics in constructive algebra is constructive algebraic structures with the relation of (tight) apartness #, the second most important relation in constructive mathematics. The principal novelty in treating basic algebraic structures constructively is that (tight) apartness becomes a fundamental notion. (Consider the reals: we cannot assert that x^{-1} exists unless we know that x is apart from zero, i.e., $|x| > 0$—constructively that is not the same thing as $x \neq y$. Furthermore, in fields x^{-1} exists only if x is apart from 0 (Beeson, 1985)).

In some books and papers, such as Troelstra and van Dalen (1988)'s book (1988), the term "preapartness" is used for an apartness relation, while "apartness" means tight apartness. The tight apartness on the real numbers was introduced by Brouwer in the early 1920s. Brouwer introduced the notion of apartness as a positive intuitionistic basic concept. A formal treatment of apartness relations began with A. Heyting's formalization of elementary intuitionistic geometry in Heyting (1927). The intuitionistic axiomatization of apartness is given in Heyting (1956). The study of algebraic structures in the presence of *tight* apartness was started by Heyting (1925). Heyting gave the theory a firm base in 1941 (Heyting, 1941).

During the implementation of the FTA Project, Geuvers et al. (2002b), the notion of commutative constructive semigroups with tight apartness appeared. We put noncommutative constructive semigroups with "ordinary" apartness in the center of our study, proving first, of course, that such semigroups do exist. Starting our work on constructive semigroups with apartness, as pointed out above, we faced an algebraically completely new area. What we had in "hand" at that moment were the experience and knowledge coming from classical semigroup theory, other constructive mathematics disciplines, and computer science.

The notion of co-quasiorder first appeared in Romano (1996). However, let us mention that the results reported from (Romano, 1996): Theorem 0.4, Lemma 0.4.1, Lemma 0.4.2, Theorem 0.5, and Corollary 0.5.1 (pages 10–11 in Romano, 2002) are not correct. Indeed, the filled product mentioned in Romano (1996, 2002) is not associative in general. The notion of co-equivalence, i.e., a symmetric co-quasiorder, first appeared in Božić and Romano (1987).

The Quotient Structure Problem (**QSP**) is one of the very first problems which has to be considered for any structure with apartness. The solutions of the **QSP** problem for sets and semigroups with apartness were given in Crvenković et al. (2013). Those results are improved in Mitrović et al. (2019).

More background on constructive mathematics can be found in the following books: Beeson (1985), Bishop (1967), Bridges and Vîţă (2011), and Troelstra and van Dalen (1988). The standard references for constructive algebra are Mines et al. (1988), Ruitenburg (1982).

4 CLASS and BISH: A Comparative Analysis

From the classical mathematics (**CLASS**) point of view, mathematics consists of a preexisting mathematical truth. From a constructive viewpoint, the judgement φ *is true* means that *there is a proof of* φ. "What constitutes a proof is a social construct, an agreement among people as to what is a valid argument. The rules of logic codify a set of principles of reasoning that may be used in a valid proof. Constructive (intuitionistic) logic codifies the principles of mathematical reasoning as it is actually practiced", Harper (2013). In constructive mathematics, the *status of an existence statement* is much stronger than in **CLASS**. The classical interpretation is that an object exists if its non-existence is contradictory. In constructive mathematics when the existence of an object is proved, the proof also demonstrates how to find it. Thus, following further Harper (2013), the *constructive logic* can be described as the logic of people matter, as distinct from the classical logic, which may be described as the logic of the mind of God. One of the main features of constructive mathematics is that the concepts that are equivalent in the presence of **LEM** need not be equivalent any more. For example, we distinguish nonempty and inhabited sets, several types of inequalities, two complements of a given set, etc.

Contrary to the classical case, a set exists only when it is defined. To define a set S, we have to give a property that enables us to construct members of S and to describe the equality $=$ between elements of S—which is a matter of convention, except that it must be an equivalence. There is another problem to face when we consider families of sets that are closed under a suitable operation of complementation. Following Bishop and Bridges (1985), "we do not wish to define complementation in the terms of negation; but on the other hand, this seems to be the only method available. The way out of this awkward position is to have a very flexible notion based on the concept of *a set with apartness*".

In **CLASS**, equivalence is the natural generalization of equality. A theory with equivalence involves equivalence and functions and relations respecting this equivalence. In constructive mathematics the same works without difficulty, Ruitenburg (1991).

Many sets come with a binary relation called inequality satisfying certain properties and denoted by $\neq$, #, or $\not\leq$. In general, more computational information is required to distinguish elements of a set S, than to show that elements are equal. Comparing with **CLASS**, the situation for inequality is more complicated. There are different types of inequalities (denial inequality, diversity, apartness, and tight apartness—to mention a few), some of them completely independent, which only in **CLASS** are equal to one standard inequality. So, in **CLASS** the study of the equivalence relation suffices, but in constructive mathematics, an inequality becomes a "basic notion in intuitionistic axiomatics". Apartness, as a positive version of inequality, in the words of Jacobs (1995), "is yet another fundamental notion developed in intuitionism which shows up in computer science".

The statement that every equivalence relation is the negation of some apartness relation is equivalent to the excluded middle. The statement that the negation of an equivalence relation is always an apartness relation is equivalent to the nonconstructive de Morgan law.

For a tight apartness, the two complements are constructive counterparts of the classical complement. In general, we have $\sim Y \subseteq \neg Y$. However, even for a tight apartness, the converse inclusion entails the Markov principle (**MP**). This result illustrates a main feature of constructive mathematics: classically equivalent notions could be no longer equivalent constructively. For which type of subset of a set with apartness do we have equality between its two complements? It turns out that the answer initiated a development of *order theory* for sets and semigroups with apartness we develop. Constructive mathematics brings to the light some notions that are invisible to the classical eye (here, the three notions of detachability).

In the constructive order theory, the notion of co-transitivity, which is the property that for every pair of related elements, any other element is related to one of the original elements in the same order as the original pair is a constructive counterpart to classical transitivity (Crvenković et al., 2013).

A relation defined on a set with apartness S is

- *weak co-quasiorder* if it is irreflexive and co-transitive
- *co-quasiorder* if it is strongly irreflexive and co-transitive.

Even if the two classically (but not constructively) equivalent variants of a co-quasiorder are constructive counterparts of a quasiorder in the case of (a tight) apartness, the stronger variant, co-quasiorder, is, of course, the most appropriate for a constructive development of the theory of semigroups with apartness we develop. The weaker variant, that is, weak co-quasiorder, could be relevant in the analysis. "One might expect that the splitting of notions leads to an enormous proliferation of results in the various parts of constructive mathematics when compared with their classical counterparts. In particular, usually only very few constructive versions of a classical notion are worth developing since other variants do not lead to a mathematically satisfactory theory", Troelstra and van Dalen (1988).

Following Bishop, every classical theorem presents the challenge: find a constructive version with a constructive proof. Within **CLASS**, the semigroups can be viewed, historically, as an algebraic abstraction of the properties of the composition of transformations on a set. Cayley Theorem for semigroups (which can be seen as an extension of the celebrated Cayley Theorem on groups) stated that every semigroup can be embedded in a semigroup of all self-maps on a set. As a consequence of Theorem 19, we can formulate *the constructive Cayley Theorem for semigroups with apartness*.

Following the standard literature on constructive mathematics, the term "constructive theorem" refers to a theorem with a constructive proof. A classical theorem that is proven in a constructive manner is a constructive theorem. This constructive version can be obtained by strengthening the conditions or weakening the conclusion of the theorem. Although constructive theorems might look like the corresponding classical versions, they often have more complicated hypotheses and proofs. Theorems and their proofs given in

- Sections 2.1.2, 2.1.3, 2.2.2, and 2.2.3 for classical case
- Sections 3.1.4, 3.1.5, 3.2.3, and 3.2.4 for constructive case

are evidence for that.

There are, often, several constructively different versions of the same classical theorem. Some classical theorems are neither provable nor disprovable, that is, they are independent of **BISH**. For some classical theorems it is shown that they are not provable constructively. More details about nonconstructive principles and various classical theorems that are not constructively valid can be found in Ishihara (2004).

5 Concluding Remarks

There are many decisions a mathematician must make when deciding to replace classical logic with intuitionistic logic. Let us mention some of them, first of all, **the choice of variant of constructive mathematics.** Constructive mathematics is not a unique notion. Our choice was the Errett Bishop-style constructive mathematics, **BISH**. The cornerstones for **BISH** include the notion of positive integers, sets, and functions. Contrary to the classical case, a set exists only when it is defined.

Going through the literature there are several variants of what is considered to be a set with apartness—depending on the relations between equality and apartness defined on a set. Our choice –our starting structure– is a **set with apartness** $(S, =, \#)$ where

- equality and apartness are basic notions
- equality and apartness are independent of each other
- apartness is not, in general, tight.

Such our choice –fully justified in Darpö and Mitrović (arXiv:2103.07105)– was and is a novelty within constructive circles.

When working constructively one has to choose definitions with some care. Heyting pointed out that it is very important to have experience of where the intuitionistic pitfalls lie. Classically equivalent notions often split into a certain number of inequivalent constructive ones: nonempty and inhabited set, several inequivalent constructive definition of inequality; two complements of a given subset of a set—just to mention a few. Thus, it is very important to be aware of the phenomena of splitting notions.

Is it easier to work within constructive mathematics than within classical? One does not develop constructive mathematics for simplification. There is something else, something so important worth that simplicity can sometimes be sacrificed. So WHY to work constructively? Is it usefulness in question? Or, something else? Instead of giving answers, let us cite Heyting, *Intuitionism—An Introduction* (1956) again.

"It seems quite reasonable to judge a mathematical system by its usefulness [...] in my eyes its chances of being useful for **philosophy, history and the social sciences** are better. In fact, mathematics, from the intuitionistic point of view, is a study of certain functions of the human mind, and as such it is akin to these sciences. But is usefulness really the only measure of value? You know how philosophers struggle with the problem of defining the concept of value in art; yet every educated person feels this value. The case is analogous for the value of intuitionistic mathematics".

Acknowledgments The authors are grateful to anonymous referees for careful reading of the manuscript and helpful comments. M. M. was financially supported by the Ministry of Science, Technological Development and Innovation of the Republic of Serbia (Contract No. 451-03-47/2023-01/ 200109). M. N. H. is supported by the ICMPA-UNESCO Chair, which is in partnership with Daniel Iagolnitzer Foundation (DIF), France, and the Association pour la Promotion Scientifique de l'Afrique (APSA), supporting the development of mathematical physics in Africa. P. C. also thanks the Portuguese Foundation for Science and Technology (FCT) under the projects UIDB/00013/2020 and UIDP/00013/2020.

6 Appendix

6.1 The CLASS Case

Proof **—Lemma 1**

Proof is straightforward as it is based on the well-known facts that each element of S belongs to a unique equivalence class, and each equivalence class is nonempty. □

***Proof* —Theorem 1**

(i) Let $x(ker\ f) = y(ker\ f)$. Then $(x, y) \in ker\ f$, thus $f(x) = f(y)$ and φ is well defined, i.e., the definition of $\varphi(x(kef\ f))$ is independent of the choice of element form $x(ker\ f)$.

Let $x \in S$. By the assumption

$$(\varphi \circ \pi_{nat})(x) = \varphi(\pi_{nat}(x)) = \varphi(x(ker\ f)) = f(x).$$

If $\varphi(x(ker\ f)) = \varphi(y(ker\ f))$, then $f(x) = f(y)$, which, further, means that $(x, y) \in ker\ f$. Thus $x(ker\ f) = y(ker\ f)$, and φ is injective.

If there is another function $\varphi' : S/ker\ f \to T$ such that $f = \varphi' \circ \pi_{nat}$, then for any $x(ker\ f) \in S/ker\ f$ we have

$$\varphi'(x(ker\ f) = \varphi'(\pi_{nat}(x)) = \varphi' \circ \pi_{nat}(x) = f(x).$$

Thus $\varphi = \varphi'$.

(ii) If f is an onto mapping, then for any $y \in T$ there is an $x \in S$ such that $y = f(x)$. But then $y = \varphi(x(ker\ f))$ and so φ is surjective. We have that φ is bijection.

(iii) Let φ be a surjective mapping. By Lemma 1, π_{nat} is surjective too. Thus $\varphi \circ \pi$ is also surjective, i.e., $f = \varphi \circ \pi_{nat}$ is surjective. □

***Proof* —Theorem 2**

Let $x\varepsilon = y\varepsilon$, $x, y \in S$. Then $(x, y) \in \varepsilon \subseteq ker\ f$. Thus $f(x) = f(y)$ and φ is well defined. It is a routine to verify that $f = \varphi \circ \pi_\varepsilon$. The rest follows by the Theorem 1. □

***Proof* —Lemma 2**

(i) Let f be an isotone mapping. If $(x, y) \in \rho$, then, by the assumption, $(f(x), f(y)) \in \sigma$, which, by the definition of η_f given above, means that $(x, y) \in \eta_f$.

Conversely, from the assumption and the definition of η_f we have

$$(x, y) \in \rho \Rightarrow (x, y) \in \eta_f \Leftrightarrow (f(x), f(y)) \in \sigma.$$

(ii) Similar to (i).

(iii) Consequence of (i) and (ii). □

***Proof* —Lemma 3**

Straightforward. □

***Proof* —Lemma 5**

(i) It follows immediately by Lemma 4

(ii) By Lemma 3, $(S/\varepsilon_\rho; \hat{\Theta}_\rho)$ is a quasiordered set. Let $(x\varepsilon_\rho, y\varepsilon_\rho), (y\varepsilon_\rho, x\varepsilon_\rho) \in \hat{\Theta}_\rho$, which, by the definition, gives $(x, y), (y, x) \in \rho$, i.e., $(x, y) \in \rho \cap \rho^{-1} = \varepsilon_\rho$. Thus $x\varepsilon_\rho = y\varepsilon_\rho$. So, $\hat{\Theta}_\rho$ is an order.

(iii) Let $x, y \in S$. Then

$$(x, y) \in \rho \Leftrightarrow (x\varepsilon_\rho, y\varepsilon_\rho) \in \hat{\Theta}_\rho$$
$$\Leftrightarrow (\pi_{nat}(x), \pi_{nat}(y)) \in \hat{\Theta}_\rho.$$

Thus π_{nat} is isotone and reverse isotone, and by the Lemma 1, π_{nat} is surjective. □

***Proof* —Lemma 6**

(i) ⇒(ii) From $(x, x) \in \varepsilon_\rho$, we have $x \in x\varepsilon_\rho$. Also $y \in y\varepsilon_\rho$. Thus, by (i) we have $(x, y) \in \rho$.

(ii) ⇒ (i) It follows immediately from $x \in x\varepsilon_\rho$ and $y \in y\varepsilon_\rho$.

(i) ⇒ (iii) Let $x_1 \in x\varepsilon_\rho$ and $y_1 \in y\varepsilon_\rho$, i.e., we have $(x_1, y_1) \in \rho$. Let $a \in x\varepsilon_\rho$ and $b \in y\varepsilon_\rho$. Since $(a, x), (x_1, x) \in \varepsilon_\rho$, we have $(a, x_1) \in \varepsilon_\rho$. Since $(y, y_1), (y, b) \in \varepsilon_\rho$, we have $(y_1, b) \in \varepsilon_\rho$. Then $(y_1, b) \in \rho$. Now, from $(a, x_1), (x_1, y_1), (y_1, b) \in \rho$ we have $(a, b) \in \rho$.

(iii) ⇒ (i) It follows immediately. □

***Proof* —Theorem 3**

(i) Reflexivity of σ gives $(f(x), f(x)) \in \sigma$, $x \in S$, and, by the definition of η_f, we have $(x, x) \in \eta_f$.

Transitivity of η_f follows by the definition of η_f and transitivity of σ, i.e., we have

$$(x, y), (y, z) \in \eta_f \Leftrightarrow (f(x), f(y)), (f(y), f(z)) \in \sigma$$
$$\Rightarrow (f(x), f(z)) \in \sigma$$
$$\Leftrightarrow (x, z) \in \eta_f.$$

(ii) By (i) and Lemma 5(i), $\varepsilon_f = \eta_f \cap \eta_f^{-1}$ is an equivalence on S.
Using the reflexivity of σ, we have

$$(x, y) \in ker\ f \Leftrightarrow f(x) = f(y)$$
$$\Rightarrow (f(x), f(y)) \in \sigma$$
$$\Leftrightarrow (x, y) \in \eta_f,$$

i.e., $ker\ f \subseteq \eta_f$. In a similar manner we can prove $ker\ \subseteq \eta_f^{-1}$. Thus $ker\ f \subseteq \varepsilon_f$. Conversely, let

$$(x, y) \in \varepsilon_f = \eta_f \cap \eta_f^{-1} \Leftrightarrow (x, y), (y, x) \in \eta_f$$
$$\Leftrightarrow (f(x), f(y)), (f(y), f(x)) \in \sigma$$
$$\Rightarrow f(x) = f(y)$$
$$\Leftrightarrow (x, y) \in ker\ f,$$

so $\varepsilon_f \subseteq ker\ f$. Thus, $ker\ f = \varepsilon_f$.

(iii) By Theorem 1(i), the mapping $\varphi : S/\varepsilon_f \to T$ defined by $\varphi(x\varepsilon_f)) = f(x)$ is the unique injective mapping such that $f = \varphi \circ \pi_{nat}$.

By Lemma 5, $(S/\varepsilon_f, \Theta_f)$ is an ordered set, and

$$\begin{aligned}(x\varepsilon_f, y\varepsilon_f) \in \Theta_f &\Leftrightarrow (x, y) \in \eta_f \\ &\Leftrightarrow (f(x), f(y)) \in \sigma.\end{aligned}$$

Therefore, φ is isotone and reverse isotone mapping such that $f = \varphi \circ \pi$.

(iv) It follows by (ii) and Theorem 1(iii). □

***Proof* —Theorem 4**

(i) From $\rho \subseteq \eta_f$, we have $\rho^{-1} \subseteq \eta_f^{-1}$. Then

$$\varepsilon_\rho = \rho \cap \rho^{-1} \subseteq \eta_f \cap \eta_f^{-1} = \varepsilon_f.$$

Now, by Theorem 3, we have $\varepsilon_\rho \subseteq \varepsilon_f = ker\ f$.

(ii) By Theorem 2 there is the unique injective mapping $\varphi : S/\varepsilon_\rho \to T$, defined by $\varphi(x\varepsilon_\rho) = f(x)$, such that $f = \varphi \circ \pi_\rho$. By Lemma 5, $(S/\varepsilon_\rho, \hat{\Theta}_\rho)$ is an ordered set.

Let $(x\varepsilon_\rho, y\varepsilon_\rho) \in S/\varepsilon_\rho$. Then

$$\begin{aligned}(x\varepsilon_\rho, y\varepsilon_\rho) \in \hat{\Theta}_\rho &\Leftrightarrow (x, y) \in \rho \subseteq \eta_f \\ &\Rightarrow (f(x), f(y)) \in \sigma.\end{aligned}$$

So, φ is an isotone mapping. □

***Proof* —Lemma 7**

Let ρ be a compatible quasiorder on S. Let $(x, y) \in \rho$ and $z \in S$. By the reflexivity of ρ, we have $(z, z) \in S$. Now, by the compatibility, it follows that $(zx, zy) \in S$, and ρ is left compatible. Right compatibility can be proved in a similar manner.

Conversely, let ρ be a left and right compatible quasiorder on S. Let $(x, y), (s, t) \in \rho$. By the left and right compatibility, it follows that $(xs, ys), (ys, yt) \in \rho$, which, by the transitivity of ρ, gives us $(xs, yt) \in \rho$. □

***Proof* —Theorem 8**

(i) All the set-theoretic results are as in Theorem 1. It only remains to prove that $ker\ f$ is compatible with multiplication as well as that φ is a homomorphism.

Let $(x, y), (s, t) \in ker\ f$. Then $f(x) = f(y)$ and $f(s) = f(t)$. It follows that

$$f(xs) = f(x)f(s) = f(y)f(t) = f(yt),$$

and hence $(xs, yt) \in ker\ f$. Therefore $ker\ f$ is a congruence.

Let $x(ker\ f), y(ker\ f) \in S/ker\ f$. Then

$$\begin{aligned}\varphi(x(ker\ f)y(ker\ f)) &= \varphi((xy)(ker\ f)) \\ &= f(xy) = f(x)f(y) \\ &= \varphi(x(ker\ f))\varphi(y(ker\ f)).\end{aligned}$$

So, φ is a homomorphism.

(ii) It follows immediately by the assumptions and Theorem 2. □

***Proof* —Lemma 8**

(i) Let $(x, y), (s, t) \in \rho^{-1}$, or, equivalently $(y, x), (t, s) \in \rho$. Then, as ρ is compatible, $(yt, xs) \in \rho$, which further means that $(xs, yt) \in \rho^{-1}$. Thus ρ^{-1} is a compatible quasiorder, and the equivalence $\varepsilon_\rho = \rho \cap \rho^{-1}$ is compatible too, i.e., it is a congruence on S.

(ii) Compatibility of $\hat{\Theta}_\rho$ follows by its definition and compatibility of ρ.

(iii) By Theorem 7 π is an epimorphism. By Lemma 5(iii), it is isotone and reverse isotone. □

***Proof* —Theorem 9**

It follows by Theorems 3 and 7. □

***Proof* —Theorem 10**

It follows by Theorems 4 and 9. □

6.2 *Within BISH*

***Proof* —Theorem 11**

See Mines et al. (1988), Theorem I 2.2. □

***Proof* —Theorem 12**

(i) Let Y be an sd-subset of S. Then, applying the definition and logical axiom, we have

$$\begin{aligned}\forall_{x\in S}\,(x \in Y \vee x \in {\sim}Y) &\Leftrightarrow \forall_{x\in S}\,(x \in Y \ \vee\ \forall_{y\in Y}(x\#y)) \\ &\Rightarrow \forall_{x\in S}\,\forall_{y\in Y}\,(x \in Y \vee x\#y).\end{aligned}$$

In order to prove the second part of this statement, we consider the real number set $\mathbb{R}$ with the usual (tight) apartness and the subset $Y = \widetilde{0}$. Then, for each real number x and for each $y \in Y$, it follows, from the co-transitivity of #, either $y\#x$ or $x\#0$, that is, either $x \in Y$ or $x\#y$. Consequently, Y is a qd-subset of $\mathbb{R}$. On the other hand, if Y is an sd-subset of $\mathbb{R}$, then for each $x \in \mathbb{R}$, either $x \in Y$ or $x \in{\sim} Y$. In the former case, $x\#0$ and in the latter $x = 0$, and hence **LPO** holds.

(ii) Let Y be a qd-subset, and let $a \in \neg Y$. By assumption, we have

$$\forall_{x \in S} \forall_{y \in Y} \, (x \in Y \vee x \# y),$$

so substituting a for x, we get $\forall_{y \in Y} \, (a \in Y \vee a \# y)$, and since, by assumption, $\neg(a \in Y)$, it follows that $a \# y$ for all $y \in Y$. Hence $a \in \sim Y$. See also Mitrović et al. (2019).

(iii) Let S be the real number set $\mathbb{R}$ with the usual apartness #. As in the proof of (i), consider the qd-subset $\widetilde{0}$ of $\mathbb{R}$. If $\widetilde{0}$ is a d-subset of $\mathbb{R}$, then $x \in \widetilde{0}$ or $\neg(x \in \widetilde{0})$, for all real numbers x. In the latter case $\neg(x \# 0)$, which is equivalent to $x = 0$. Thus we obtain the property $\forall_{x \in \mathbb{R}} \, (x \# 0 \vee x = 0)$ which, in turn, is equivalent to **LPO**.

(iv) Consider a real number a with $\neg(a = 0)$ and let S be the set $\{0, a\}$ endowed with the usual apartness of $\mathbb{R}$. For $Y = \{0\}$, since $0 \in Y$ and $a \in \neg Y$, it follows that Y is a d-subset of S. On the other hand, if Y is a qd-subset of S, then $a \# 0$. It follows that for any real number with $\neg(a = 0)$, $a \# 0$ which entails the Markov Principle (**MP**).

(v) The first part follows immediately from (i), (ii) and the definition of d-subsets. The converse follows from (i) and (iv).

(vi) Consider again $\mathbb{R}$ with the usual apartness and define $Y = \{0\}$. If Y is a qd-subset of $\mathbb{R}$, then for all $x \in \mathbb{R}$, we have $x = 0$ or $x \# 0$, and hence **LPO** holds. □

***Proof* —Lemma 9**

If $\alpha \subseteq \gamma$ and $\beta \subseteq \delta$, then

$$\begin{aligned} (x, y) \in \alpha * \beta &\Leftrightarrow \forall_{z \in S} \, ((x, z) \in \alpha \vee (z, y) \in \beta) \\ &\Leftrightarrow \forall_{z \in S} \, ((x, z) \in \gamma \vee (z, y) \in \delta) \\ &\Leftrightarrow (x, y) \in \gamma * \delta. \end{aligned}$$

□

***Proof* —Lemma 10**

(i) If α is strongly irreflexive, then

$$\begin{aligned} (x, y) \in \alpha * \beta &\Leftrightarrow \forall_{z \in S} \, ((x, z) \in \alpha \vee (z, y) \in \beta) \\ &\Rightarrow (x, x) \in \alpha \vee (x, y) \in \beta \\ &\Rightarrow (x, y) \in \beta. \end{aligned}$$

The case when β is strongly irreflexive is analogous.

(ii) By (iv), we have $\alpha * \alpha \subseteq \alpha \subseteq \#$. Thus, $\alpha * \alpha$ is strongly irreflexive. □

***Proof* —Lemma 12**

(i) Let α be a strongly irreflexive relation on S. For each $a \in S$, it can be easily proved that $(a, a) \# (x, y)$ for all $(x, y) \in \alpha$.

Let $\sim \alpha$ be reflexive, that is, $(x, x) \in\sim \alpha$, for any $x \in S$. On the other hand, the definition of the a-complement implies $(x, y)\#(x, x)$ for any $(x, y) \in \alpha$. So, $x\#x$ or $x\#y$. Thus, $x\#y$, that is, α is strongly irreflexive.

(ii) If α is symmetric, then

$$\begin{aligned}(x, y) \in\sim \alpha &\Leftrightarrow \forall_{(a,b)\in\alpha} ((x, y)\#(a, b))\\ &\Rightarrow \forall_{(b,a)\in\alpha} ((x, y)\#(b, a))\\ &\Rightarrow \forall_{(b,a)\in\alpha} (x\#b \ \vee \ y\#a)\\ &\Rightarrow \forall_{(a,b)\in\alpha} ((y, x)\#(a, b))\\ &\Leftrightarrow (y, x) \in\sim \alpha.\end{aligned}$$

(iii) If $(x, y) \in\sim \alpha$ and $(y, z) \in\sim \alpha$, then, by the definition of $\sim \alpha$, we have that $(x, y) \bowtie \alpha$ and $(y, z) \bowtie \alpha$. For an element $(a, b) \in \alpha$, by co-transitivity of α, we have $(a, x) \in \alpha$ or $(x, y) \in \alpha$ or $(y, z) \in \alpha$ or $(z, b) \in \alpha$. Thus $(a, x) \in \alpha$ or $(z, b) \in \alpha$, which implies that $a\#x$ or $b\#z$, that is, $(x, z)\#(a, b)$. So, $(x, z) \bowtie \alpha$ and $(x, z) \in \sim \alpha$. Therefore, $\sim \alpha$ is transitive.

(iv) Let $(x, y), (y, x) \in\sim \alpha$. Assume $x\#y$. Then, by the co-antisymmetry of α, we have $(x, y) \in \alpha$ or $(y, x) \in \alpha$, which is impossible. Thus, $\neg(x\#y)$, i.e., $x = y$ as apartness is tight. □

Proof **—Proposition 2**

(i) Let $(x, y) \in S \times S$. Then, for all $(a, b) \in \tau$,

$$\begin{aligned}a\tau x \vee x\tau b &\Rightarrow a\tau x \vee x\tau y \vee y\tau b\\ &\Rightarrow a\#x \vee x\tau y \vee y\#b\\ &\Rightarrow (a, b)\#(x, y) \vee x\tau y,\end{aligned}$$

that is, τ is a qd-subset.

(ii) It follows from (i) and Theorem 12(ii).

(iii) This is a consequence of Lemma 12(i), (iii). □

Proof **—Lemma 13**

From the strong irreflexivity of τ and σ, i.e., from $\tau \subseteq \#$ and $\sigma \subseteq \#$, we have $\tau \cup \sigma \subseteq \#$. Thus, $\tau \cup \sigma$ is strongly irreflexive.

From $\tau \subseteq \tau \cup \sigma$ and $\sigma \subseteq \tau \cup \sigma$, by Lema 9 we have

$$\begin{aligned}\tau * \tau &\subseteq (\tau \cup \sigma) * (\tau \cup \sigma)\\ \sigma * \sigma &\subseteq (\tau \cup \sigma) * (\tau \cup \sigma).\end{aligned}$$

By the assumption and by the Proposition 2(i), we have

$$\tau \subseteq (\tau \cup \sigma) * (\tau \cup \sigma)$$
$$\sigma \subseteq (\tau \cup \sigma) * (\tau \cup \sigma),$$

which gives us $\tau \cup \sigma \subseteq (\tau \cup \sigma) * (\tau \cup \sigma)$. Thus $\tau \cup \sigma$ is co-transitive. □

***Proof* —Lemma 14**

Indeed, let α be a co-quasiorder and β an equivalence on S such that $\boldsymbol{\alpha}$ defines an apartness on $\mathbf{S}/\boldsymbol{\beta}$. Let $(x, a), (y, b) \in \beta$, i.e., $a \in x\beta$ and $b \in y\beta$, which, by the assumption, gives $a\beta = x\beta$ and $b\beta = y\beta$. If $(x, y) \in \alpha$, then, by (Ap6), $x\beta \# y\beta$, which, by (Ap5'), gives $a\beta \# b\beta$. By (Ap6) we have $(a, b) \in \alpha$. In a similar manner, starting from $(a, b) \in \alpha$, we can conclude $(x, y) \in \alpha$. □

***Proof* —Theorem 13**

(i) By the Proposition 2(iii), κ^c is a quasiorder on S, and, by the Lemma 12(ii), it is symmetric. Thus, κ^c is an equivalence.

Let $(x, y) \in \kappa$ and $(y, z) \in \kappa^c$. Thus $(x, y) \in \kappa$ and $(y, z) \bowtie \kappa$. By the co-transitivity of κ, we have $(x, z) \in \kappa$ or $(y, z) \in \kappa$. Thus $(x, z) \in \kappa$, and $\kappa \looparrowleft \kappa^c$.

(ii) The strong irreflexivity of # is implied by its definition and by the strong irreflexivity of κ.

Let $a\kappa^c \# b\kappa^c$. Then $(a, b) \in \kappa$ implies that $(b, a) \in \kappa$, that is, $b\kappa^c \# a\kappa^c$.

Let $a\kappa^c \# b\kappa^c$ and $u\kappa^c \in S/\kappa^c$. Then $(a, b) \in \kappa$, and, by the co-transitivity of κ, we have $(a, u) \in \kappa$ or $(u, b) \in \kappa$. Finally we have that $a\kappa^c \# u\kappa^c$ or $u\kappa^c \# b\kappa^c$, so the relation # is co-transitive.

(iii) Let $\pi(x) \# \pi(y)$, i.e., $x\kappa^c \# y\kappa^c$, which, by what we have just proved, means that $(x, y) \in \kappa$. Then, by the strong irreflexivity of κ, we have $x \# y$. So π is an se-mapping.

Let $a\kappa^c \in S/\kappa^c$ and $x \in a\kappa^c$. Then $(a, x) \in \kappa^c$, i.e., $a\kappa^c = x\kappa^c$, which implies that $a\kappa^c = x\kappa^c = \pi(x)$. Thus π is an se-surjection. □

***Proof* —Theorem 14**

(i) The strong irreflexivity of coker f is easy to prove: if $(x, y) \in$ coker f, then $f(x) \# f(y)$ and therefore $x \# y$.

If $(x, y) \in$ coker f, then, by the symmetry of apartness in T, $f(y) \# f(x)$; so $(y, x) \in$ coker f.

If $(x, y) \in$ coker f and $z \in S$, i.e., $f(x) \# f(y)$ and $f(z) \in T$, then either $f(x) \# f(z)$ or $f(z) \# f(y)$; that is, either $(x, z) \in$ coker f or $(z, y) \in$ coker f. Hence coker f is a co-equivalence on S.

(ii) Let $(x, y) \in$ coker f and $(y, z) \in$ ker f. Then $f(x) \# f(y)$ and $f(y) = f(z)$. Hence $f(x) \# f(z)$, that is, $(x, z) \in$ coker f, and coker $f \looparrowleft$ ker f.

Now let $(x, y) \in$ ker f, so $f(x) = f(y)$. If $(u, v) \in$ coker f, then, by the co-transitivity of coker f, it follows that $(u, x) \in$ coker f or $(x, y) \in$ coker f or $(y, v) \in$ coker f. Thus either $(u, x) \in$ coker f or $(y, v) \in$ coker f, and, by the strong irreflexivity of coker f, either $u\#x$ or $y\#v$; whence we have $(x, y)\#(u, v)$. Thus $(x, y) \bowtie \text{coker}\, f$, or, equivalently $(x, y) \in (\text{coker}\, f)^c$.

(iii) This follows from the definition of # in $S/\ker f$ and (i).

(iv) Let us first prove that φ is well defined. Let $x(\ker f), y(\ker f) \in S/\ker f$ be such that $x(\ker f) = y(\ker f)$, that is, $(x, y) \in \ker f$. Then we have $f(x) = f(y)$, which by the definition of φ, means that $\varphi(x(\ker f)) = \varphi(y(\ker f))$.

Now let $\varphi(x(\ker f)) = \varphi(y(\ker f))$; then $f(x) = f(y)$. Hence $(x, y) \in \ker f$, which implies that $x(\ker f) = y(\ker f)$. Thus φ is an injection.

Next let $\varphi(x(\ker f))\#\varphi(y(\ker f))$; then $f(x)\#f(y)$. Hence $(x, y) \in \text{coker}\, f$, which, by (iii), implies that $x(\ker f)\#y(\ker f)$. Thus φ is an se-mapping.

Let $x(\ker f)\#y(\ker f)$; that is, by (iii), $(x, y) \in \text{coker}\, f$. So we have $f(x)\#f(y)$, which, by the definition of φ means $\varphi(x(\ker f))\#\varphi(y(\ker f))$. Thus φ is a-injective.

By the definition of composition of functions, Theorem 13, and the definition of φ, for each $x \in S$, we have

$$(\varphi\pi)(x) = \varphi(\pi(x)) = \varphi(x(\ker f)) = f(x).$$

(v) Taking into account (iv), we have to prove only that φ is a surjection. Let $y \in T$. Then, as f is onto, there exists $x \in S$ such that $y = f(x)$. On the other hand $\pi(x) = x(\ker f)$. By (iv), we now have

$$y = f(x) = (\varphi\pi)(x) = \varphi(\pi(x)) = \varphi(x(\ker f)).$$

Thus φ is a surjection. □

***Proof* —Theorem 15**

(i) Let $x, y \in S$ and assume that $(x, y) \in \varepsilon \cap \kappa$. Then $(x, y) \in \varepsilon$ and $(y, y) \in \varepsilon$, which, by Lemma 14, i.e., (Ap6') and $(x, y) \in \kappa$, gives $(y, y) \in \kappa$, which is impossible. Thus, $\varepsilon \cap \kappa = \emptyset$.

Conversely, let $(x, a), (y, b) \in \varepsilon$ and $(x, y) \in \kappa$. Then, by the co-transitivity of κ and by the assumption, we have

$$\begin{aligned}(x, y) \in \kappa \;&\Rightarrow\; (x, a) \in \kappa \vee (a, y) \in \kappa\\ &\Rightarrow\; (x, a) \in \kappa \vee (a, b) \in \kappa \vee (b, y) \in \kappa\\ &\Rightarrow\; (a, b) \in \kappa.\end{aligned}$$

(ii) Let $\pi(x)\#\pi(y)$, that is, $x\varepsilon\#y\varepsilon$, which, by (i), means that $(x, y) \in \kappa$. Then, by the strong irreflexivity of κ, we have $x\#y$. So π is an se-mapping.

Let $a\varepsilon \in S/\varepsilon$ and $x \in a\varepsilon$. Then $(a, x) \in \varepsilon$, i.e., $a\varepsilon = x\varepsilon$, which implies that $a\varepsilon = x\varepsilon = \pi(x)$. Thus π is an se-surjection. □

***Proof* —Theorem 16**

(i) It follows from Theorem 15(i).

(ii) It follows from Theorem 15(ii).

(iii) This was shown in Theorem 14(iv).

(iv) Let φ be an se-mapping. Let $(x, y) \in \operatorname{coker} f$ for some $x, y \in S$. Then, by the definition of coker f and φ, the assumption, and (i), we have

$$\begin{aligned} f(x)\#f(y) &\Leftrightarrow \varphi(x(\ker f))\#\varphi(y(\ker f)) \\ &\Rightarrow x(\ker f)\#y(\ker f) \\ &\Leftrightarrow (x, y) \in \kappa. \end{aligned}$$

Conversely, let $\operatorname{coker} f \subseteq \kappa$. By assumption, (i), and the definitions of φ and coker f, we have

$$\begin{aligned} \varphi(x(\ker f))\#\varphi(y(\ker f)) &\Leftrightarrow f(x)\#f(y) \\ &\Leftrightarrow (x, y) \in \operatorname{coker} f \\ &\Rightarrow (x, y) \in \kappa \\ &\Leftrightarrow x(\ker f)\#y(\ker f). \end{aligned}$$

(v) Let φ be a-injective, and let $(x, y) \in \kappa$. Then, by (i), we have

$$\begin{aligned} x(\ker f)\#y(\ker f) &\Rightarrow \varphi(x(\ker f))\#\varphi(y(\ker f)) \\ &\Leftrightarrow f(x)\#f(y) \\ &\Leftrightarrow (x, y) \in \operatorname{coker} f. \end{aligned}$$

Conversely, let $\kappa \subseteq \operatorname{coker} f$. Then

$$\begin{aligned} x(\ker f)\#y(\ker f) &\Leftrightarrow (x, y) \in \kappa \\ &\Rightarrow (x, y) \in \operatorname{coker} f \\ &\Leftrightarrow f(x)\#f(y) \\ &\Leftrightarrow \varphi(x(\ker f))\#\varphi(y(\ker f)). \end{aligned}$$

(vi) If φ is an se-mapping then, by (iv), we have that $\operatorname{coker} f \subseteq \kappa$. So, the strong irreflexivity of κ implies f is an se-mapping. □

***Proof* —Lemma 15**

(i) It follows by Lemma 11 and Lemma 13.

(ii) This is a consequence of Theorem 13(ii).

(iii) Let $x\kappa_\tau^c, y\kappa_\tau^c \in S/\kappa_\tau^c$ such that $(x\kappa_\tau^c, y\kappa_\tau^c) \in \Upsilon_\tau$, i.e., $(x, y) \in \tau$, which, by (ii), gives $x\kappa_\tau^c \# y\kappa_\tau^c$. Thus, Υ_τ is strongly irreflexive.

Let $x\kappa_\tau^c, y\kappa_\tau^c \in S/\kappa_\tau^c$ such that $(x\kappa_\tau^c, y\kappa_\tau^c) \in \Upsilon_\tau$ and let $z\kappa_\tau^c \in S/\kappa_\tau^c$. Then $(x, y) \in \tau$. By the co-transitivity of τ, we have $(x, z) \in \tau$ or $(z, y) \in \tau$, which, in turn, gives $(x\kappa_\tau^c, z\kappa_\tau^c) \in \Upsilon_\tau$ or $(z\kappa_\tau^c, y\kappa_\tau^c) \in \Upsilon_\tau$. So, Υ_τ is co-transitive.

Let $x\kappa_\tau^c, y\kappa_\tau^c \in S/\kappa_\tau^c$ such that $x\kappa_\tau^c \# y\kappa_\tau^c$. Then, by (ii), $(x, y) \in \kappa_\tau = \tau \cup \tau^{-1}$. Thus $(x, y) \in \tau$ or $(y, x) \in \tau$, which implies $(x\kappa_\tau^c, y\kappa_\tau^c) \in \Upsilon_\tau$ or $(y\kappa_\tau^c, x\kappa_\tau^c) \in \Upsilon_\tau$. Therefore, Υ_τ is co-antisymmetric. We have proved Υ_τ is a co-order.

(iv) By Theorem 13(iii), π_τ is an se-surjection. By the definition of Υ_τ, π_τ is isotone and reverse isotone. □

***Proof* —Theorem 17**

(i) Let $x, y \in S$ such that $(x, y) \in \mu_f$, that is $(f(x), f(y)) \in \sigma$. Now we have $f(x) \# f(y)$, and, finally, as f is an se-mapping, $x \# y$ follows. Thus, μ_f is strongly irreflexive.

Let $x, y \in S$ such that $(x, y) \in \mu_f$, and let $z \in S$. Then, by the definition of μ_f, $(f(x), f(y)) \in \sigma$. By co-transitivity of σ, we have $(f(x), f(z)) \in \sigma$ or $(f(z), f(y)) \in \sigma$, that is, $(x, z) \in \mu_f$ or $(z, y) \in \mu_f$. So, μ_f is co-transitive.

(ii) By Lemma 15(i) $\kappa_f = \mu_f \cup \mu_f^{-1}$ is a co-equivalence on S. Let $(x, y) \in \kappa_f$, i.e., $(x, y) \in \mu_f \cup \mu_f^{-1}$, that is, $(x, y) \in \mu_f$ or $(y, x) \in \mu_f$. So we have $(f(x), f(y)) \in \sigma$ or $(f(y), f(x)) \in \sigma$. Strong irreflexivity of σ implies $f(x) \# f(y)$, i.e., $(x, y) \in \operatorname{coker} f$. Thus, $\kappa_f \subseteq \operatorname{coker} f$.

(iii) Let $x, y \in S$ such that $(x, y) \in \operatorname{coker} f$, i.e., $f(x) \# f(y)$. Now, by the co-antisymmetry of σ, we have $(f(x), f(y)) \in \sigma$ or $(f(y), f(x)) \in \sigma$, that is, by the definition of μ_f, $(x, y) \in \mu_f$ or $(y, x) \in \mu_f$, i.e., $(x, y) \in \mu_f \cup \mu_f^{-1} = \kappa_f$.

(iv) It follows immediately by Lemma 15(ii),(iii).

(v) Let us, first, prove that ψ is well defined. Let $x\kappa_f^c = y\kappa_f^c$. Then

$$\begin{aligned}
x\kappa_f^c = y\kappa_f^c &\Leftrightarrow (x, y) \in \kappa_f^c \\
&\Leftrightarrow \neg((x, y) \in \kappa_f) \\
&\Leftrightarrow \neg((x, y) \in \mu_f \vee (y, x) \in \mu_f) \\
&\Rightarrow \neg((f(x), f(y)) \in \sigma \vee (f(y), f(x)) \in \sigma) \\
&\Leftrightarrow \neg((f(x), f(y)) \in \sigma \cup \sigma^{-1} = \kappa_\sigma) \\
&\Leftrightarrow (f(x), f(y)) \in \kappa_\sigma^c \\
&\Leftrightarrow (f(x))\kappa_\sigma^c = (f(y))\kappa_\sigma^c \\
&\Leftrightarrow \psi(x\kappa_f^c) = \psi(y\kappa_f^c).
\end{aligned}$$

Let $\psi(x\kappa_f^c) \# \psi(y\kappa_f^c)$. Then

$$\begin{aligned}
\psi(x\kappa_f^c) \# \psi(y\kappa_f^c) &\Leftrightarrow (f(x))\kappa_\sigma^c \# (f(y))\kappa_\sigma^c \\
&\Leftrightarrow (f(x), f(y)) \in \kappa_\sigma^c \\
&\Leftrightarrow (f(x), f(y)) \in \sigma \vee (f(y), f(x)) \in \sigma
\end{aligned}$$

$$\Rightarrow (x, y) \in \mu_f \vee (y, x) \in \mu_f$$

$$\Leftrightarrow (x, y) \in \kappa_f$$

$$\Leftrightarrow x\kappa_f^c \# y\kappa_f^c,$$

so ψ is an se-mapping.

Let $x \in S$. Then

$$(\psi\pi_f)(x) = \psi(\pi_f(x)) = \psi(x\kappa_f^c) = (f(x))\kappa_\sigma^c = \pi_\sigma(f(x)) = (\pi_\sigma f)(x).$$

□

Proof **—Theorem 18**

Let us first prove that φ is well defined. Let $x\kappa_\tau^c, y\kappa_\tau^c \in S/\kappa_\tau^c$ be such that $x\kappa_\tau^c = y\kappa_\tau^c$, i.e., $(x, y) \bowtie \kappa_\tau$. Assume $\varphi(x\kappa_\tau^c)\#\varphi(y\kappa_\tau^c)$, i.e., by the definition of φ, $f(x)\# f(y)$, that is, $(x, y) \in \operatorname{coker} f$. Now, by Theorem 17(iii), we have $(x, y) \in \kappa_f$, which is a contradiction. Thus, $\neg(f(x)\# f(y))$, which, as apartness on T is tight one, implies $f(x) = f(y)$.

Let $x\kappa_\tau^c, y\kappa_\tau^c \in S/\kappa_\tau^c$ be such that $\varphi(x\kappa_\tau^c)\#\varphi(y\kappa_\tau^c)$, i.e., $f(x)\# f(y)$. Therefore, $(x, y) \in \operatorname{coker} f = \kappa_f$. From the assumption $\mu_f \subseteq \tau$, it follows that $\kappa_f \subseteq \kappa_\tau$, which, in turn, implies $(x, y) \in \kappa_\tau$. Now, by Lemma 15(ii), $x\kappa_\tau^c \# y\kappa_\tau^c$, and φ is an se-mapping.

Finally, let $x \in S$. Then

$$\varphi\pi_\tau(x) = \varphi(\pi_\tau(x)) = \varphi(x\kappa_\tau^c) = f(x).$$

□

Proof **—Theorem 19**

As in the classical case, composition of functions is associative. By Theorem 11, $(\mathcal{T}_X^{se}, =, \#)$ is a set with apartness.

Let $f, g \in \mathcal{T}_X^{se}$ and suppose that $(fg)(x)\#(fg)(y)$ for some $x, y \in X$. Then, by the definition of the composition, $f(g(x))\# f(g(y))$, and, as f is an se-mapping, we have $g(x)\#g(y)$. Finally, as g is an se-mapping as well, we have $x\#y$. Thus, fg is an se-mapping and $fg \in \mathcal{T}_X^{se}$.

Let $f, g, h, w \in \mathcal{T}_X^{se}$ and $fh\#gw$. Then, by the definition of apartness in S, there is an element $x \in X$ such that $(fh)(x)\#(gw)(x)$, i.e., $f(h(x))\#g(w(x))$. Now we have

$$f(h(x))\#f(w(x)) \vee f(w(x))\# g(w(x)),$$

which, further, implies $h(x)\#w(x)$ (because f is an se-mapping) or $f\#g$ (by the definition of the apartness relation on $\mathcal{T}_X^{se}$). Thus $f\#g \vee h\#w$, that is, composition is an se-operation and $(\mathcal{T}_X^{se}, =, \#)$ is a semigroup with apartness. □

Proof **—Theorem 20**

Let $(S, =, \#; \cdot)$ be a semigroup with apartness. The semigroup S embeds into the monoid with apartness $S^1 = (S \cup \{1\}, =_1, \#_1, \cdot)$ with equality $=_1$ which consists of all pairs in $=$ and the pair $(1, 1)$, and with apartness $\#_1$, which consists of all pairs in $\#$ and the pairs $(a, 1), (1, a)$ for each $a \in S$.

Let f_a be a left translation of S^1, i.e., $f_a(x) = ax$, for all $x \in S^1$. Then f_a is an se-function. Indeed, $f_a(x)\#_1 f_a(y)$ is equivalent to $ax\#_1 ay$. The strong extensionality of multiplication implies $x\#_1 y$.

Denote by $\mathcal{T}^{se}_{S^1}$ the set of all se-functions from S^1 to S^1. In a pretty much similar way as in **CLASS**, define a mapping $\varphi : S^1 \to \mathcal{T}^{se}_{S^1}$ to be $\varphi(a) = f_a$, for each $a \in S^1$. It is routine to verify that

$$\varphi(ab) = f_{ab} = f_a\, f_b = \varphi(a)\varphi(b),$$

as well as

$$\varphi(a)\#_T\varphi(b) \Rightarrow a\#_1 b.$$

Thus, φ is an se-homomorphism. Also, $\varphi(a) =_T \varphi(b)$ iff $ax =_1 bx$ for all $x \in S^1$, and, for $x = 1$, we have $a =_1 b$. Therefore, φ is an embedding. □

Proof **—Lemma 16**

Let τ be a co-compatible co-quasiorder w.r.t. multiplication on S, and let $x, y, z \in S$. Then $(zx, zy) \in \tau$ implies $(x, y) \in \tau$ or $(z, z) \in \tau$. The latter is impossible because of strong irreflexivity of τ. Thus $(x, y) \in \tau$, i.e., τ is left co-compatible.

Conversely, let τ be a left and a right co-compatible co-quasiorder on S. Let $x, y, z, t \in S$ be such that $(xz, yt) \in \tau$. By the co-transitivity of τ, it follows either $(xz, yz) \in \tau$ or $(yz, yt) \in \tau$. Now, by the assumption, we have $(x, y) \in \tau$ or $(z, t) \in \tau$, as required. □

Proof **—Theorem 21**

By Theorem 13, $(S/\kappa^c, =, \#)$ is a set with apartness. The associativity of multiplication in S/κ^c follows from that of multiplication on S.

Let $a\kappa^c x\kappa^c \# b\kappa^c y\kappa^c$. Then $(ax)\kappa^c\#(by)\kappa^c$. By Theorem 13, we have that $(ax, by) \in \kappa$. But κ is a co-congruence, so either $(a, b) \in \kappa$ or $(x, y) \in \kappa$. Thus, by the definition of $\#$ in S/κ^c, either $a\kappa^c\#b\kappa^c$ or $x\kappa^c\#y\kappa^c$. So $(S/\kappa^c, =, \#, \cdot)$ is a semigroup with apartness. Using that fact and the definition of π, we have

$$\pi(xy) = (xy)\kappa^c = x\kappa^c\, y\kappa^c = \pi(x)\pi(y).$$

Hence π is a homomorphism, and, by Theorem 13, π is an se-surjection. □

Proof **—Theorem 22**

(i) Taking into account Theorem 14, it is enough to prove that coker f is co-compatible with multiplication in S. Let $(ax, by) \in$ coker f, i.e., $f(ax)\#f(by)$. Since f is a homomorphism, we have $f(a)f(x)\#f(b)f(y)$. The strong extensionality of multiplication implies that either $f(a)\#f(b)$ or $f(x)\#f(y)$. Thus either $(a, b) \in$ coker f or $(x, y) \in$ coker f, and therefore coker f is a co-congruence on S.

(ii) This is Theorem 14(ii).

(iii) This follows by Theorem 14 and Theorem 21.

(iv) Using (iii) and the assumption that f is a homomorphism, we have

$$\begin{aligned}\varphi(x(ker\ f)\, y(ker\ f)) &= \varphi((xy)(ker\ f))\\ &= f(xy)\\ &= f(x)f(y)\\ &= \varphi(x(ker\ f))\, \varphi(y(ker\ f)).\end{aligned}$$

Now, by Theorem 14, φ is an apartness embedding.

(v) This follows by Theorem 14 and (iv). □

Proof **—Theorem 23**

(i) If κ defines an apartness on S/μ, then, by Theorem 15(i), $\mu \cap \kappa = \emptyset$.

Let μ be a congruence and κ a co-congruence on a semigroup with apartness S such that $\mu \cap \kappa = \emptyset$. Then, by Theorem 15(i), κ defines apartness on S/μ.

Let $a\mu\, x\mu \# b\mu\, y\mu$, then $(ax)\mu\#(by)\mu$ which further, by the definition of apartness on S/μ, ensures that $(ax, by) \in \kappa$. But κ is a co-congruence, so either $(a, b) \in \kappa$ or $(x, y) \in \kappa$. Thus, by the definition of apartness in S/μ again, either $a\mu\#b\mu$ or $x\mu\#y\mu$. So $(S/\mu, =, \#, \cdot)$ is a semigroup with apartness.

(ii) Straightforward. □

Proof **—Lemma 17**

(i) By Lemma 15(i), κ_τ is a co-equivalence. Let $(ax, by) \in \kappa_\tau$. Then, by the definition of κ_τ, we have $(ax, by) \in \tau$ or $(by, ax) \in \tau$. By the co-compatibility of τ, we have $(a, b) \in \tau$ or $(x, y) \in \tau$ or $(b, a) \in \tau$ or $(x, y) \in \tau$, i.e., $(a, b) \in \kappa_\tau$ or $(x, y) \in \kappa_\tau$. So, κ_τ is co-compatible and, therefore, co-congruence.

(ii) Follows by (i) and Theorem 21.

(iii) By Lemma 15(ii), it follows that Υ_τ is a co-order. Let us prove that it is co-compatible. Let $x\kappa_\tau^c, y\kappa_\tau^c, a\kappa_\tau^c, b\kappa_\tau^c \in S/\kappa_\tau^c$ be such that $(x\kappa_\tau^c\, a\kappa_\tau^c, y\kappa_\tau^c\, b\kappa_\tau^c) \in \Upsilon_\tau$. Then $((xa)\kappa_\tau^c, (yb)\kappa_\tau^c) \in \Upsilon_\tau$, i.e., $(xa, yb) \in \tau$. By the co-compatibility of τ, we now have $(x, y) \in \tau$ or $(a, b) \in \tau$. By the definition of Υ_τ, it follows that $(x\kappa_\tau^c, y\kappa_\tau^c) \in \Upsilon_\tau$ or $(a\kappa_\tau^c, b\kappa_\tau^c) \in \Upsilon_\tau$; that is Υ_τ is co-compatible.

(iv) It follows by Theorem 21 and Lemma 15(iii). □

References

Baroni, M., & Bridges, D. S. (2008). Continuity properties of preference relations. *Mathematical Logic Quarterly, 54(5)*, 454–459.

Beeson, M. J. (1985). *Foundations of Constructive Mathematics*. Springer-Verlag.

Birkhoff, G. (1967). *Lattice Theory*. American Mathematical Society, Providence, Rhode Island, 1940 (First Edition), 1948 (Second (Revised) Edition), 1967 (Third Edition).

Birkhoff, G. (1971). The role of modern algebra in computing. Pp. 1–47 in *Computers in Algebra and Number Theory, Proceedings Sympos. Appl. Math. (New York, 1970)*, SIAM-AMS Prec., AMS, Providence, R. I. 4.

Bishop, E. (1967). *Foundations of Constructive Analysis*, McGraw-Hill, New York.

Bishop, E., & Bridges, D. S. (1985). *Constructive Analysis*, Grundlehren der mathematischen Wissenschaften 279, Springer, Berlin.

Bloom L. S. (1976). Varieties of ordered algebras. *J. Comput. System Set., 13*, 200–212.

Blyth, T. S. (2005). *Lattices and ordered algebraic structures*. Springer, 2005.

Boyd, J. P. (1991). *Social semigroups a unified theory of scalling and blockmodelling as applied to social networks*. George Mason University Press, Fairfax, Virginia.

Božić, M., & Romano, D. A. (1987). Relations, functions and operations in the constructive mathematics. *Publ. VTS, Ser. A: Math., 2*, 25–39.

Bras-Amorós, M. (2019). Tempered Monoids of Real Numbers, the Golden Fractal Monoid, and the Well-Tempered Harmonic Semigroup. *Semigroup Forum, Springer, 99(2)*, 496–516.

Bras-Amorós, M. (2020). Increasingly Enumerable Submonoids of R: Music Theory as a Unifying Theme. *The American Mathematical Monthly, Mathematical Association of America, 127(1)*, 33–44.

Bridges, D. S., & Havea, R. (2001). A Constructive Version of the Spectral Mapping Theorem. *Math. Log. Quart., 47, 3*, 299–304.

Bridges, D. S., & Vîţă, L. S. (2011). *Apartness and Uniformity - A Constructive Development*. CiE series on Theory and Applications of Computability, Springer.

Brouwer, L. E. J. (1975). The unreliability of the logical principles. Pp. 107–111 in Heyting, A. (Ed.), *Collected Works, Vol. 1*. North-Holland, Amsterdam.

Calderón, G. (2017). Formalizing constructive projective geometry in Agda. In *The Proceedings of The 12th Workshop on Logical and Semantic Frameworks, with Applications, LSFA 2017 (Brasilia, 23–24 September 2017* - http://lsfa2017.cic.unb.br/)

Chajda, I. (2004). Congruences in transitive relational systems. *Miskolc Mathematical Notes Vol. 5, No. 1*, 19–23.

Chajda, I., & Snášel, V. (1998). Congruences in ordered sets. *Mathematica Bohemica, Vol. 123, No. 1*, 95–100.

Crvenković, S., Mitrović, M., & Romano, D. A. (2013). Semigroups with Apartness. *Mathematical Logic Quarterly, 59 (6)*, 407-414.

Crvenković, S., Mitrović, M., & Romano, D. A. (2016). Basic Notions of (Constructive) Semigroups with Apartness. *Semigroup Forum, Volume 92, Issue 3*, 659–674.

Davey, A, & Priestley, H. A. (2002). *Introductions to lattices and orders*. Cambridge University Press.

Darpö, E., & Mitrović, M. Some results in constructive semigroup theory. arXiv:2103.07105.

Duffus, D., & Rival, I. (1981). A structure theory for ordered sets. *Discrete Mathematics 35*, 53–118, North-Holland Publishing Company.

Geuvers, H., Wiedijk, F., & Zwanenburg, J. (2002a). A constructive proof of the Fundamental theorem of algebra without using the rationals. Pp. 96–111 in Callaghan, P., Luo, Z., McKinna, J., & Pollack, R. (Eds.), *Proceedings of TYPES 2000 Workshop Durham Springer-Verlag, LNCS 2277*.

Geuvers, H., Pollack, R., Wiedijk, F., & Zwanenburg, J. (2002b). A Constructive algebraic hierarchy in Coq. *J. Symb. Comput. 34*, 271–286.

Grzegorczyk, A. (1964). A philosophically plausible formal interpretation of intuitionistic logic. *Indagationes mathematicae, vol. 26*, 596-601. (Pp. 596–601 in *Koninklijke Nederlandse Akademie van Wetenschappen, Proceedings, series A, vol. 67 (1964)*).

Halmos, P. R. (1960). *Naive set theory*. Princeton, N.J., Van Nostrand, 1960.

Harper, R. (2013). *Practical Foundations for Programming Languages*. Cambridge University Press.

Heyting, A. (1925). *Intuitionistische axiomatick der projectieve meetkunde*. Thesis, P. Noordhoof.

Heyting, A. (1927). Zur intuitionistischen Axiomatik der projektiven Geometrie. *Mathematische Annalen, 98*, 491–538.

Heyting, A. (1941). *Untersuchungen über intuitionistische Algebra*. Nederl. Akad. Wetensch. Verh. Tweede Afd. Nat. 18/2.

Heyting, A. (1956). *Intuitionism, an Introduction*. North - Holland.

Hille, E. (1948). *Functional Analysis and Semi-groups*. New York, American Mathematical Society, Providence.

Howie, J. M. (1995). *Fundamentals of Semigroup Theory*, London Mathematical Society Monographs, New Series, Oxford; Clarendon Press.

Ishihara, H. (2004). Informal Constructive Reverse Mathematics. *CDMTCS Research Report 229*, University of Auckland.

Jacobs, B. (1995). *Bisimulation and Apartness in Coalgebraic Specification*. 1995 - Notes of lectures given in January 1995 at the joint TYPES/CLICS workshop in Gothenburg and BRICS seminar in Aarhus.

Kehayopulu, N., & Tsingelis, M. (1995a). On subdirectly irreducible ordered semigroups. *Semigroup Forum Vol. 50*, 161–177.

Kehayopulu, N., & Tsingelis, M. (1995b). Pseudoorder in ordered semigroups. *Semigroup Forum Vol. 50*, 389–392.

Kline, M. (1967). *Mathematics for nonmathematicians*. Dover Publications, INC., New York.

Kórtesi, P., Radeleczki, S., & Szilágyi, S. (2005). Congruences and isotone maps on partially ordered sets. *Mathematica Pannonica 16/1*, 39–55.

Martin-Löf, P. (1982). Constructive mathematics and computer programming. Pp. 153–175 in Kohen, L. J., Los, J., Pfeiffer, H. & Podewski, K-P. (Eds.), *Proceedings of the Sixth International Congress of Logic, Methodology and Philosophy of Science*. North-Holland Publishing Company, Amsterdam.

Mines, R., Richman, F., & Ruitenburg, W. (1988). *A Course of Constructive Algebra*. New York: Springer-Verlag.

Mitrović, M., & Hounkonnou, M. N. *Constructive binary structures with apartness vs. classical ones*, Book in progress.

Mitrović, M., Hounkonnou, M. N., & Baroni, M. A. (2021). Theory of constructive semigroups with apartness - foundations, development and practice. *Fundamenta Informaticae 184(3)*, 233–271.

Mitrović, M., & Silvestrov, S. (2020). Apartness Isomorphism theorems for basic constructive algebraic structures with special emphasize on constructive semigroups with apartness - an overview. Pp. 653–686 in Silvestrov, S., Malyarenko, M., & Rančić, M. (Eds.), *Stochastic Processes and Algebraic Structures -From Theory Towards Applications, Volume II: Algebraic Structures and Applications, Västerås and Stockholm, Sweden, October 4–6*, Springer.

Mitrović, M., Silvestrov, S., Crvenković, S., & Romano, D. A. (2019). Constructive semigroups with apartness: towards new algebraic theory. *Journal of Physics: Conference Series (JPCS), Volume 1194*, 012076, https://doi.org/10.1088/1742-6596/1194/1/012076.

Moshier, M. A. (1995). A rational reconstruction of the domain of feature structures. *J. of Logic, Language and Information, 4* 2, 111–143.

Newman, J. R. (1956). *The World of Mathematics*, 4 Volume set, New York.

Partee, B. H., ter Meulen, A., & Wall, R. E. (1990). *Mathematical methods in linguistics*. Kluwer Academic Publishers, Dordrecht/Boston/London.

Pattison, P. (1993). *Algebraic models for social networks*. Cambridge University Press 1993.

Reddy, U. S. (2014). *Notes on Semigroups*. University of Birmingham. https://www.cs.bham.ac.uk/~udr/notes/semigroups.pdf

Romano, D. A. (1996). Coequality relations, a survey. *Bull. Soc. Math. Banja Luka, 3*, 1–36.

Romano, D. A. (2002). Some relations and subsets generated by principal consistent subset of semigroup with apartness. *Univ. Beograd, Publ. Elektrotehn. Fak. Ser. Mat. 13*, 7–25.

Romano, D. A. (2005). A note on family of quasi-antiorder on semigroup. *Kragujevac J. Math., 27*, 11–18.

Romano, D. A. (2007). A note on quasi-antiorder on semigroup. *Novi Sad J. Math., 37*, 3–8.

Ruitenburg, W. B. G. (1982). *Intuitionistic Algebra, Theory and Sheaf Models.* Utrecht, Ph.D. Thesis, 143 pages, D. van Dalen (adviser).

Ruitenburg, W. (1991). Inequality in Constructive Mathematics. *Notre Dame Journal of Formal Logic, Volume 32*, 533–553.

Schein, B. M. (1970). Relation algebras and/unction semigroups. *Semigroup Forum 1, no. 1*, 1–62.

Schein, B. M. (1997). Book Review - Social semigroups a unified theory of scaling and blockmodelling as applied to social networks (by J. P. Boyd). *Semigroup Forum, 54*, 264–268.

Schröder, B. S. W. (2003). *Ordered Sets - An Introduction*. Springer Science+Business Media, LLC.

Szédli, G., & Lenkehegyi, A. (1983a). On classes of ordered algebras and quasiorder distributivity. *Acta Sci. Math., 46*, 41–54.

Szédli, G., & Lenkehegyi, A. (1983b). On congruence n-distributivity of ordered algebras. *Acta Math. Hungar., 41*, 17–36.

Troelstra, A. S., van Dalen, D. (1988). *Constructivism in Mathematics, An Introduction, (two volumes)*, North - Holland, Amsterdam.

Wells, R. B. (2010). *Introduction to Biological Signal Processing and Computational Neuroscience*, https://webpages.uidaho.edu/rwells/techdocs/Biological%20Signal%20Processing/Chapter%2000.pdf

Woźny, J. (2018). *How We Understand Mathematics*, Springer.

Algebraic Approaches to the Analysis of Social Networks

Philippa Pattison

1 Introduction

This chapter reviews the use of algebraic approaches for understanding the interconnected nature of social relationships. It outlines the algebraic approaches that have been taken, the types of questions they have sought to address, and some of their successes and shortcomings. These approaches include analyzing observed binary social relations among a set of actors to identify *social positions* of actors and constructing the partially ordered semigroup of the binary relations to represent their *role structure*. Analyses of the role structure using the lattice of order-preserving homomorphisms and corresponding partial functions on the actor set are then discussed. The chapter concludes with a discussion of three broad challenges for this line of inquiry.

The idea of using algebraic approaches to understand social relational forms dates back to at least the pioneering work in the 1940s of mathematician André Weil and anthropologist Claude Lévi-Strauss who collaborated to characterize certain Australian Aboriginal kinship structures as mathematical groups (Lévi-Strauss 1969). Building on this work, Robert R. Bush and Harrison White described group structures for some other Australian Aboriginal peoples, representing these structures as permutation groups of two generator relations on a set of classes (termed *clans* by White), one describing the clan to which a man's wife belonged and one describing the clan of the man's children (White 1963; see also Kemeny et al. 1957). Although it was recognized that the permutation group structures were

P. Pattison (✉)
University of Sydney, Sydney, NSW, Australia

Melbourne School of Psychological Sciences, University of Melbourne, Parkville, VIC, Australia
e-mail: philippa.pattison@sydney.edu.au

© The Author(s), under exclusive license to Springer Nature Switzerland AG 2023

M. N. Hounkonnou et al. (eds.), *Mathematics for Social Sciences and Arts*, Mathematics in Mind, https://doi.org/10.1007/978-3-031-37792-1_8

particular to certain peoples and idealized rather than exact descriptions, these early efforts served as an important foundation for further generalizations. White saw this work as a first step toward a broader set of insights and tools for "penetrating beneath the everyday perceptions of our own social structure" (White 1963, p. 2). In an insightful review of White's book, *An Anatomy of Kinship*, John Boyd also articulated the value of broadening the representation to more general social relations and defining an algebraic system as a homomorphism from a finitely generated free semigroup into a regular monoid (Boyd 1964). His recognition that a broader framework might prove fruitful is sympathetic with a similar observation by Schein that "irreversible processes are much more common than reversible ones" and "one meets functions and transformation semigroups much more often than groups" (Schein 1970, p. 54).

Boyd (1969) also extended White's approach for group representations of certain Australian Aboriginal kinship systems to analyze the relationships *among* the group representations, highlighting the role of homomorphic mappings in understanding these relationships. He argued "that if a group G_1 evolves into a group G_2, then G_1 will be a homomorphic image of G_2 but that the remnants of G_1 will be revealed in the coding of G_2 in its everyday use" (Boyd 1969, p. 139). Boyd was able to realize this intention in several specific instances by linking (idealized) group structures, kinship grammars, and componential (i.e., semantic) analysis of kinship terms. Importantly, his approach demonstrated how an algebraic formulation could reflect a strong theoretical claim – in this case, about the nature of change in kinship systems – and additional evidence could provide a supporting interpretive context.

Boyd and colleagues also studied other kinship systems for which inverse semigroups provide a compelling representational form (Boyd et al. 1972). More recently, scholars of kinship systems have turned to other tools such as expert systems (Read 2006) and agent-based models (Itao and Kaneko 2020) to understand the emergence of various characteristics of kinship systems.

White (1963) foreshadowed a broader agenda for his study of kinship systems, envisaging the study of many different types of social relations in contemporary society. In doing so, he appealed to the theoretical insights of Siegfried Nadel who also sought the mathematical means of understanding the "interlocking of relationships, whereby the interactions implicit in one determine those occurring in others" (Nadel 1957, p. 16). Nadel had a particular conception of this interlocked system of relationships as a "network," comprising not just the relationships or links between pairs of actors (with actors and their links construed as "knots"), but also "the further linkage of the links themselves and the important consequence that, what happens so-to-speak between one pair of 'knots', must affect what happens between other adjacent ones" (Nadel 1957, p. 16). Crucial to this characterization was the expectation that the interlocking of relationships had a bearing on "'what happens' in 'neighboring' relationships and hence on their effective interdependence" (Nadel 1957, p. 17). In Nadel's formulation, social structure can be understood by "abstracting from the concrete population and its behavior the pattern or network (or system) of relationships" (Nadel 1957, p. 12), with roles reflecting the abstracted ways in which people act toward one another. Nadel thus construed a network of interlocking relationships as a system with coherence and closure and a resulting patterning of social relational forms.

In a linked pair of papers, White et al. (1976), and Boorman and White (1976), proposed an algebraic approach to the characterization of this interlocking in response to Nadel's theoretical framing. Their approach yielded an empirical means of identifying social *roles* and social *positions*. It begins with a fixed collection N of n actors and a fixed set $R = \{R_1, R_2, \ldots, R_p\}$ of $n \times n$ binary (social) relations observed among them. The types of relations are not specified theoretically but rather assumed to be relevant to the context of a particular setting and limited only by the assumption that "all ties of a given observed type share a common signification (whatever their content might be)" (White et al. 1976, p. 734). Importantly, the relation set R comprises a set of relational *data*. We refer to R as the set of *generator* relations for the actor set N.

White and colleagues proposed two key constructions. The first was that actors have the same *social position*, or are *structurally equivalent*, if and only if they each stand in the same relationship to any other member of the set N. That is, two actors, k and l, are *structurally equivalent* (we write $(k,l) \in \text{SE}$) whenever $(k, m) \in R_s$ if and only if $(l, m) \in R_s$, and $(m, k) \in R_s$ if and only if $(m, l) \in R_s$, for all $m \in N$ and all relations R_s in R. Structural equivalence is associated with a partition on the actor set N. Denoting by $c_k = \{m \in N : (k, m) \in \text{SE}\}$ the class of actors structurally equivalent to actor k, and by C the set of distinct such classes, then SE can be associated with a *relational* or *graph* homomorphism of R, namely, a mapping $\psi : N \to C$ and a set of relations B_s on C, defined by $(c_k, c_l) \in B_s$ if $(k, l) \in R_s$, for c_k, c_l, $\in C$; $s = 1$, 2, ..., p. By virtue of the definition of *SE*, $(k, l) \in R_s$ if and only if $(k', l') \in R_s$ for all $k' \in c_k$, $l' \in c_l$, so that, in this special case, the mapping ψ and the relation set $B = \{B_1, B_2, \ldots, B_p\}$ are sufficient to re-construct the original relation set R. The relation set B is one instance of a *blockmodel* for the relation set R, namely, a mapping of the actor set N to a set of *social positions*, and a specification of the social relations among those abstracted positions.

The second construction proposed by White and colleagues was to represent the *role structure* implicit in the relation set R by the semigroup of binary relations generated by $R = \{R_1, R_2, \ldots, R_p\}$ under relational composition. The *composition* of two binary relations, U and V on the set N is the relation UV in which $(i,j) \in UV$ if and only if $(i,k) \in U$ and $(k,j) \in V$ for some $k \in N$. The composition operation applied to a finite set of relational terms leads to a free semigroup whose elements correspond to all possible finite strings of elements from that set of relational terms. Invoking what Boorman and White (1976) termed the *Axiom of Quality* – the proposition that two compound relations should be regarded as equal if they link exactly the same pairs of actors in N – then yields a finite semigroup as Boyd (1964) indicated. Semigroups of binary relations have been well studied by mathematicians (e.g., McKenzie and Schein 1997; Plemmons and West 1970; Schein 1970; Schwarz 1970).

An appealing feature of these constructions is that the role structure of R on the actor set N is identical to the derived role structure defined on positions of structurally equivalent actors. The role structure can therefore be seen, in principle, as a property of the relations among social positions, rather than being tied specifically to the actors who occupy those positions.

In practice, the analysis of actors into a set of social positions, each of which comprises actors who are approximately structurally equivalent, often precedes the development of a blockmodel among those positions and the construction of its role structure. As an example of the construction of a role structure, we consider Breiger's (1974) re-analysis of data from a study of community elites by Laumann and Pappi (1976). Franz and Pappi had identified members of the community elite in a German town with the pseudonym Altneustadt and interviewed members of the elite to determine those with whom each had the closest business/professional ties most often discussed community affairs and most frequently engaged in social activity. Breiger (1974) re-analyzed these relational data to identify a partition of the elite members into four classes of approximately structurally equivalent elite members. He was able to demonstrate the meaningfulness of the distinctions among the four classes by establishing their differentiation according to influence ranks and religious and political affiliations. He then inferred from the relational data the relations among these four classes or positions on the basis of the density of ties among and between the elite members in the four classes. The classes are labelled from 1 to 4 and the resulting blockmodel of relations among the classes is depicted graphically in Fig. 1. Classes 1 and 3 contain the more influential members of the community, but the two are sharply distinguished by political (and also religious) affiliation. Classes 1, 2, and 4 share some political affiliations, and class 4 has a distinctive cultural involvement in the affairs of the community. We use the blockmodel in Fig. 1 to illustrate a number of approaches described in the chapter.

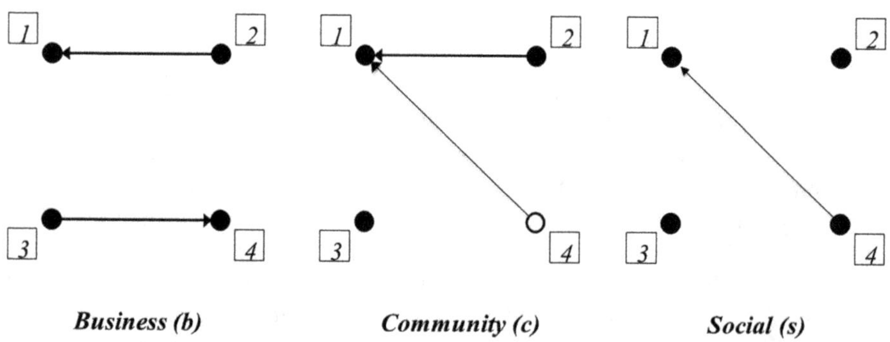

Fig. 1 Blockmodel for Altneustadt community elite: business (*b*), community affairs (*c*), and social (*s*) relations among 4 classes (note that a node for class *i* is solid if (*i*,*i*) is included in the relation)

2 Partially Ordered Semigroups from Binary Relations

The set $R = \{R_1, R_2, \ldots, R_p\}$ of $n \times n$ binary relations gives rise to a partially ordered semigroup under relational composition and the natural partial order on binary relations, defined by:

- $U \leq V$ whenever $(i,j) \in U$ implies $(i,j) \in V$ for all $i, j \in N$.

Following Pattison (1993), we focus here on the construction of partially ordered semigroups but note that a number of other useful algebraic structures can also be defined on $\{R_1, R_2, \ldots, R_p\}$ (see Pattison, 2009, for some examples). Key definitions are included in the text; others can be found in the glossary of terms included in the Appendix.

Formally, a *semigroup* $[S,\circ]$ can be defined as a set S with a binary operation, $\circ$, that satisfies:

- $(a \circ b) \circ c = a \circ (b \circ c)$ for all $a, b, c \in S$ (associativity).

A *quasi-order* on S is a relation $\leq$ that is both reflexive and transitive. A quasi-order that is also *anti-symmetric* is a *partial order.*

A *partially ordered semigroup* $[S,\circ,\leq]$ is a semigroup with a partial order $\leq$ for which the binary operation $\circ$ is *isotone* in each of its variables; that is, $a \leq b$ and $c \leq d$ implies $a \circ c \leq b \circ d$, for any $a, b, c, d \in S$.

To obtain the partially ordered semigroup $S_R = [R^*, \circ, \leq]$ associated with the generator relation set R, we create the finite set of distinct binary relations generated by R under relational composition. More precisely, we define the closure R^* of R under relational composition as the set R^* of binary relations that is *closed, inclusive of the generator set R* and *minimal*; that is, R^* satisfies:

- For all $a, b \in R^*$, $a \circ b \in R^*$;
- $R \subseteq R^*$; and
- For any set W that includes the generator set R and also satisfies closure, then $R^* \subseteq W$.

Implicit in the creation of R^* is the notion that any two networks in R^* are regarded as equal if they link precisely the same pairs of actors in N. As noted earlier, Boorman and White referred to this assumption as the *Axiom of Quality*, observing that it captures equations among direct and indirect (compound) relations on N.

The equations of the role structure semigroup arguably serve to describe the interlocking of social relations, as Nadel had sought. If the eq. $T = UV$ holds, for example, then actors i and j are linked directly by a relation T if and only if they are also indirectly linked through at least one other actor k by an (indirect) relation UV. As White later put it, a social tie "exists in, and only in, a relation between actors that *catenates*, that is, that entails (some) compound relation through other such ties of those actors. ... A social tie generates and is warranted by other such ties in one or another network. This is most evident in what are presumably the oldest and most basic social ties, those of kinship" (White 2002, p. 312). In other words, the equations of the semigroup are seen as reflecting the interlocking (and hence the interdependence) of relations and their resulting patterning.

If sets of relations of the same type on different actor sets give rise to isomorphic semigroups, there is a strong case for inferring a common form of relational interlocking in the two actor sets (Boorman and White 1976). More broadly, homomorphic mappings among semigroups of binary relations offer the means for comparing them, as Boyd demonstrated for group representations of kinship systems. In the sections to follow, we review how this potential has been utilized to analyze and compare role structures.

The partially ordered semigroup S_R can be conveniently represented by:

- An $m \times p$ right multiplication table indicating the composition, $s^\circ g$ *of* each of the m distinct binary relations $s \in S_R$ and each of the p generator relations $g \in R$; and
- An $m \times m$ array capturing the quasi-order on S_R, with a unit entry in the array for (t,s) if $s \leq t$, with $s, t \in S_R$.

The partially ordered semigroup generated by the blockmodel of Fig. 1 is displayed in Fig. 2. The closure R^* of the relational set $R = \{b,c,s\}$ includes the three generator relations and five additional distinct compound relations, $\{bc, bs, cb, sb, cbs\}$ (note that the binary operation is implicit in these expressions; that is, we write $b^\circ c$ as bc). Equations such as $bb = b$ and $cs = c$ can be read from the right multiplication table; the partial ordering can be read from the partial order diagram in which one element s is below and connected by an ascending path to another element t if $s \leq t$ (e.g., in the semigroup in Fig. 2, $c \leq sb$, $cbs \leq bs$).

Many interesting aspects of the interlocking of relations can be read from the right multiplication table and the partial order diagram. For example, each generator relation demonstrates a form of closure, in that the composition of the relation

Elements	*b*	*c*	*s*
b	*b*	*bc*	*bs*
c	*cb*	*c*	*c*
s	*sb*	*c*	*s*
bc	*cbs*	*bc*	*bc*
bs	*bs*	*bc*	*bs*
cb	*cb*	*bc*	*cbs*
sb	*sb*	*bc*	*bs*
cbs	*cbs*	*bc*	*cbs*

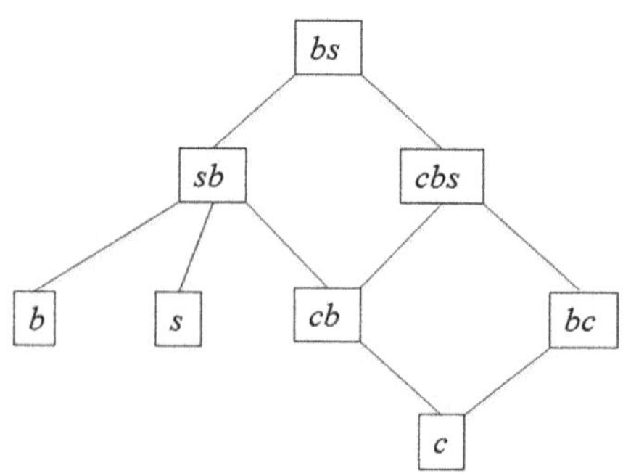

Fig. 2 The right multiplication table (left) and the partial order diagram (right) for the partially ordered semigroup of the Altneustadt blockmodel of Fig. 1

with itself yields the original relation (in other words, $bb = b$, $cc = c$ and $ss = s$ and the three elements are *idempotent*). Moreover, compound relations create new connections among the four classes only where they comprise two or more types of relations one of which is a business relation. Social ties also act as an identity for community ties ($cs = sc = c$, $ss = s$) indicating that social ties are "stronger" (i.e., more tightly held) than community ties (see Breiger and Pattison 1986; also Granovetter 1973), even though both are relatively "strong."

As Boyd argued for group representations for kinship, it is of value to be able to compare semigroups. For this purpose, we introduce the concept of homomorphic mappings among partially ordered semigroups.

3 Semigroup Homomorphisms

An *(isotone) homomorphism* from a partially ordered semigroup $[S,\circ,\leq]$ onto a partially ordered semigroup $[T,\circ,\leq]$ is a mapping φ for which:

- $\varphi(s \circ t) = \varphi(s) \circ \varphi(t)$ for all $s, t \in S$;
- $s \leq t$ implies $\varphi(s) \leq \varphi(t)$ for any $s, t \in S$; and
- For each $u \in T$, there is some $s \in S$ for which $\varphi(s) = u$.

T is termed a *homomorphic image* of S and we write $T = \varphi(S)$. Each homomorphism φ on $[S,\circ,\leq]$ can be represented by a quasi-order π_φ, termed a *π-relation*, in which $(s, t) \in \pi_\varphi$ if and only if $\varphi(t) \leq \varphi(s)$, for $s, t \in S$. It is also convenient to write $T = \varphi(S) = S/\pi_\varphi$, where π_φ is the *π-relation* corresponding to the homomorphism φ.

For example, there is a homomorphism from the semigroup S of Fig. 2 onto the semigroup T of Fig. 3.

Elements	b	c	s
b	b	b	b
c	b	c	c
s	b	c	s

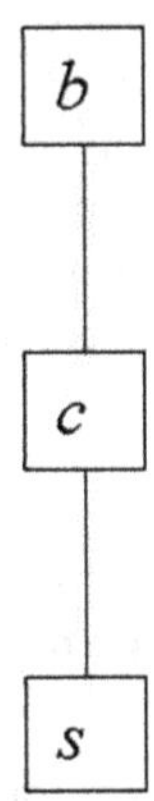

Fig. 3 Right multiplication table and partial order for the semigroup T

	b	c	s	bc	bs	cb	sb	cbs
b	1	1	1	1	1	1	1	1
c	0	1	1	0	0	0	0	0
s	0	0	1	0	0	0	0	0
bc	1	1	1	1	1	1	1	1
bs	1	1	1	1	1	1	1	1
cb	1	1	1	1	1	1	1	1
sb	1	1	1	1	1	1	1	1
cbs	1	1	1	1	1	1	1	1

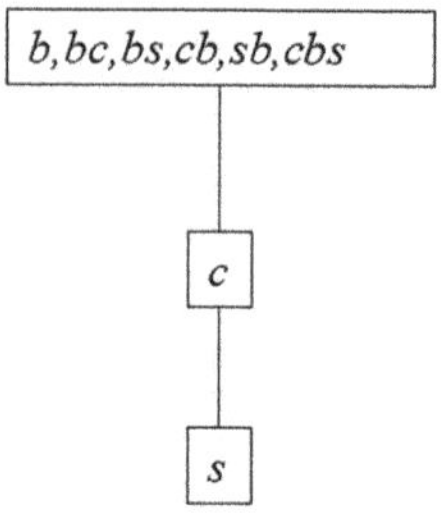

Fig. 4 The π-relation on the semigroup S of Fig. 2 associated with the isotone homomorphism from S to the semigroup T

The π-relation associated with the mapping is shown in Fig. 4 both as a quasi-order relation and in diagrammatic form; it maps all of the relations in the semigroup S that contain a business tie b to the b relation in the semigroup T, and it also maps the relations c and s in S to the c and s relations in T, respectively.

Boyd (1969) argued that the presence of a homomorphic mapping from one algebraic structure onto another suggested a relation of structural *similarity* between the systems that the structures represent, with the second being an "approximation" to the first and, as mentioned earlier, a potential step toward its evolution. He also proposed that the approximations (or homomorphic images) of an algebraic structure could themselves be ordered in the way just described. With that, he anticipated sequences of similar structures, ordered by proximity to a given structure, and argued that "a structural description of a behavioral phenomena should also account for the dynamics of its use and its changes through time" (Boyd 1969, p. 165). His insistence on explicitly accounting for the dynamics of change is important, as it encourages exploration of the context for change and potential mechanisms of change. It is also important to recognize that the idea that a homomorphic image of a semigroup is *simpler* than the original semigroup is itself a theoretical claim, as quickly became evident in the proposals discussed in Sect. 6 for representing shared role structure.

4 Relational Homomorphisms and Semigroup Homomorphisms

A *relational, or graph, homomorphism* of a relation set $R = \{R_1, R_2, \ldots, R_p\}$ on an actor set N onto a relation set $T = \{T_1, T_2, \ldots, T_{op}\}$ on an actor set M is a mapping ψ from N onto M for which:

- For each $k \in M$, $\psi(i) = k$ for some $i \in N$, and
- $(\psi(i), \psi(j)) \in T_s$, if and only if $(i,j) \in R_s$, for some $i, j \in N$; $s = 1, 2, \ldots, p$.

As observed earlier, when the relational homomorphism comprises a mapping of N onto the classes associated with a structural equivalence relation on R, two equivalence classes are related by R_s if and only if each actor in the first class is related by R_s to each actor in the second class. This means that the semigroup induced by the network homomorphism is isomorphic to the original semigroup of R. In general, though, this is not the case and two classes are linked in the induced set of relations whenever any actor in the first class is linked by R_s to any actor in the second. Indeed, the semigroup of the relational system induced by the relational homomorphism is not necessarily even a homomorphic image of the original semigroup; hence, the interest in when relational and semigroup homomorphisms are aligned.

In the case of the semigroup T, the mapping ψ defined by:

- $1, 4 \to 1, 2 \to 2, 3 \to 3$

yields the relational homomorphism of the blockmodel of Fig. 1 to that shown in Fig. 5. The partially ordered semigroup of the reduced blockmodel of Fig. 5 is T. The relational homomorphism emphasizes the roles of blocks 2 and 3 in the ordering of relations by their strength of closure that is evident in the semigroup equations and the partial order of T. Social relations are the most "closed" and indeed the social relation in Fig. 5 is an identity relation. Business relations are the least "closed" and more connective of the different classes, including to class 2 and also across the political divide to class 3. The community relations are intermediate in terms of how tightly they are held and connect class 2 to the combined classes 1 and 4; however, there are no community affairs connections to class 3.

Given that not all relational homomorphisms give rise to semigroup homomorphisms, the question naturally arises of the conditions under which they do. In fact, a number of generalizations of structural equivalence have been identified that

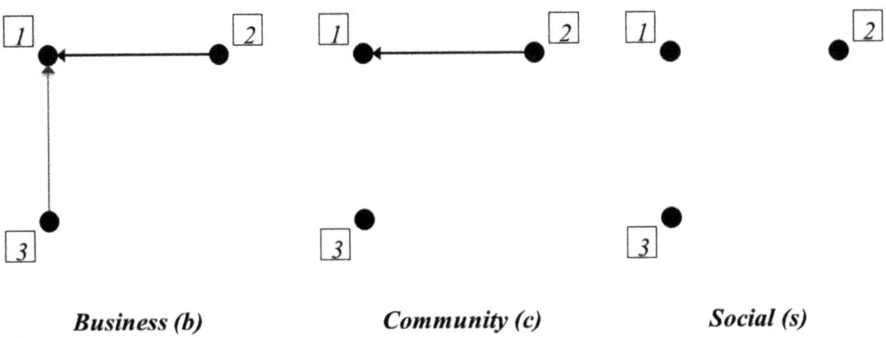

Fig. 5 A relational homomorphism of the blockmodel of Fig. 1 giving rise to the semigroup T (the mapping is defined by $1, 4 \to 1, 2 \to 2, 3 \to 3$)

guarantee that the relational homomorphism associated with a partition of the actors of N gives rise to an isotone semigroup isomorphism. The most general of these to have been identified is termed the condition G_{im} and was defined in two steps by Kim and Roush (1984).

First, recall that $c_k = \{j \in N : \psi(j) = k; k \in M\}$ is the class of actors in N mapped by ψ to the element $k \in M$. The mapping ψ on N satisfies Kim and Roush's *condition* G_i if, whenever $(j, l) \in R_{\underline{s}}$ for any $j, l \in N$ and any $s = 1, 2, \ldots, p$, then any subset of i actors in $c_{\psi(j)}$ is related by $R_{\underline{s}}$ to at least i actors in $c_{\psi(l)}$ (or to all actors in $c_{\psi(l)}$ if $|c_{\psi(l)}| < i$).

Second, the mapping ψ satisfies the more general *condition* G_{im} if, for each class c_k, there is a subset c_k' of c_k such that $(j, l) \in R_{\underline{s}}$ implies $(j', l') \in R_{\underline{s}}$ for some $j' \in c_{\psi(j)}'$ and $l' \in c_{\psi(l)}'$ and ψ satisfies the condition G_i when restricted to relations involving actors in the set $N' = \cup_k c_k'$.

The conditions G_i and G_{im} cleverly capture a level of consistency in the relations between actors in two classes of a relational homomorphism that guarantees the existence of a semigroup homomorphism.

Two interesting cases of the condition G_i occur for $i = 1$ and $i = n$. When G_1 holds, if some actor in an equivalent class c_k is linked by a relation R_s to some actor in the equivalence class c_l, then *every* actor in c_k is linked by R_s to some actor in c_l. Conversely, when G_n holds, if some actor in an equivalent class c_k is linked by a relation R_s to an actor in the equivalence class c_l, then there is a relation of type R_s from some actor in c_k to *every* actor in c_k. These two conditions, originally termed the *outdegree* and *indegree* conditions, respectively (e.g., see Pattison 1993), generalize structural equivalence by requiring, respectively, consistency within each class and for each relation of outgoing *or* incoming ties.

If both conditions hold, then the partition associated with the relational homomorphism is termed a *regular equivalence* (White and Reitz 1983). Regular equivalence is an appealing form of generalization of structural equivalence, since it requires a qualitative consistency in the relations of actors sharing the same class. Audenaert et al. (2019) have recently described an algorithm for identifying regular equivalences in very large networks. A comprehensive treatment of regular and other equivalences can be found in Doreian et al. (2004).

5 Comparing and Analyzing Role Structures

In order to compare the semigroups of two relation sets with the same set of relational terms but distinct actor sets, N and M, Boorman and White (1976) defined the *joint* homomorphic image as the largest semigroup that was a homomorphic image of both. They argued that this joint homomorphic image reflected features common to both, seeing each semigroup reflected by the set of its "simpler" homomorphic images. Bonacich (1980) proposed an alternative representation, the *common structure semigroup*, the semigroup generated by the disjoint union of the two relational systems on the actor set $N \cup M$ and the smallest semigroup having a set of equations present in each of the constituent semigroups and also *containing* each of the two semigroups as a homomorphic image. Bonacich (1980) was viewing each semigroup in terms of its set of equations and was drawn as a result to the view that the set of shared equations reflected their common structure. Both constructions are potentially useful and these different propositions for shared structure illustrate that care is needed in translating mathematical relationships into propositions in the substantive domain of application. This point is further elaborated below.

Indeed, there is value in understanding how homomorphisms might inform a more systematic analysis of the partially ordered semigroups of one or more relation sets. In this section, therefore, we turn to some algebraic approaches for comparing and analyzing algebraic structures.

We note first that the collection of all (isotone) homomorphisms on a partially ordered semigroup $A = [S,^\circ,\leq]$ can themselves be partially ordered by defining:

- $\varphi \leq \rho$ for two homomorphisms φ, ρ on A if, for all $s, t \in S$, $\rho(t) \leq \rho(s)$ implies $\varphi(t) \leq \varphi(s)$.

This partial ordering gives rise to a lattice $L(A)$ of homomorphisms of A (e.g., Pattison 1993).

A lattice $L_\pi(A)$ can also be defined by a partial ordering on the corresponding π-relations:

- $\pi_\varphi \leq \pi_\rho$ if and only if $(s, t) \in \pi_\varphi$ implies $(s, t) \in \pi_\rho$ for any $s, t \in S$.

The partial ordering in $L_\pi(A)$ is *dual* to that in $L(A)$; that is, $\varphi \leq \rho$ in $L(A)$ if and only if $\pi_\rho \leq \pi_\varphi$ in $L_\pi(A)$. As a result, the lattice $L_\pi(A)$ is dual to $L(A)$.

Pattison (1993) proposed analyzing the semigroup of a set of generator relations by deriving certain minimal subdirect representations of the semigroup (see glossary for definitions).

Well-known theorems in universal algebra (e.g., Birkhoff 1963) show that the collection of direct and subdirect representations of a semigroup $A = [S,^\circ,\leq]$ (and indeed of algebras, in general) depend on the lattice of homomorphisms of the semigroup (or, equivalently, on the lattice $L_\pi(A)$ of π-relations of A). We refer here to $L_\pi(A)$ but an equivalent (and dual) account can be given by referring to the lattice $L(A)$.

If $\{\pi_1, \pi_2, \ldots, \pi_f\}$ is a set of π-relations in $L_\pi(A)$ for which $\pi_1 \wedge \pi_2 \wedge \ldots \wedge \pi_f = \pi_{min}$, the minimal element of $L_\pi(A)$, then A can be expressed as the subdirect product of the semigroups A/π_1, A/π_2, $\ldots$, A/π_f (Birkhoff 1963). If $\{\pi_1, \pi_2, \ldots, \pi_f\}$ is a minimal irredundant set in $L_\pi(A)$, the subdirect representation to which it gives rise is termed a *factorization* of A. By virtue of its definition, a factorization of a semigroup is an efficient form of subdirect representation, identifying a minimal collection of irreducible homomorphic images whose direct product includes the semigroup as a subsemigroup. Each component semigroup associated with a factorization is, by definition, irreducible.

Factorizations are not necessarily unique and depend to some extent on the structure of $L_\pi(A)$. If $L_\pi(A)$ is distributive, then a unique factorization exists. Where

$L_\pi(A)$ is modular, then each factorization includes the same number of elements. Furthermore, if A can be represented as a direct product of semigroups, then the factorization of A will yield the irreducible components of the direct product representation.

Pattison (1993) proposed factorization as a useful means of analyzing the partially ordered semigroup of a relational system R into smaller, arguably "simpler," irreducible components. Each irreducible semigroup in this representation has a unique maximal homomorphic image and that unique image is itself irreducible or possesses a factorization. A useful analysis of the structure of the semigroup can therefore be represented by a directed graph beginning with a node representing the original semigroup $[S_R, \circ, \leq]$ and directing an edge from that node to each of the semigroups associated with its factorization or, alternatively if $[S_R, \circ, \leq]$ is irreducible, directing a single edge to its unique maximal homomorphic image. This same strategy can be applied to each of the new nodes in the diagram – multiple directed edges to semigroups associated with factorization of a node representing a semigroup or a single directed edge to a unique maximal homomorphic image – with the process continuing until each new node is a one-element semigroup. The resulting directed acyclic graph, termed a *reduction diagram*, summarizes the lattice-based analysis of the semigroup.

What value does the analysis represented by the reduction diagram offer? Pattison (1993) argued that a factorization of the semigroup of a relational system summarizes the elemental relational features of the semigroup in terms of smaller, simpler semigroups from which the full semigroup of the relational system R could be constructed. The definition of factorization makes precise the nature of this representation. Like Boyd (1969), Pattison argued that these more elementary component semigroups could have an historical significance in the emergence of the relation set under consideration but more generally could also represent the relational structure of potentially different facets of the role structure as we describe below. This consideration again raises the question considered in the previous section of the circumstances under which a homomorphic image of the semigroup of a set R of generator relations correspond to a (relational) homomorphism of R itself.

Pattison (1993) proposed an "empirical" approach of identifying maximal partial mappings on N for which there existed an induced graph homomorphism *corresponding to* a target semigroup homomorphism (specifically, containing the target semigroup as a homomorphic mapping). A partial mapping ψ of the actor set N onto a set M maps a potentially proper subset N' of N onto M (and N' is equal to the union of the equivalence classes associated with ψ). Partial mappings are potentially useful in this context because we know that the disjoint union of two relational sets give rise to the common structure semigroup of their individual semigroups and that the semigroup of each constituent relation set is a homomorphism of the common structure semigroup. As a result, functions on subsets of actors are a legitimate setting of interest for a component of the role structure. These empirical correspondences with partial mappings of N are particularly useful for the semigroup homomorphisms associated with a factorization. The resulting analysis

A1	*b*	*c*
b	*b*	*bc*
c=*s*	*cb*	*c*
bc	*bc*	*bc*
cb	*cb*	*bc*

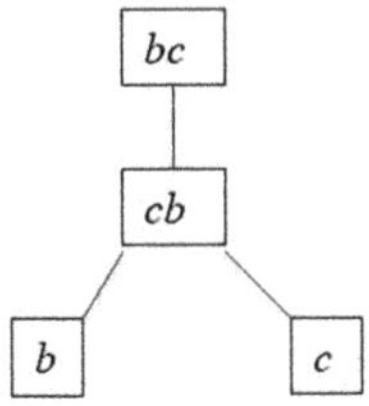

Actor set partial function:
1→1, 3→3, 4→4*

A2	*b*	*c*	*s*
b	*b*	*c*	*b*
c	*b*	*c*	*c*
s	*b*	*c*	*s*

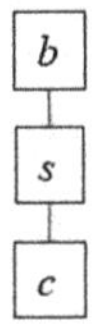

Actor set partial function:
3→3, 4→4*

A3	*b*	*c*
b=*s*	*b*	*b*
c	*b*	*c*

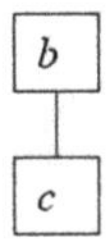

Actor set partial function:
4→4

A4	*b*	*s*
b=*c*	*b*	*b*
s	*b*	*s*

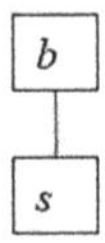

Actor set partial function:
1→1, 2→2

*not exact

Fig. 6 Factorization of the partially ordered semigroup of the Altneustadt blockmodel (Fig. 1), showing the right multiplication table, partial order, and corresponding partial function on the actor set

identifies not just an efficient description of relational structure in terms of structural building blocks but can also point to one or more maximal partial functions of the actor set with which each of these structural building blocks is associated.

For example, the factorization of the partially ordered semigroup for the Altneustadt blockmodel in Fig. 1 is shown in Fig. 6. The images are labelled A1 to A4, and each emphasizes some feature of the semigroup and can, in most cases, be associated with a facet of the Altneustadt blockmodel identified in the right panel of Fig. 6. The image A1 reflects relations among the two more influential classes (1 and 3) and members of the primarily cultural class (4). Indeed, the mapping that aggregates classes 1 and 2 is also consistent with A1. The role structure associated with A1 emphasizes the role of compound relations involving business and community ties in providing greater connectivity among these members of the elite and especially the role of class 4 in linking the politically distinct but influential class 3 to the most influential class 1 through compound community affairs and business ties. Interestingly, the community affairs partners of business associates provide more extensive connections than the business associates of community affairs contacts.

The image A2 in Fig. 6 is also an image of the blockmodel of classes 3 and 4; indeed, it is also an image of the blockmodel that aggregates classes 1, 2, and 3 and considers their aggregated connections to class 4. Its role structure features social ties as an identity relation reflective of strong internal closure, community affairs ties as a right zero given the absence of internal community affairs ties for class 4, and business ties as maximally connective among the two classes. The image A3 can be associated with the internal ties of class 4 alone, and hence the absence of internal community affairs ties within this primarily cultural sector class. The image A4 reflects the relations among classes 1 and 2 with social ties again serving as an identity and reflecting strong internal closure, and business and community ties linking the less influential class 2 to the more influential class 1 and equal to all compound ties involving business or community affairs ties.

This approach of associating partial functions of the actor set with homomorphic images is also potentially useful for comparing role structures in distinct relation sets. This is because the relational setting corresponding to a shared component can be identified in each relation set, allowing for a detailed understanding of similarities and differences across the two relation sets. Importantly, also, it provides a way of pulling the analysis conducted at the algebraic level – namely, the factorization of the partially ordered semigroup – back to the relational level, allowing the structural building blocks to also be described in terms of reductions or components of the relation set itself. In other words, this analytic approach yields a decomposition at the level of the algebra into components of the factorization *and* a decomposition at the relational level into relational components defined by corresponding partial functions on the actor set.

6 Open Challenges and Opportunities

It is worth noting that the constructions presented above have been generalized to other forms of relational data and other perspectives. These have included, for example, multi-mode relations in which the relations link distinct types of actors (such as people and organizations); a further generalization to multi-level configurations, in which relations describe linkages among pairs of three or more distinct sets (e.g., among people, teams and organizations); valued and signed relations; time-stamped relations; and relational data that is augmented with actor attributes (see, e.g., Doreian et al. 2004; Kontoleon et al. 2013; Ostoic 2021). They have also included constructions that focus on the perspective of a single actor or a subset of actors (e.g., Breiger and Pattison 1986; Pattison 1993). Ostoic (2021) has developed a suite of R functions to carry out many of the algebraic analyses described above as well a number of the generalizations just mentioned; Doreian et al. (2004) also provide a comprehensive account of generalized block modelling approaches.

Interestingly, the positional analysis spawned by the introduction of the concept of structural and more general equivalences in a relation set has been taken up by scholars in the social sciences much more enthusiastically than the role structure analysis enabled by the construction of the semigroup of the relation set. This is despite some quite strong interest in the algebraic developments from a theoretical point of view. To some degree, this can be attributed to the relative unfamiliarity of algebraic methods to many social scientists and to the lack of ready access until recently to user-friendly computational tools. Nonetheless, this is unlikely to be the full explanation and we consider in this section the further challenges and opportunities associated with progress to date.

Here I focus on three implicit and interrelated tensions in the work I have just described. Each poses a key mathematical challenge as well as a substantive one and each offers potential opportunities for further work.

The challenges, arguably, are these:

1. *Attention to sociocultural dynamics*. Boyd (1969) advocated making strong and compelling social theoretical claims for relationships and operations at the algebraic level, not just at the relational level. The first challenge we identify is therefore to develop a sufficiently rich understanding through case analysis or other means of how algebraic analytic distinctions relate to the interpretive context of application and changes through time. It is likely that the greater uptake of positional analysis is due in part to the ease with which the positional analysis can be assessed relative to characteristics of the actors as well as the observed relational data and also viewed as exploratory in mode. By contrast, algebraic mappings often operate on derived structures which have been "lifted above" the level of the relational data. As Nadel (1957) proposed, there is significant potential theoretical value in this level of description, and as described earlier, we can – with some effort – pull back the analysis to the relational level by linking homomorphic images, including components of a factorization, to partial functions on the actor set. But is that information sufficient or should we be finding other empirical ways of testing the relevance and sociocultural coherence of distinctions important to an algebraic analysis, and hence of formulating stronger claims in compelling, precise, and testable terms?
2. *Joint analysis of roles and positions*. Second, can we develop a coherent, simultaneous analysis of positions and roles, one that addresses effectively the *duality*, or mutually constitutive (Breiger 1974), nature of *positions* and *roles*? In the approach developed by White and colleagues and in many applications to date, positional and role analyses have been undertaken sequentially. The first, positional, analysis has often been seen as sufficient, perhaps because – quite reasonably – interest has primarily been in the macro-structure of the network rather than in the interlocked character of relations and hence in its role structure. Yet the challenge of a joint analysis of what are, in principle, mutually interdependent concepts has been noted by both Boyd (1992) and Otter and Porter (2020), and we explore this important challenge further below.
3. *Taking account of variability in relational data*. Third, can we take account of and understand the uncertainty implied by variability in the relational data? Relational data, like most data, may be subject to fluctuations and errors in

reporting, and this variability is known to have a potentially significant impact on derived algebraic relations (e.g., Pattison 1993). Is it possible to propose and assess positional and role models for relational data that accommodate this potential variability? As we describe below, some substantial efforts have been made in this direction but can we build models that yield fruitful, ideally joint, models for both positions and role structures?

The three challenges are considered in turn despite their overlap, because each has a distinctive characteristic. Nonetheless, an approach that addresses all three challenges at once is arguably the ultimate goal.

6.1 Attention to Sociocultural Dynamics

The value of providing a richer interpretive context for an algebraic operation or relation and of understanding changes through time is not unique to algebraic representations of the interlocking of social relations or indeed to any specific mathematical representation of social phenomena. For example, a number of dimension reduction techniques commonly used in the social sciences, such as principal components analysis, embed data in low dimensional "spaces" and seek to understand each component of that space in terms of the contrasts among data elements that it emphasizes. This is exactly the approach proposed by Pattison (1993) in linking each component of a factorization to one or more partial functions on the actor set. The partition emphasizes some distinctions among actors but not others, and the corresponding component of the factorization expresses the relational feature with which the distinctions captured in the partial function are associated. The fact that the components of the factorization are as independent as possible makes the description of relational features and associated distinctions among actors as "efficient" as possible. This approach is nonetheless representational rather than interpretive in form: the analysis represents the data faithfully – in that the distinctions made are present in the data – but it does not draw in any contextual dynamic or interpretive information.

Boyd's (1969) challenge was to go beyond this representational form and seek evidence for social meaning in the distinctions emphasized in an algebraic relation, operation, or mapping. For him, the link was to a semantic analysis of kinship terms and hence to the possible dynamics of change. Importantly, this challenge pushes us in a direction that many social scientists have been advocating, namely, to integrate an understanding of the relational and cultural dynamics of social life (e.g., Emirbayer 1997; Pachuki and Breiger 2010). Cultural data could include, for example, beliefs or practices of the actors that might be expected to align meaningfully with distinctions emphasized in an analysis and in changes through time.

A good example of the type of data that might be helpful in this regard is provided by Tasselli et al. (2020), who demonstrate an alignment between the vocabularies used by actors in their day-to-day interactions and their social relationships. If such data are also aligned with algebraic distinctions derived from the relational data, then this would provide a helpful interpretive context for these distinctions. Meaningful correspondences between algebraic and contextual distinctions would suggest a coherent embedding of the algebraic distinctions in this context, and this would add interpretive value to both sets of distinctions. Importantly, this approach to integrating relational and cultural perspectives requires careful attention to the up-front design of social observational schemes, since there is a need to anticipate the kind of data that would provide this helpful interpretive context. The mathematical challenge entailed by this approach is to develop joint analyses of the more complex longitudinal relational and cultural data structures to which the approach would give rise, as Ostoic (2021), for example, has anticipated.

6.2 Simultaneous Analysis of Positions and Roles

Positions and roles are interdependent concepts, and there is evident value in exploring options for their joint analysis (e.g., see Boyd 1992). Recently, Otter and Porter (2020) have also argued for a viable joint analysis of positions and roles and recommended adopting a category theory perspective. As well, they proposed paying more attention to shorter compound relations in this process, where the *length* of a compound relation $U_1U_2 \ldots U_h$ in which each U_j ($j = 1,2, \ldots h$) is a generator relation in R, is h. Their argument is compelling and reinforces a number of earlier efforts to focus analysis on length-constrained compound relations (e.g., Pattison et al. 2000; Winship and Mandel 1983).

Let $S_{R,h}$ be the set of all compound relations generated by relations in R of length $\leq h$ and regard a generator relation as a compound relation of length 1. Also let s_h be the total number of relations in $S_{R,h}$. The relations in $S_{R,h}$ can be represented in an $n \times n \times s_h$ array of relations of length up to h and give rise to a *partial* partially ordered semigroup (see, e.g., Pattison 2009; Pattison et al. 2000). In general, h would be expected to be quite small in practical applications, taking values such as 2, 3, or 4. Otter and Porter (2020) suggested simply setting compound relations of length greater than h to a zero element z in the semigroup (satisfying $sz = zs = z$, for all $s \in S_{R,h}$). An appropriate form of dual clustering of the n actors in X and the s_h relations in $S_{R,h}$ would then yield the required joint analysis.

As an alternative to setting compound relations that exceed a threshold length to zero, Pattison (2009) also demonstrated that all of the algebraic tools introduced in earlier sections can be applied to the partial algebra defined on generator relations and compound relations of length no greater than h, with a compound relation regarded as undefined if it is distinct from any relations in $S_{R,h}$ (see also Pattison and Wasserman 1995; Pattison et al. 2000).

In either case, the three-way $n \times n \times s_h$ relational array for relations of length up to h is a promising focus for the joint analysis of positions (mappings ψ on N) and roles (mappings φ on $S_{R,h}$).

A further option for consideration is to re-construe the three-way $n \times n \times s_h$ relational array just described as a two way $n^2 \times s_h$ relational array. The n^2 rows of this array correspond to the *tie bundles* of $S_{R,\,h}$ (White and Reitz 1983), one for each ordered pair of actors (k,l). Each tie bundle is a binary vector describing whether the ordered pair is linked by each relation in $S_{R,\,h}$. The columns of the two-way array are the s_h relations in $S_{R,\,h}$. The advantage of this representation is that it casts the individual potential links – that is, the potential relations of each type from each actor to each other actor – as the elementary units of analysis. Each such potential connection is a constituent of both a tie bundle and a relation and the representation expresses the *duality* (Breiger 1974) or mutually constitutive nature of tie bundles and relations.

Analysis of this two-way array could then also serve as a valuable focus of analysis. For example, the *concept lattice* (Ganter and Wille 1999) of the array could be analyzed to represent the interrelationships of the tie bundles and relationships. In the concept lattice, each tie bundle and relation are mapped to a node in the lattice in such a way that every tie bundle sits below exactly those relations that are present in the bundle, and every relation sits above exactly those tie bundles for which it is present. The concept lattice is computed by constructing all possible intersections of tie bundles, adding them as rows to the array, adding also the *universal* tie bundle (each entry of which is a 1), and constructing the partial order on this collection of tie bundles and their intersections. This lattice has a unique factorization and a unique reduction diagram and each component of the factorization is associated with:

- A homomorphism of the lattice;
- A partition on the tie bundles;
- A partition on the relations; and
- A unique reduction of the two-way tie bundle by relation array (Pattison and Breiger 2002).

An appealing aspect of this analytic approach is that each partition on the set of tie bundles may imply a partition on the actor set itself, but it also may not. In this way, the existence of a social equivalence among actors into social positions is not assumed but can emerge from the analysis.

An alternative and closely related approach would be to begin with a lattice-ordered semigroup by including the intersection operation as well as composition in the operation set for the algebra. Indeed, as White (2008) recommended, it could also include a unary converse operation $'$ defined for a relation U by $(k,l) \in U'$ if and only if $(l,k) \in U$.

These approaches and those proposed by Otter and Porter (2020) all appear to offer fruitful lines of enquiry for a genuinely joint analysis of positions and roles.

6.3 Taking Account of Uncertainty in Relational Data

One other consideration that is repeatedly raised in the algebraic analysis of relation sets is the issue of data accuracy: what impact does variability in the underlying relational data have on the analysis made of it? In general, of course, the impact is potentially quite significant.

In practice, the impact of data variability may be argued to be moderated by the two-stage approach often taken to data analysis: first, aggregation of actors into compelling social positions and second, algebraic analysis of the relations among these inferred positions. Certainly, this approach has yielded interesting analysis in many applications (e.g., Lazega 2001; Ostoic 2021) and, in any case, serves as a useful form of exploratory analysis. However, more robust approaches would clearly be of value.

In relation to models for social positions, a number of statistical models have, in fact, been successfully developed (e.g., Hoff et al. 2002; Nowicki and Snijders 2001; Schweinberger and Snijders 2003; for a recent review of further developments, see Lee and Wilkinson 2019). These models propose a probability model for the relational data conceptualized as a three-way $n \times n \times p$ matrix $X = [X_{\mathrm{kls}}]$ of binary random variables X_{kls}, with $k, l \in N$, $s = 1, 2, \ldots, p$. The variables X_{kls} are often termed *tie variables*, given that each specifies a potential tie in a (multi-relational) network. A probability distribution on the set of all possible realizations $x = [x_{\mathrm{kls}}]$ of X in which $x_{\mathrm{kls}} = 1$ if $(k, l) \in R_{\mathrm{s}}$ and $x_{\mathrm{kls}} = 0$, otherwise, gives rise to a *random (multi) graph model*, $\Pr(X = x)$.

The most direct stochastic generalization of structural equivalence is arguably that of Nowicki and Snijders (2001). Nowicki and Snijders developed a random graph model $\Pr(X_{\mathrm{kls}} = x_{\mathrm{kls}}; k, l \in N; s = 1, 2, \ldots, p)$ in terms of an unobserved, or *latent*, set of social positions and an assumption of stochastic equivalence. Two actors k and l are defined to be *stochastically equivalent* if exchanging k and l in the network leaves the probability distribution $\Pr(X_{\mathrm{kls}} = x_{\mathrm{kls}}; k, l \in N; s = 1, 2, \ldots, p)$ unchanged. The model is parameterized in terms of the probability with which each actor belongs to each of c assumed latent classes, and the probability of a tie of each type for members of each of the c^2 ordered pair of classes. Given these latent values, the binary relational variables are assumed to be conditionally independent. Nowicki and Snijders (2001) developed an algorithm for estimating model parameters, and deriving the posterior probabilities of characteristics such as shared class memberships. The model has been widely used and provides an excellent stochastic approach to position analysis that generalizes the concept of structural equivalence.

A range of other latent structure forms have also been proposed, including a latent embedding of the actors in a metric (Hoff et al. 2002) or ultrametric space (Schweinberger and Snijders 2003), with the probability of a relationship between two actors a function of their separation in the latent space.

While these developments provide a compelling stochastic approach to the analysis of social positions, they do not address directly the interdependence of relations. A different line of statistical development has focused more directly on the interdependence of relations. Consistent with the insights of Nadel and White cited earlier, that "what happens so-to-speak between one pair of 'knots', must affect what happens between other adjacent ones" (Nadel 1957, p. 16), exponential random graph models allow for dependencies to be specified among the collection of random tie variables, $X = [X_{kls}]$, associated with potential relational links.

Frank and Strauss (1986) first proposed the use of a dependence structure in this context, defining a dependence graph on the collection of tie variables $\{X_{kls}\}$ as a specification of the pairs of tie variables assumed to be conditionally dependent, given the values of all other variables. They demonstrated the way in which the assumed dependence structure constrained the form of an associated exponential random graph model. For example, the so-called Markov assumption that two relational tie variables X_{kls} and X_{abt} are conditionally independent unless $\{k, l\} \cap \{a, b\} \neq \varnothing$, for k, l, a, $b \in N$, allows for conditional dependencies among relations that "catenate" through third parties (but not over longer chains of catenation). For example, whereas an algebraic equation of the form $T = UV$ could represent a form of interlock among the relations, U, V and T, a related form of interlock could be captured in a Markov random graph model by assuming dependencies among all pairs of the variables in the triple $\{T(k,l), U(k,m), V(m,l)\}$, for k, l, $m \in N$. Under this Markov assumption, Frank and Strauss demonstrated that model parameters corresponded to statistics associated with two types of relational configurations: *triangles*, such as x_{kls}, x_{lat}, x_{aku}; and *w-stars*, comprising w observed relational ties each with a common actor k (for $w = 1, 2, \ldots, n$-1). A homogeneity assumption (that parameters corresponding to isomorphic relational configurations are equal) entails that the parameters of the resulting model correspond to sufficient statistics that are counts of these relational configurations in x. The approach is not described in full detail here, but as this case of the Markov random graph model suggests, there is an appealing directness to the way in which interdependencies among relations might be represented in a statistical model.

Markov random graph models are an instance of exponential random graph models. The topic of exponential and indeed other random graph models has generated a substantial literature and a much deeper understanding of the properties of various model forms. Although the Markov assumption of Frank and Strauss (1986) proved to be less useful than anticipated in its original homogenous form, various elaborations have proved extremely useful (e.g., Snijders et al. 2006; Hunter and Handcock 2006; Schweinberger 2020). Three considerations have contributed to the success of these elaborations.

The first has been to specify *realization-dependent* model forms (Pattison and Robins 2002) in which the conditional dependence of two relational tie variables could depend on the realized value of other tie variables; a key step, in particular, was to allow for the conditional dependence of X_{kls} and X_{abt} whenever $(k, a) \in R_u$ and $(b, l) \in R_v$; for some R_u, $R_v \in R$, in other words, when the tie variables had the potential to create a four cycle in the relational structure. Pattison and Snijders (2013) went on to propose a more general hierarchy of dependence assumptions for exponential-family random graph models based on a graph theoretical analysis of potential proximity among tie variables. The second construction was to propose model parameters that were a non-linear function of certain common relational substructures and hence avoid the so-called *degenerate* model forms in which very high probability was attached to a small number of instances of the possible relational structures (e.g., Hunter and Handcock 2006; Snijders et al. 2006). The third has been to limit the dependence among tie variables to subsets of actors within local settings, that is, within blocks of a block structure that may be unobserved (Schweinberger 2020). Schweinberger and Luna (2018) have defined a more general class of hierarchical exponential-family random graph models. Their models are of interest in the current context because they jointly seek to represent a latent partition of the actor set *and* relational interdependence among ties (and are clearly of value where there are likely to be known or unknown barriers to dependence among ties, for example, a hidden opportunity structure limiting interaction among certain actors and hence the emergence of ties among them). Although they are not joint analyses of positions and roles as currently formulated, the question of whether they could be adapted for such use is clearly of interest to this third challenge. For example, could a latent positional structure be assumed at the same time as dependencies among tie variables, with ties between distinct pairs of actors more likely if each pair links actors from the same positions? Could statistics inspired by exponential random graph models or by a stochastic blockmodel be used to explore unexplained network features for a model of the other type? More generally, how should we conceptualize unobserved social positions in the context of a random graph model with dependencies among "proximal" tie variables? In fact, Koskinen (2009) has taken initial steps in this direction by proposing and developing an algorithm to estimate a model with two latent classes and certain exponential random graph model parameters. Further generalization of his approach would be very valuable.

Another approach would be to include within a longitudinal analysis relational or attribute variables likely to be indicative of social positions and thereby address all three challenges simultaneously: a joint analysis of positions and role structure; a rich interpretive context for the dynamics of change; and a theoretically defensible and estimable stochastic model. This approach is simpler from a mathematical perspective since the modelling machinery already exists for such an endeavor, but, of course, it is much less straightforward from a social science perspective, since it requires having a good understanding of the likely indicators of social positions and their sociocultural dynamics.

7 Concluding Remarks

The application of algebra to our understanding of social relations has been a theoretically engaging endeavor, one that has led to a wealth of promising new concepts and modelling approaches. Uptake of these applications has nonetheless been partial and/or primarily exploratory to date, and the development of methods, particularly for the analysis of role structure, is ongoing. This chapter has reviewed progress to date in applying algebraic concepts to our understanding of social positions and social roles from social relational data. Three evident challenges and opportunities for further development have been identified, and some prospective leads for the interrogation of these challenges have also been sketched.

As in the development of most modelling applications, regular interplay between richly theorized observational designs and careful data gathering, on the one hand, and the application and development of new mathematical concepts, on the other, will help to advance and refine the ambitious theoretical agendas of those who first saw the potential of mathematical representation for rigorous analysis of social relational phenomena and took the first, important steps toward their development.

Appendix: Glossary of Notation and Definitions

Actor set. $N = \{1,2,\ldots,n\}$ is a fixed set of actors.

Binary relation. A binary relation on N is a subset of $N \times N = \{(k,l)\text{: } k, l \in N\}$. We write $(k, l) \in R_s$ if there is a relation of type s from actor k to actor l, for $k, l \in N$; $s = 1, 2, \ldots, p$.

Relation set and generator relations. $R = \{R_1, R_2, \ldots, R_p\}$ is a set of binary relations on N of a fixed set of p relation types; each relation R_s is termed a *generator relation*.

Boolean matrix representation of a relation set. A relation set can be represented by a three-way $n \times n \times p$ Boolean array $x = [x_{kls}]$, with $x_{kls} = 1$ if $(k, l) \in R_s$ and $x_{kls} = 0$ otherwise.

Composition of binary relations. The *composition UV* of two binary relations, U and V on N is the binary relation on N in which $(k,l) \in UV$ if and only if $(k,m) \in U$ and $(m,l) \in V$ for some $m \in N$.

Length of compound relation. The *length* of a compound relation $U_1 U_2 \ldots U_h$ in which each U_j ($j = 1,2, \ldots h$) is a generator relation in R, is h.

Partial order on binary relations. $U \leq V$ for binary relations U, V on N whenever $(k,l) \in U$ implies $(k,l) \in V$ for all $k, l \in N$.

Closure of a relational set under composition. The closure R^* of a relational set R on N under composition is the set R^* of distinct binary relations for which:

- For all $U, V \in R^*$, $UV \in R^*$ (*closure*);
- $R \subseteq R^*$ (*inclusive of the generator set R*); and
- For any set W that includes the generator set R and also satisfies closure, then $R^* \subseteq W$ (*minimal*).

Axiom of Quality. The *Axiom of Quality* is the proposition that two binary relations U and V are *equal* if they link exactly the same pairs of actors in N; that is, $(k,l) \in U$ if and only if $(k,l) \in V$; $k, l \in N$.

Semigroup. A *semigroup* $[S,\circ]$ is a set S with a binary operation, $\circ$, that satisfies:

- $(a^\circ b)^\circ c = a^\circ(b^\circ c)$ for all $a, b, c \in S$ (associativity).

Quasi-order. A *quasi-order* on a set A is a relation $\leq$ for which:

- $a \leq a$, for all $a \in A$ (reflexivity); and
- $a \leq b$ and $b \leq c$, implies $a \leq c$, for all $a, b, c \in A$ (*transitivity*).

Partial order. A *partial order* on a set A is a quasi-order on A for which:

- $a \leq b$ and $b \leq a$, implies $a = b$, for all $a, b \in A$ (*anti-symmetry*).

Equivalence relation. An *equivalence relation* E on a set A is a relation for which:

- $(a,a) \in E$ for all $a \in A$ (reflexivity);
- $(a,b) \in E$ implies $(b,a) \in E$ for all $a, b \in A$ (symmetry); and
- $(a,b) \in E$ and $(b,c) \in E$ implies $(a,c) \in E$ for all $a, b, c \in A$ (transitivity).

Structural equivalence. Two actors, k and l, are *structurally equivalent* (that is, $(k,l) \in SE$) whenever $(k, m) \in R_s$ if and only if $(l, m) \in R_s$, and $(m, k) \in R_s$ if and only if $(m, l) \in R_s$, for all $m \in N$ and all relations R_s in R.

A *relational, or graph, homomorphism* of a relation set $R = \{R_1, R_2, \ldots, R_p\}$ on an actor set N onto a relation set $T = \{T_1, T_2, \ldots, T_{op}\}$ on an actor set M is a mapping ψ from N onto M for which:

- For each $k \in M$, $\psi(i) = k$ for some $i \in N$, and
- $(\psi(i), \psi(j)) \in T_s$, if and only if $(i,j) \in R_s$, for some $i, j \in N$; $s = 1, 2, \ldots, p$.

Partially ordered semigroup. A *partially ordered semigroup* $[S,^\circ,\leq]$ is a semigroup with a partial order $\leq$ for which the binary operation $^\circ$ is *isotone* in each of its variables; that is, $a \leq b$ and $c \leq d$ implies $a^\circ c \leq b^\circ d$, for any $a, b, c, d \in S$.

Partially ordered semigroup of a relation set. The *partially ordered semigroup* S_R of *the relation set* R is the semigroup with element set R^*, relational composition as its binary operation and the partial order on binary relations as its partial order. We write $S_R = [R^*, {}^\circ, \leq]$.

Subsemigroup of a partially ordered semigroup. A semigroup $B = [W,^\circ,\leq]$ is a *subsemigroup* of $A = [S,^\circ,\leq]$ if W is a subset of S, W is closed under relational composition and $s \leq t$ for $s, t \in W$ if and only if $s \leq t$ for $s, t \in S$.

Right multiplication table of a semigroup. The *right multiplication table* for a semigroup $S_R = [R^*, {}^\circ, \leq]$ is an $n_R \times p$ table indicating the composition of each of the n_R distinct binary relations in R^* and each of the p generator relations R.

Quasi-order table of a partially ordered semigroup. The *quasi-order table* on a partially ordered semigroup $S_R = [R^*, {}^\circ, \leq]$ is an $n_R \times$ table with a unit entry in the array for (V,U) if $U \leq V$, with $U, V \in R^*$.

Length-restricted subset $S_{R,h}$ *of the semigroup* $S_R = [R^*, {}^\circ, \leq]$. Define $S_{R,h}$ to be the set of all compound relations generated by relations in R of length $\leq h$, with a generator relation regarded as a compound relation of length 1. The relations in $S_{R,h}$ can be represented in an $n \times n \times s_h$ array, x_h, where s_h is the total number of relations in $S_{R,h}$ and give rise to a *partial* partially ordered semigroup in which compound relations of length greater than h are regarded as undefined. The tie bundle $(x_h)_{kl}$ is a binary vector of length s_h describing whether the ordered pair (k,l) is linked by each relation in $S_{R,\,h}$.

Isotone homomorphism. An *(isotone) homomorphism* from a partially ordered semigroup $[S,°,\leq]$ onto a partially ordered semigroup $[T,°,\leq]$ is a mapping φ for which:

- $\varphi(s°t) = \varphi(s)°\varphi(t)$ for all s, t $\in$ S;
- $s \leq t$ implies $\varphi(s) \leq \varphi(t)$ for any s, t $\in$ S; and
- For each u $\in$ T, there is some $s \in S$ for which $\varphi(s) = u$.

T is termed a *homomorphic image* of S and we write $T = \varphi(S)$.

The π-relation π_φ for an isotone homomorphism φ. Each homomorphism φ on $[S,°,\leq]$ can be represented by a quasi-order π_φ, termed a *π-relation*, in which $(s,t) \in \pi_\varphi$ if and only if $\varphi(t) \leq \varphi(s)$, for s, $t \in S$. It is also convenient to write $T = S/\pi_\varphi$ $(= \varphi(S))$ where π_φ is the *π-relation* corresponding to the homomorphism φ.

Partial ordering and lattice of isotone homomorphisms. The collection of all (isotone) homomorphisms on a partially ordered semigroup $A = [S,°,\leq]$ can be partially ordered by defining:

- $\varphi \leq \rho$ for two homomorphisms φ, ρ on A if, for all $s, t \in S$, $\rho(t) \leq \rho(s)$ implies $\varphi(t) \leq \varphi(s)$.

This partial ordering gives rise to a *lattice L(A) of homomorphisms* of A.

Partial ordering and lattice of π-relations. The π-relations associated with isotone homomorphisms can also be partially ordered by defining:

- $\pi_\varphi \leq \pi_\rho$ if and only if $(s,t) \in \pi_\varphi$ implies $(s,t) \in \pi_\rho$ for any $s, t \in S$.

The π-relations also give rise to the lattice $L_\pi(A)$, with a partial ordering *dual* to that in $L(A)$; that is, $\varphi \leq \rho$ in $L(A)$ if and only if $\pi_\rho \leq \pi_\varphi$ in $L_\pi(A)$.

Direct product of partially ordered semigroups. Let $A_q = [S_q, °, \leq]$, $q = 1, 2, \ldots, f$ be a collection of (partially ordered) semigroups. The *direct product* $A_1 \times A_2 \times \ldots \times A_f$ of the semigroups $A_1, A_2, \ldots A_f$ is the semigroup with element set $S_1 \times S_2 \times \ldots \times S_f$, relational composition defined by $(s_1, s_2, \ldots, s_h) ° (t_1, t_2, \ldots, t_h) = (s_1 ° t_1, s_2 ° t_2, \ldots, s_h ° t_h)$; and the partial order defined by $(s_1, s_2, \ldots, s_f) \leq (t_1, t_2, \ldots, t_f)$ if and only if $s_q \leq t_q$, for each $q = 1, 2, \ldots f$.

Subdirect product of partially ordered semigroups. A subsemigroup $B = [W,°,\leq]$ of the direct product $A_1 \times A_2 \times \ldots \times A_f$ of semigroups $A_1, A_2, \ldots A_f$ is a *subdirect product* of $A_1, A_2, \ldots A_f$ if for each $s_q \in S_q$ $(q = 1, 2, \ldots, f)$ there is an element $t \in W$ having s_q as its component in S_q.

Lattice. A lattice $L = [X,\wedge,\vee,\leq]$ is a set of elements with two binary operations, meet $\wedge$ and join $\vee$, satisfying:

- $s \wedge t = t \wedge s$ and $s \vee t = t \vee s$ for all s, t $\in$ X (commutativity);
- $(s \wedge t) \wedge u = s \wedge (t \wedge u)$ and $(s \vee t) \vee u = s \vee (t \vee u)$ for all s, t, u $\in$ X (associativity); and

- $s \vee (s \wedge t) = s$ and $s \wedge (s \vee t) = s$ for all $s, t \in X$ (absorption).

A partial order in L is defined by setting $s \leq t$ for any $s, t \in X$ if $s \wedge t = s$ or, equivalently, if $s \vee t = t$.

Distributive lattice. A lattice L is distributive if the identity $s \wedge (t \vee u) = (s \wedge t) \vee (s \wedge u)$ *holds for all* $s, t, u \in S$*).*

Modular lattice. A lattice L is modular if $s \leq u$ *implies* $s \vee (t \wedge u) = (s \vee t) \wedge u$ *for all* $t \in S$*, and for any* $s, u \in S$*).*

Minimal and maximal elements of a lattice. Each lattice L has a maximal element s_{max} *for which* $s \vee s_{max} = s_{max}$ *for each* $s \in L$ *and a minimal element* s_{min} *for which* $s \wedge s_{min} = s_{min}$ *for each* $s \in X$*.*

Meet-irreducible element of a lattice. An element $s \in L$ *is termed meet-irreducible if* $s = t \wedge u$ *implies that* $s = t$ *or* $s = u$*.*

Irredundant subset of elements of a lattice. A subset $s_1, s_2, \ldots, s_f$ *of elements of S is termed irredundant if each* s_i *is meet-irreducible and* $s_1 \wedge s_2 \wedge \ldots \wedge s_{i-1} \wedge s_{i+1} \wedge \ldots \wedge s_f \neq s_{min}$*.*

Ordering on irredundant subsets of a lattice. Define $\{s_1, s_2, \ldots, s_f\} \leq \{t_1, t_2, \ldots, t_g\}$ *if for each* $j = 1, 2, \ldots g$*, there exists some* $s_i \in \{s_1, s_2, \ldots, s_f\}$ *such that* $t_j \leq s_i$ *in L.*

Factorization of a partially ordered semigroup. A factorization of a partially ordered semigroup is s subdirect representation associated with a minimal irredundant subset $\{\pi_1, \pi_2, \ldots, \pi_f\}$ *of* $L_\pi(A)$*.*

The condition G_i *for a relational homomorphism.* Let ψ be a relational homomorphism from the relation set $R = \{R_1, R_2, \ldots, R_p\}$ on the actor set N onto a relation set $T = \{T_1, T_2, \ldots, T_{op}\}$ on the actor set M and define $c_k = \{j \in N : \psi(j) = k; k \in M\}$ be the class of actors in N mapped by ψ to $k \in M$. The mapping ψ on N satisfies Kim and Roush's *condition* G_i if, whenever $(j, l) \in R_{\underline{s}}$ for any $j, l \in N$ and any $s = 1, 2, \ldots, p$, then any subset of i actors in $c_{\psi(j)}$ is related by $R_{\underline{s}}$ to at least i actors in $c_{\psi(l)}$ (or to all actors in $c_{\psi(l)}$ if $|c_{\psi(l)}| < i$). The conditions G_1 and G_n are also termed the *outdegree* and indegree conditions, respectively.

Regular equivalence on a relation set. A relational homomorphism of a relation set R on actor set N is a *regular equivalence* if it satisfies the conditions G_1 and G_n.

The condition G_{im} *for a relational homomorphism.* The mapping ψ satisfies the *condition* G_{im} if, for each class c_k, there is a subset c_k' of c_k such that $(j, l) \in R_{\underline{s}}$ implies $(j', l') \in R_{\underline{s}}$ for some $j' \in c_{\psi(j)}'$ and $l' \in c_{\psi(l)}'$ and ψ satisfies the condition G_i when restricted to relations involving actors in the set $N' = \cup_k c_k'$.

Random (multi) graph model. A random (multi) graph model is a probability distribution $\Pr(X = x)$ over a specified set of possible (multi) graph realizations x of a random (multi) graph X on a set N of actors. Two actors k and l in N are defined to be *stochastically equivalent* if exchanging k and l in the network leaves the probability distribution $\Pr(X = x)$ unchanged.

References

Audenaert, P., Colle, D., & Pickavet, M. (2019). Regular equivalences for social networks. *Applied Sciences*, *9*, 117: https://doi.org/10.3390/app9010117.

Birkhoff, G. (1963). *Lattice Theory* (3rd edition). Providence, Rhode Island: American Mathematical Society.

Boorman, S. A., & White, H. C. (1976). Social structure from multiple networks, II. Role structures. *American Journal of Sociology*, *81*, 1384–1446.

Bonacich, P. (1980). The "common structure semigroup", a replacement for the Boorman and White "joint reduction". *American Journal of Sociology*, *86*, 159–166.

Boyd, J. P. (1964). Review of H. C. White (1863). *An Anatomy of Kinship* Englewood Cliffs, NJ: Prentice-Hall. *Behavioral Science*, *9*, 372–376.

Boyd, J. P. (1969). The algebra of group kinship. *Journal of Mathematical Psychology*, *6*, 139–167.

Boyd, J. P. (1992). Relational homomorphisms. *Social Networks*, *14*, 163–186.

Boyd, J. P., Haehl, J. H., & Sailer, L. D. (1972). Kinship systems and inverse semigroups. *The Journal of Mathematical Sociology*, *2*, 37–61.

Breiger, R. L. (1974). The duality of persons and groups. *Social Forces*, *53*, 181–190.

Breiger, R. L., & Pattison, P. E. (1986). Cumulated social roles: the duality of persons and their algebras. *Social Networks*, *8*, 215–256.

Doreian, P., Batagelj, V., & Ferligoj, A. (2004). *Generalized blockmodeling*. New York, NY: Cambridge University Press.

Emirbayer, M. (1997). Manifesto for a relational sociology. *American Journal of Sociology*, *103*, 281–317.

Frank, O., & Strauss, D. (1986). Markov graphs. *Journal of the American Statistical Association*, *81*, 832–842.

Ganter, B., & Wille, R. (1999). *Formal Concept Analysis: Mathematical Foundations*. Berlin: Springer.

Granovetter, M. S. (1973). The strength of weak ties. *American Journal of Sociology, 78*, 1360–1380.

Hoff, P. D., Raftery, A. E. & Handcock, M. S. (2002). Latent space approaches to social network analysis. *Journal of the American Statistical Association*, *97*, 1090–1098.

Hunter, D. R., & Handcock, M. S. (2006). Inference in curved exponential family models for networks. *Journal of Computational and Graphical Statistics*, *15*, 565–583.

Itao, K., & Kaneko, K. (2020). Evolution of kinship structures driven by marriage tie and competition. *PNAS*, *117*, 2378–2384.

Kemeny, J. G., Snell, J. L., & Thompson, G. L. (1957). *Introduction to Finite Mathematics*. Englewood Cliffs, NJ: Prentice-Hall.

Kim, K. H., & Roush, F. W. (1984). Group relationships and homomorphisms of Boolean matrix semigroups. *Journal of Mathematical Psychology*, *28*, 448–452.

Kontoleon, N., Falzon, L., & Pattison, P. E. (2013). Algebraic structures for dynamic networks. *Journal of Mathematical Psychology*, *57*, 310–319.

Koskinen, J. (2009). Using latent variables to account for heterogeneity in exponential family random graph models. Pp. 845–849 in S.M. Ermakov, V.B. Melas & A.N. Pepelyshev (Eds.), *Proceedings of the 6th St. Petersburg Workshop on Simulation Vol II*.

Laumann, E. O., & Pappi, F. U. (1976). Networks of Collective Action: *A Perspective on Community Influence Systems*. London: Academic Press.

Lazega, E. (2001). The Collegial Phenomenon: *The Social Mechanisms of Cooperation Among Peers in a Corporate Law Firm*. Oxford: Oxford University Press.

Lee, C., & Wilkinson, D. J. (2019). A review of stochastic blockmodels and extensions for graph clustering. *Applied Network Science*, *4*, 122. https://doi.org/10.1007/s41109-0232-2.

Lévi-Strauss, C. (1969). *The Elementary Structures of Kinship* (Rev. ed., translated from the French by J. H. Bell, R. von Sturmer & R. Needham). Boston: Beacon Press.

McKenzie, R., & Schein, B. M. (1997). Every semigroup is isomorphic to a transitive semigroup of binary relations. *Transactions of the American Mathematical Society*, *349*, 271–285.

Nadel, S. F. (1957). *The Theory of Social Structure*. Melbourne: Melbourne University Press.

Nowicki, K., & Snijders, T. (2001). Estimation and prediction for stochastic blockstructures. *Journal of the American Statistical Association*, *96*, 1077–1087.

Ostoic, J. A. R. (2021). *Algebraic Analysis of Social Networks: Models, Methods and Applications Using R*. Hoboken, NJ: John Wiley & Sons.

Otter, N., & Porter, M. (2020). A unified framework for equivalences in social networks. arXiv:2006.10733**[cs.SI]**.

Pachuki, M. A., & Breiger, R. L. (2010). Cultural holes: Beyond relationality in social networks and culture. *Annual Review of Sociology*, *36*, 205–224.

Pattison, P. E. (1993). *Algebraic Models for Social Networks*. New York: Cambridge University Press.

Pattison, P. E. (2009). Algebraic models for social networks. Entry 00199 in Meyers, R. A. (Editor-in-chief), *Encyclopedia of Complexity and System Science*. New York, NY: Springer.

Pattison, P. E., & Breiger, R. L. (2002). Lattices and dimension representations: matrix decompositions and ordering structures. *Social Networks*, *24*, 423–444.

Pattison, P. E., & Robins, G. L. (2002). Neighborhood-based models for social networks. *Sociological Methodology*, *32*, 300–337.

Pattison, P., & Snijders, T. A. B. (2013). Modelling social networks: next steps. Pp. 287–302 in Lusher, D., Koskinen, J., and Robins, G. (Eds.), *Exponential random graph models for social networks: Theory, method and applications*. Cambridge University Press.

Pattison, P. E., & Wasserman, S. (1995). Constructing algebraic models for local social networks using statistical methods. *Journal of Mathematical Psychology*, *39*, 57–72.

Pattison, P. E., Wasserman, S., Robins, G. L., & Kanfer, A. (2000). Statistical evaluation of algebraic constraints for networks using statistical methods. *Journal of Mathematical Psychology*, *44*, 536–568.

Plemmons, B. J., & West, M. T. (1970). On the semigroup of binary relations. *Pacific Journal of Mathematics*, *35*, 743–753.

Read, D. W. (2006). Kinship Algebra Expert System (KAES): A software implementation of a cultural theory. *Social Science Computer Review*, *24*, 1–25.

Schein, B. M. (1970). Relation algebras and function semigroups. *Semigroup Forum*, *1*, 1–62.

Schwarz, S. (1970). On the semigroup of binary relations on a finite set. *Czechoslovak Mathematical Journal*, *20*, 632–679.

Schweinberger, M. (2020). Consistent structure estimation of exponential-family random graph models with block structure. *Bernoulli*, *26*, 1205–1233.

Schweinberger, M., & Luna, P. (2018). hergm: Hierarchical exponential-family random graph models. *Journal of Statistical Software*, *85*, 1–39.

Schweinberger, M., & Snijders, T. A. B. (2003). Settings in social networks: A measurement model. *Sociological Methodology*, *33*, 307–341.

Snijders, T. A. B., Pattison, P. E., Robins, G. L., & Handcock, M. S. (2006). New specifications for exponential random graph models. *Sociological Methodology*, *36*, 99–153.

Tasselli, S., Zappa, P., & Lomi, A. (2020). Bridging cultural holes in organizations: The dynamic structure of social networks and organizational vocabularies within and across subunits. *Organization Science*, *31*, 1292–1312.

White, D. R., & Reitz, K. P. (1983). Graph and semigroup homomorphisms on networks of relations. *Social Networks*, *5*, 193–234.

White, H. C. (1963). *An Anatomy of Kinship*. Englewood Cliffs, N.J.: Prentice-Hall.

White, H. C. (2002). *Markets from Networks: Socioeconomic Models of Production*. Princeton: Princeton University Press.

White, H. C. (2008). *Identity and Control: How Social Formations Emerge – Second Edition*. Princeton, N.J.: Princeton University Press.

White, H. C. Boorman, S. A., & Breiger, R. L. (1976). Social structure from multiple networks, I. Blockmodels of roles and positions. *American Journal of Sociology*, *81*, 730–780.

Winship, C., & Mandel, M. (1983). Roles and positions: a critique and extension of the blockmodeling approach. *Sociological Methodology*, *14*, 314–344.

Relational Systems of Transport Network and Provinces in Ancient Rome

J. Antonio Rivero Ostoic

1 Introduction

Algebraic models have been used for many years for the study of social phenomena. An early application of permutation groups to the study of kinship systems was made in the 1940s by André Weil, as described in a book by Lévi-Strauss (1949). In this application, algebraic groups represented the intertwining of marriage and descent rules as permutation relations.

While kinship systems are now a specific branch of network analysis, the employment of group objects is inadequate for analyzing social networks with directed relations due to their symmetry. In most networks related to people, including undirected systems, the invertibility property that characterizes symmetry does not hold. Semigroups (Clifford and Preston 1967; Howie 1996; Suschkewitsch 1928) provide a solution by allowing us to model directed social networks with multiple relations, specifically their relational system associated with a partial order structure.

Relational systems define algebraic constraints on network connections of different types and explain the relational interlock of the network structure in substantial terms. Networks can have one set of related entities in a one-mode structure, or different sets of entities cross-tied, such as in multimodal systems. Formal concept analysis is a mathematical setting for modeling cross-domain relations in affiliation networks, and it plays a role in reducing multilevel structures, which are complex arrangements where sets of entities in one- and two-mode networks stand for different levels in the system.

J. A. R. Ostoic (✉)
School of Culture and Society, Aarhus University, Aarhus, Denmark
e-mail: jaro@cas.au.dk

© The Author(s), under exclusive license to Springer Nature Switzerland AG 2023
M. N. Hounkonnou et al. (eds.), *Mathematics for Social Sciences and Arts*, Mathematics in Mind, https://doi.org/10.1007/978-3-031-37792-1_9

This chapter uses algebraic representations of a spatial network from ancient Rome as multilevel systems, incorporating both maritime and terrestrial routes used for trade and military purposes, as well as the politico-administrative affiliations of the provinces to the Roman Empire at that time. Positional analyses consider relations within and between domains among classes of provinces, revealing role structures that can be interpreted using different methods for a substantial interpretation of the role interlock.

2 RE Transport Network and Provinces

Settings for the algebraic analyses are a transport network and political types of affiliations of provinces in ancient Rome. The transport network is made of the main communication shaping the Roman Empire (RE) that lasted between 27 BC and AD 271. One reference year for provinces is 117 when RE had its maximal expansion under Emperor Trajan and for the transport network ca. 125. The RE transport network is a one-mode multiplex system represented by X^+ with Roman provinces in N as actors and $r = 2$ for terrestrial and maritime routes as relations. Moreover, X^B stands for an affiliation network having two domains associated with X^+ where $R_{N \times M}$ are constitutive relations of N to a set M in X^B. Systems X^+ and X^B make a multilevel structure X^{mlvl} made of provinces and government types where $R_M = \varnothing$.

Figure 1a has two representation forms of the RE transport network X^+ with the main terrestrial and maritime routes, or roads and shipping routes between the provinces. The cartographical map of the transport network to the left is according to (Rodrigue 2020), and there is a corresponding multigraph representing the transport system with provinces as nodes in the graph being Ita the X Italian regions and where province abbreviations are according to (Epigraphic Database Heidelberg 2022). Edges stand for the two kinds of transportation routes, and they can carry a value to reflect transportation costs between points.

The political affiliations of provinces during the Roman Imperial period X^B according to (United Nations of Roma Victrix 2021) are a two-mode structure represented as a bipartite graph in Fig. 1b. There are two components in the affiliation network reflecting the politico-administrative division of the RE at that time with classes of ties across domains. The graph depiction of X^B has a binomial projection on the nodes, which allows applying a force-directed layout that places structurally related provinces in the two-mode structure close to each other.

2.1 *Transportation Routes*

The aggregated relations for the main terrestrial and maritime routes between Roman provinces correspond to a one-mode structure of the RE transport network

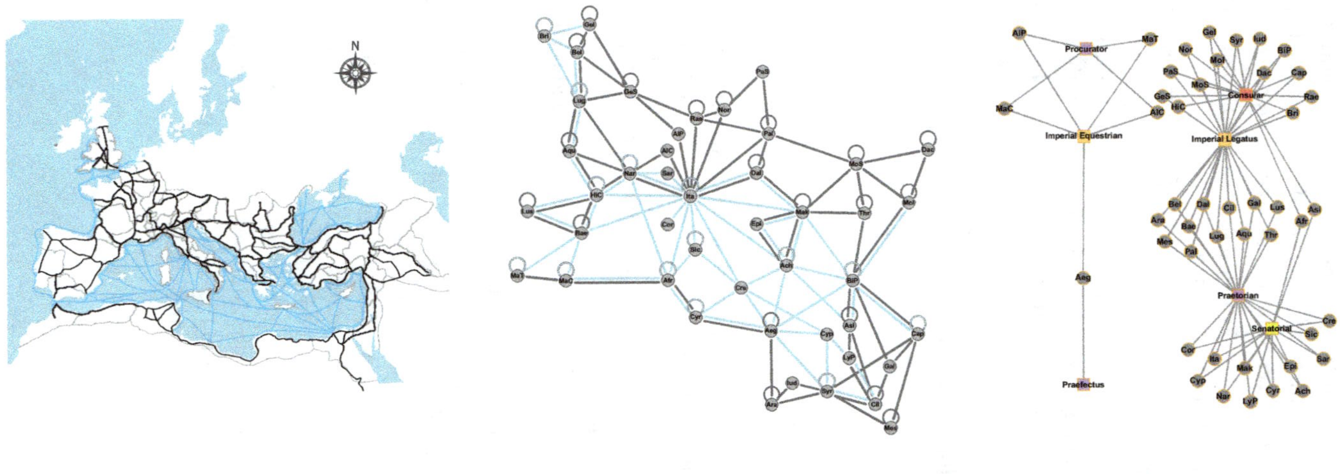

(a) (b)

Fig. 1 Roman Empire transport network with main routes ca. AD 125 and types of government affiliations of provinces in year AD 117. (**a**) Cartographical map with graph representation where circles are Roman provinces, and dark grey and light blue edges stand for terrestrial and maritime routes. (**b**) Bipartite graph with political affiliations where squares are government types

made with undirected relations. Since the different types of routes, R_{N_1} and R_{N_2}, are often between various human settlements and crossing points, the multigraph of the transport network in Fig. 1a has loops representing self-relations to indicate that main roads and maritime routes are connecting significant points within the province and not only across other provinces. The force-directed layout of the graph has a correspondence with the cartographical map because spatial networks with roads and shipping routes are physically constrained.

There are in total 45 provinces with $m = 191$ of 95 terrestrial routes and 66 main maritime routes in the RE transport network X^+ including connections between settlements and crossing points within provinces by road or by the sea. However, the network relational system of a semigroup of relations $S(R)$ with $n = 45$ and $r = 2$, the order of the semigroup will presumably be immensely large. Because large semigroups make difficult the substantial interpretation of the relational system, there is a good reason for the reduction of X^+ into a positional system with multiple role ties Y^+ with the interlock of different types of role relations within the network *role structure* Q.

2.2 Province Affiliations

Affiliations between Roman provinces add another dimension to the network arrangement and correspond to a two-mode structure X^B where edges stand for affiliation relations. X^B is related to the one-mode system in the RE transport network X^+ in an integrated multilevel system X^{mlvl}. In the binomial projection graph of the two-mode system of provinces and government types in Fig. 1b, different classes of a "set of entities" are linked through undirected edges to the type of governments each province has within the political and administrative organization during the height of the Roman Empire.

An algebraic approach for dealing with two-mode structures is found in formal concept analysis where the affiliation system constitutes the formal context. In this framework, the context bears the network cross-domain relations as concepts of intents and extents representing network actors and their affiliation types. One aspect to consider in the two-mode structure of X^B is the fact that in some cases province affiliations do not overlap, and this is clearly with senatorial/imperial territories, but it is also true for certain government types. These constrictions have implications in the network positional system and its combinatorial role structure as well, particularly if they take part in the network reduction where cross-domain associations provide the clustering information.

3 Positional Analysis

A *positional analysis* refers to the partition and reduction process of network structures into a system of social roles connecting approximately structurally equivalent entities as positions within social network analysis. There are different methods to find structurally equivalent parts in a network to produce a system made of positions and the corresponding role structure. With complex systems, the equivalence criterion should preserve the multiplicity of the ties when occur and be able to incorporate attributes of the network members as well.

In the case of the RE network X^+, the positional system Y^+ is made of classes of correspondent provinces in the transportation network and taking on their government types as political affiliations. Role structures from positional arrangements facilitate the substantial interpretation of the network relational system, either of the RE transport network as X^+ and of its multilevel version Y^{mlvl} with provinces and class affiliations.

Figure 2 has representations of two partial order configurations at different levels of the RE transport network and provinces government types as inclusion lattices. The ordered structure in Fig. 2a is the positional system product of compositional equivalence, and the other lattice in Fig. 2b is, within the formal concept analysis

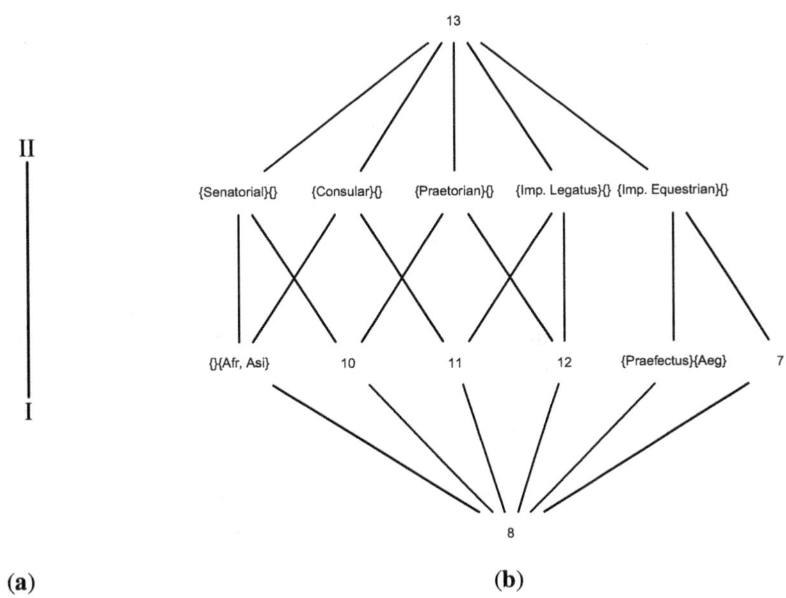

Fig. 2 Lattices of partial order structures with cross-domain relations of Roman provinces and government types from compositional equivalence and from formal concepts. (**a**) Two class inclusion lattice of the cumulated province hierarchy with administrative affiliations and compounds of length 6. (**b**) Concept diagram of provinces and government types with a reduced labelling of concept from multilevel network Imp. stands for Imperial, { } implies {∅}, and numbers replace six sets of intents and extents

framework, a concept diagram of concepts product of Galois derivations with cross-domain relations between Roman provinces and government types. The two lattice structures incorporate the province affiliations to the transport system in a different manner, and they provide sources of clustering information for the partition of the network and its reduction into a simpler configuration. This process serves to produce the system *role structure* Q, which in algebraic terms is represented by a semigroup of role relations.

3.1 Positional System from Compositional Equivalence

Composition equivalence (Breiger and Pattison 1986) is a local correspondence that seeks to find structurally equivalent entities in multiplex networks by considering both the multiplicity of network ties and their composition as a dual structure of the network members and their relations. Compositional equivalence works at both the "local" and the "global" levels and allows producing a reduced system made of positions, which can incorporate the attributes of the network members in the modeling as self-relationships. With composition equivalence, the positional system Y^+_{CE} is made of partial structural equivalent classes of provinces where the correspondences among relations and the composition of ties are equal from the standpoint of each province. Two relation types R_1 and R_2 in X^+ are indistinguishable from the perspective of a Roman province if and only if both types of relations coincide as incoming and outgoing ties to the given province.

To produce a positional system Y^+_{CE} from compositional equivalence, simple and compound relations are stacked together into a Relation-Box (Mandel 1983), which is a three-dimensional array where diagonal matrices stand for nodal attributes, one per characteristic, by the Kronecker delta function. Hence, Y^+_{CE} takes senatorial/imperial affiliations as self-relations with zeroes/ones for the presence/absence of affiliations as diagonal matrices in this device. Plane surfaces or "slices" in the Relation-Box occurring across all relations reflect role sets of provinces that are cumulated into a single arrangement with transitive closure where all surfaces are partially ordered by inclusion where the basis for network partition lies.

Figure 2a has a lattice diagram of the minimal non-trivial partial order in the cumulated "province" hierarchy from compound relations of length 6 where "structure" arises. Structure arises because there are different levels of provinces in the inclusion lattice, while the cumulated province hierarchy with shorter compounds does not have any inclusion; however, the partial ordered structure changes completely with longer compounds, which makes the positional system one option among others.

The cumulated province hierarchy constitutes the basis for a positional system for the RE network with compositional equivalence Y^+_{CE} that corresponds to inclusion levels in the lattice where provinces have implicit affiliations. In this case, the positional system Y^+_{CE} of X^+ is made of two positions represented by Roman numerals with an unambiguous ordering relation $\mathrm{I} \leq \mathrm{II}$ as the inclusion levels in the lattice of the cumulated province hierarchy.

For compounds of length 6, there are two classes of provinces:

I All provinces except in class II
II {Ach, Afr, Asi, Cyr, Epi, Ita, Mak, Nar, Sic}

The positional system with compositional equivalence Y^{+}_{CE} has explicit two administrative affiliations, which are reflected in the RE network role structure Q_{CE}. The cumulated province hierarchy reveals that there are overlapping ordered structures between the Roman territories, which means that for the analysis of the RE network, whether the province is senatorial or imperial constitutes two additional generator relations in the system that are combined with the two kinds of transportation routes in the relational system. However, adding other characteristics such as government types increases dramatically the complexity of the system, and therefore, for the analysis of this system, only administrative affiliations act as node attributes.

$$\begin{array}{c|cc} \text{I} & 1 & 1 \\ \text{II} & 1 & 1 \end{array} \qquad \begin{array}{c|cc} \text{I} & 1 & 1 \\ \text{II} & 1 & 1 \end{array} \qquad \begin{array}{c|cc} \text{I} & 1 & 0 \\ \text{II} & 0 & 1 \end{array} \qquad \begin{array}{c|cc} \text{I} & 0 & 0 \\ \text{II} & 0 & 1 \end{array}$$

terrestrial, t   maritime, m   senatorial, s   imperial, i

Positions from levels in the inclusion lattice of the cumulated province hierarchy with affiliations, Y^{+}_{CE}.

In this positional system with two categories of provinces, class I has senatorial provinces only, while class II has both senatorial and imperial provinces, which happens to be equivalent when considering their simple and compound relations in the RE network. At this aggregate level, both terrestrial and maritime routes have an absorbing character in the semigroup since the universal relation represents it, while the identity element with no structuring effect represents senatorial provinces. This leaves the imperial class of provinces to the shaping of the role structure for the RE transport network and administrative affiliations with a positional system produced by compositional equivalence.

3.2 *Positional System from Formal Concepts*

Alternative clustering information for the construction of the positional system Y^{+} with cross-domain relations in the network lies in the formal context framework. Formal contexts serve to represent the affiliations of the provinces as formal concepts with Galois derivations between objects and attributes. Formal concepts can have a full or reduced labeling of intents and extents in which the reduced format comes by discarding the repeated provinces and recurrent affiliation types from the full labeling of the formal concepts in the context.

Figure 2b is a concept diagram with reduced labeling of the formal concepts in context X^B in which both provinces and affiliations are present in the lattice structure. The fact that concepts are made of intents and extents with provinces and their political affiliations means that the clustering information to construct Y^+_{FC}, which is the positional system from formal concepts for the RE network, comes from the set of partially ordered formal concepts in the concept diagram with reduced labeling straightforwardly.

A grouping of non-recurrent provinces and their political affiliations in the concept diagram is self-evident, and the number of province classes doubles the number of categories with compositional equivalence applied. There is also a correspondence between the concept diagram with cross-domain relations in terms of groups of provinces and the force-directed layout of the projection graph given in Fig. 1b for the political affiliation of the RE provinces network.

The other formal concepts having provinces in the concept diagram from Fig. 2b are:

$\mathbb{C}_7$ {Procurator} {AlC, AlP, MaC, MaT}
$\mathbb{C}_{10}$ { } {Ach, Cor, Cre, Cyp, Cyr, Epi, Ita, LyP, Mak, Nar, Sar, Sic}
$\mathbb{C}_{11}$ { } {BiP, Bri, Cap, Dac, Gel, GeS, HiC, Iud, Mol, MoS, Nor, PaS, Rae, Syr}
$\mathbb{C}_{12}$ { } {Aqu, Ara, Bae, Bel, Cil, Dal, Gal, Lug, Lus, Mes, Pal, Thr}

Formal concepts having provinces in the partially ordered structure of the concept lattice with reduced labeling correspond to classes of provinces of the positional system. One position has province Aeg or Aegyptus as a single member with Praefectus as a particular political affiliation, while the two alpine provinces AlC and AlP are in the same class with the Mauretania provinces, and that is the reason why the position with the alpine provinces has shipping routes, despite being landlocked. Classes IV, V, and VI group the rest of the provinces of the RE network, and they correspond to the formal context concepts $\mathbb{C}_{10}$, $\mathbb{C}_{11}$, and $\mathbb{C}_{12}$, which make a single cluster with generalized equivalence (Doreian et al. 2004).

The positional system Y^+_{FC} below produced from formal concepts includes the political affiliations of Roman provinces, and it has six classes of approximately structurally correspondent actors where terrestrial and maritime routes are collective relations.

As with the construction of Y^+_{CE}, the positional system of the RE network produced with a formal concept analysis approach reflects the affiliation of the provinces in the reduced structure. Y^+_{FC} constitutes another basis for the construction of the RE network role structure Q_{FC}, which is another reduced configuration that describes the intertwining of classes of transportation routes as role relations in the RE transport network.

I	1	0	0	1	0	1
II	0	1	1	1	1	0
III	0	1	1	1	1	1
IV	1	1	1	1	1	1
V	0	1	1	1	1	1
VI	1	0	1	1	1	1

terrestrial, t

I	0	0	0	1	1	0
II	0	1	1	0	0	1
III	0	1	1	1	1	0
IV	1	0	1	1	1	1
V	1	0	1	1	1	1
VI	0	1	0	1	1	1

maritime, m

Positions in Y^{+}_{FC} are I : Aeg. II : AlC,AlP, MaC,MaT. III : Afr, Asi. IV : $\mathbb{C}_{10}$. V : $\mathbb{C}_{11}$. VI : $\mathbb{C}_{12}$.

3.3 *Multilevel Positional Systems*

Multilevel systems can integrate the RE transport network and the political affiliations of the Roman provinces; however, combining one- and two-mode network structures can easily produce large and complex configurations that need reduction for the analysis. In this case, the multilevel structures Y^{mlvl}_{CE} and Y^{mlvl}_{FC} of X^{+} are associations between classes of structurally equivalent provinces in X^{B} with the positional systems produced with compositional equivalence and from formal concepts. Both the associated subsets in the concept diagram and the cumulated province hierarchy with political affiliations provided a set of clustering information for the reduction of the RE transport system X^{+} into Y^{+}_{CE} and Y^{+}_{FC}, and which are part of Y^{mlvl}_{CE} and Y^{mlvl}_{FC}.

Figure 3 depicts two graphs of the multilevel positional systems Y^{mlvl}_{CE} and Y^{mlvl}_{FC} of RE transportation network with the political affiliations of provinces obtained both from compositional equivalence and from formal concepts. As with the two-mode structures, the graphs are from a binomial projection with a force-directed layout algorithm to the mixed configuration. The multilevel structure from the compositional equivalence has only two types of government affiliations by design, while the multilevel system with correspondences from formal concepts has all government types related to six classes of Roman provinces.

The multilevel system Y^{mlvl}_{FC} in Fig. 3a reduces further with equivalences in the formal concepts between classes of provinces and the government types of the provinces. Imperial provinces are connected both by terrestrial and maritime routes where classes I and V are structural equivalent so do also classes II and VI if and only if the two imperial government types are regarded equivalent as well. The stringent condition of structural equivalence applies in this case even if these government types connect different classes of provinces in the multilevel structure. In the case of the multilevel system Y^{mlvl}_{CE} from compositional equivalence in Fig. 3b, any of the classes of provinces is equivalent unless the difference between imperial and senatorial provinces vanishes.

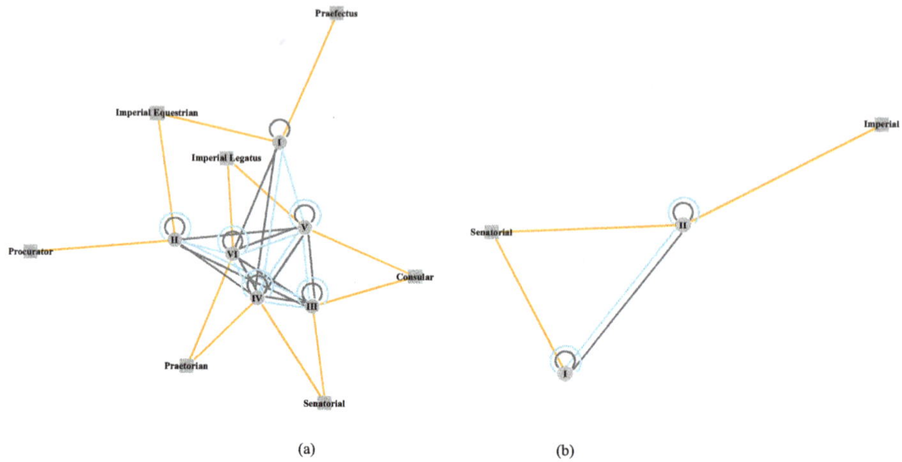

Fig. 3 Multilevel representations of the RE network positional systems Y_{FC}^{mlvl} and Y_{CE}^{mlvl} with provinces government types. Circles with Roman numerals are for classes of provinces as positions and squares are for government types. (**a**) Y_{FC}^{mlvl} with equivalence classes from formal concepts. (**b**) Y_{CE}^{mlvl} with classes from compositional equivalence

Ad hoc algebraic representations of multilevel structures with the positional systems made of correspondent provinces and class affiliations are per domain. A semigroup of relations stands for the one-mode network made of multiplex ties X^+, which incorporate cross-domain role relations as additional generators in the Relation-Box from which the cumulated province hierarchy arises for producing the positional system Y_{CE}^{mlvl}. Besides, the concept lattice includes role relations across domains in the reduced arrangement Y_{FC}^{mlvl}.

For the RE province affiliation network that is part of the multilevel system, the associated ordered structure of the semigroup is isomorphic to the concept lattice in Fig. 2b, and this is because the concept lattice provides the clustering information of the positional system with the role structure for the multilevel version of the RE network in X^+ by connecting provinces and their properties.

4 Semigroup and Role Structure

With multiplex networks like the RE transport system, it is possible to link different types of ties into a relational system that encodes the interlock of the relations occurring in the network. In the case of connections between network positions, the intertwining of these collective relations produces the network role structure Q where role relations are the semigroup objects. Even though the positional analysis reduced the complexity of the relational structure dramatically, there are still options for a simplification of the RE network role structure through different decomposition processes. The focus at this stage shifts from grouping provinces in the RE network to grouping strings in the semigroup representing its relational structure, which arrive from one of the positional systems.

4.1 Decomposition and Quotient Semigroup

Decomposition is the homomorphic reduction of a semigroup S as a mapping $\pi : S \rightarrow S\backslash \sim$ where $S\backslash \sim$ is the homomorphic image of S, which is a quotient semigroup under the equivalence relation on the semigroup that is stable under composition. Mapping π is surjective and maps x to $\tilde{x}$ or the equivalence class of S containing x where relationships in the quotient semigroup between elements in set $\tilde{x}$ represent the role structure as an image matrix. In this sense, image matrices are reduced representations of relational systems belonging to multiplex network structures that define their role interlock.

In the case of the RE network, the associations between the classes of provinces in the two positional systems constitute different versions of its role structure where partially ordered semigroups serve to represent these reduced systems in a similar way as $S(R)$ and associated partial order do for individual relations. Partially ordered semigroups encode a set of inclusions that, together with the set of equations and the relational structure itself, comprise three algebraic constraints of the network relational structure.

Figure 4 has graphical representations of partially ordered semigroups for the two role structures produced from the positional systems with formal concepts and with compositional equivalence for the clustering information. Each panel in the figure has a Cayley graph and the associated inclusion lattice of the role structures with equations in different notations among the shortest strings. The force-directed layout of the Cayley graphs evidences role structures with different topologies where both systems are symmetric because the network is undirected.

The role structure from compositional equivalence in Fig. 4b, in which terrestrial and maritime routes are equated, has more generators than the role structure from formal concepts in Fig. 4a that has just the two transportation types. In this latter role structure, the fact that generators are equal implies that whenever a statement is true for t it is also for m. The semigroup of the role structure from compositional equivalence has the senatorial role relationship s acting as the identity element, which tends to be placed at the center of the graph with a force-directed layout algorithm, while the semigroup of the role structure from formal concepts is without an identity element.

Semigroups related to social networks are typically large and complex, which can make difficult its interpretation in substantial terms. There are different semigroup decomposition strategies to reduce these relational systems into homomorphic images or aggregated role tables while preserving their essential structure. Some strategies establish equivalences among strings in a given semigroup representing a relational system, by looking at their Green's relations, by finding classes of

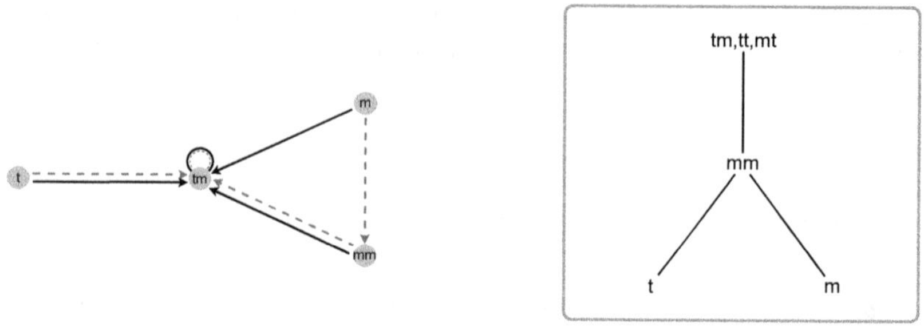

(a)

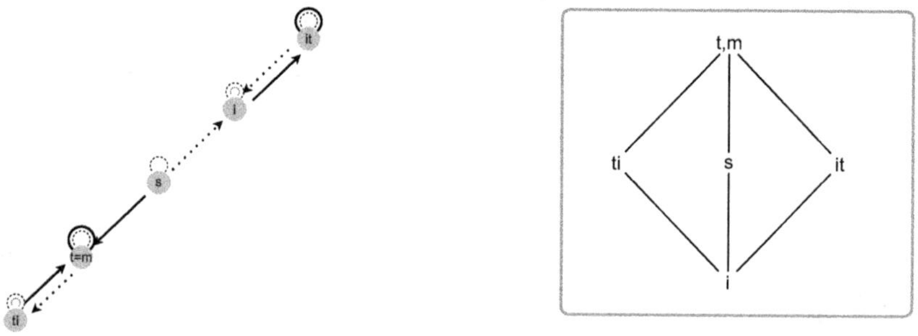

(b)

Fig. 4 Graphical representations of two role structures of RE transport network with province affiliations as partially ordered semigroups. Cayley graphs and inclusion lattice diagrams in framed boxes with equations among strings of length 2. (**a**) Role structure Q_{FC} from formal concepts. (**b**) Role structure Q_{CE} from composition equivalence

congruent semigroup elements through factorization or another technique that applies the substitution property.

4.1.1 Green's Relations

Green's relations (Green 1951) are five equivalence relations that characterize the semigroup elements in terms of the principal ideals they generate, or ideals generated by one element of the semigroup. The relationship between Green's relations and ideals is given in (Boyd 1991; Degenne and Forsé 1999) with applications in social networks. To define this type of equivalence relations among semigroup elements, take a monoid T or a semigroup with identity adjoined where relations $a, b \in T$ have a correspondence to $\mathcal{R}$ or $a\mathcal{R}b$ if and only if $aT = bT$. Similarly, $a\mathcal{L}b$ exists if and only if $Ta = Tb$.

Equivalence $\mathcal{R}$ is left compatible, and equivalence $\mathcal{L}$ is right compatible, and both the intersection and the union of these produce correspondences $\mathcal{H}$ and $\mathcal{D}$. For non-finite semigroups, there is also an equivalence $\mathcal{J}$ that covers $\mathcal{D}$ in the ordered configuration where the $\mathcal{D}$-class structure is the *egg-box diagram* where

∘	t	tm	mm	m
t	tm	tm	tm	tm
tm	tm	tm	tm	tm
mm	tm	tm	tm	tm
m	tm	tm	tm	mm

(a)

∘	t	it	s	i	ti
t	t	t	t	ti	ti
ti	t	t	ti	ti	ti
s	t	it	s	i	ti
i	it	it	i	i	i
it	it	it	it	i	i

(b)

Fig. 5 Permuted and partitioned multiplication tables of role structures for the RE transport network with affiliations according to $\mathcal{R}$ and $\mathcal{L}$ equivalences. (**a**) Role structure from formal concepts Q_{FC} where tm=mt=tt. (**b**) Role structure from composition equivalence Q_{CE} where t=m=tt=ts=st, and i=ii=si=is

a representative element typically characterizes each "block" or "egg-cell" of aggregated role relations.

Figure 5 has the two role structures of the RE network permuted and partitioned according to their Green's relations; that is, these are $\mathcal{D}$-class representations of Q_{FC} and Q_{CE}. The symmetry of $\mathcal{H}$-classes across the diagonal in both multiplication tables is because the network is undirected and the order of the semigroup with compositional equivalence is higher than that of the semigroup generated by formal concepts that have only two classes of equivalent string relations. Role structure Q_{FC} has a generator in each class, while both types of transportation routes belong to the same category in role structure Q_{CE} with senatorial provinces as the identity of the semigroup with no structuring effect in the role structure.

The two classes of strings in the role structure from the formal concepts in Fig. 5a are examples of left-zero band and right-zero band with representative compounds of length two. The same is for the $\mathcal{H}$-classes in Fig. 5b that are made of compounds in the role structure product of compositional equivalence.

4.1.2 Congruence Relations

Congruence relations are equivalences that are left and right compatible with the semigroup elements and are used to simplify its relation structure into a homomorphic image. Congruences preserve the operation of the algebraic structure and have the properties of the equivalence relation, reflexive, symmetric, and transitive, together with the substitution property, which served to decompose abstract semigroups like the role structures from positional systems produced with different strategies. Applications of congruence relations for the decomposition of abstract semigroup structures were for the partition of sequential machines (Hartmanis and Stearns 1966), and in the context of social networks (Pattison 1993).

In theory, there are several possibilities for additional equations among strings in the semigroup of role relations, which means that potentially there are different logics of role interlock in both Q_{FC} and Q_{CE} of the RE transport network with provinces affiliations. However, during the decomposition process, some homomorphic images, either from the formal concepts or from the compositional equivalence approaches, are trivial reductions of the original structure like the one-element quotient semigroup Q where all strings make a single class. Another typical aggregated role tables have two classes with a core-periphery prototype structure with the identity element as a single class of the quotient semigroup. Sometimes the identity element is associated with a "weak" kind string in the reduced structure where the "strong" relation is the one with the absorbing character in Q.

In the case of role structure Q_{FC} from formal concepts, the decomposition of its semigroup representing through congruence relations brings a set of equations that yields the one-element semigroup with a class of all string relations. This means that the most aggregated configuration of RE transport network that is not trivial is the $\mathcal{D}$-class of the role structure Q_{FC}. The only congruence relation that produces a larger homomorphic image than the one-element semigroup is on the role structure Q_{CE} from compositional equivalence where two out of three generators have an additional equation between strings to produce a homomorphic image of order three:

t	s	i	ti	it	*Equation*
1	2	3	3	1	t=it, i=ti

o	t	s	i
t	t	t	i
s	t	s	i
i	t	i	i

The aggregated configuration of the RE transport network with administrative affiliations through congruence relations has the two types of transportation routes equated, and where the class of senatorial provinces is the identity with no structuring effect in the role structure Q_{CE}. With compositional equivalence, there is also a close relationship between decomposition through congruence and Green's relations, and this is because the aggregated role structure implies two further equations on role relations that are within the equivalences conforming to the $\mathcal{D}$-class of Q_{CE}.

4.1.3 Factorization

Further versions of the RE network role interlock come from the set of inclusions among string relations in the relational structure that is part of the decomposition process; that is to say strings of aggregated routes and political affiliations of provinces in the role structure having a hierarchy. *Factorization* (Pattison and Bartlett 1982) is a decomposition method for partially ordered semigroups aimed to find factors or simpler constituent components that are the homomorphic images of the semigroup algebra as quotient semigroups. The partition relations are from induced inclusions to the partial order that satisfy the substitution property. One advantage of having partial order structures associated with the homomorphic images is that the substantial interpretation of the role structure is also in terms of hierarchies among string inclusions reflected in the lattice diagram.

Because each atom leads to different clustering information, having more than one atom implies different image matrices that are sub-products of the role semigroup, which reflect superimposed role interlocks in the reduced system. In case factorization brings more than one atom, the decomposition of the partial order semigroup of the role structure as factors is represented by different quotient semigroups where the partitions from atoms are closer to the original structure. The smallest non-trivial factors come from the partition relations produced by the maximal meet complements of the atoms when they do not equal the atoms themselves because these factors are closer to the maxima or the universal relation in the congruence lattice, which is the 1-element semigroup where all strings belong to the same class. Hence, the meet complements of the atoms will produce a coarser configuration than the ones produced by the atoms themselves.

The aggregated role structure Q_{FC} of the RE transport network and province affiliations from formal concepts that is a product of factorization has an additional equation that is tm=mm, which in this case corresponds to the first $\mathcal{R}$- and $\mathcal{L}$-class of the semigroup structure given in Fig. 5a. As a result, the order of the quotient semigroup reduces to three where all products in the semigroup of this aggregated role structure correspond to a single-element rectangular band made of compounds belonging to this equation.

For role structure Q_{CE} produced with compositional equivalence in Fig. 5b, there are two atoms in the congruence lattice where each atom has its meet complement. Based on aggregations to the partial order structure in the two meet complements of the atoms, there are partition relations on the partially ordered semigroup that produce two factors with an identical hierarchy of representative strings role relations. Aggregated role structures of Q_{CE} from the positional system product of compositional equivalence have homomorphic images with a unique partial order structure. The additional equations on generators through the factorization process are t=ti and i=it for the first factor and t=it and i=ti for the second factor that are dual structures reflecting the symmetry of undirected relations in the system.

∘	t	s	i
t	t	t	t
s	t	s	i
i	i	i	i

∘	t	s	i
t	t	t	i
s	t	s	i
i	t	i	i

$\leq$	t	s	i
t	1	0	0
s	1	1	0
i	1	1	1

In the case of the relational system of the RE network, there is a strong connection between the reduced partial order structure and quotient semigroup with congruence classes with factors because the homomorphic image produced with congruence relations is isomorphic to the second factor of the factorization process.

5 Shared Structure

Shared structures are a common set of equations and a common set of inclusions representing equality and hierarchy constraints. Role tables encode algebraic constraints as well as the associations among representative role relations in quotient semigroups. The shared structures between two or more semigroup role structures lie somewhere in the lattice of homomorphisms of the semigroup with immense size having the universal semigroup as the supremum and free semigroup of relations as its infimum. Because the quotient semigroups of Q_{FC} from formal concepts and Q_{CE} from compositional equivalence are the product of substantially the same generator relations, it is possible to compare these configurations and look at their shared structure in terms of these algebraic constraints.

The shared configuration of factors and congruence classes from role structures Q_{FC} and Q_{CE} leads to another aggregated representation of the RE network role structure. In the case of factorization, the decomposition of the RE network role structure brought overlapping aggregated configurations, which means that there are competing role interlocks in the image matrices of the positional systems with transportation routes and province affiliations. However, there is any set of shared inclusions or aggregated role structures made of congruence classes because the decomposition by congruence relations is for abstract semigroups.

An alternative for a shared structure is Q_{JNT} for the *joint homomorphic reduction* (Boorman and White 1976), which is the joint reduction $S \vee T$ or lattice union that is the least upper bound of semigroups S and T in the lattice of homomorphisms of the semigroup. The joint homomorphic reduction implies that relations that are congruent in either semigroup must be congruent in their joint reduction as well, and this means that the joint reduction of S and T produces a least refined structure than the two constitutive semigroups or a smaller data algebra. For the RE network positional systems Y^+_{FC} and Y^+_{CE}, the two role structures, Q_{CE} from compositional equivalence and Q_{FC} from formal concepts, have a joint homomorphic reduction of a quotient semigroup Q_{JNT} of order 2 where the identity or senatorial provinces distinguish from the rest of the string role relations that make a single class where the imperial provinces lie. This is prototypical core-periphery configuration or a pattern between strong and weak relationships.

5.1 *Common Structure Semigroup*

The intersection of role systems S and T in the lattice of homomorphisms of the semigroup is the *Common Structure Semigroup* (Bonacich 1980) that is the other alternative to joint homomorphic reduction in finding a shared structure. In this lattice of homomorphisms, the common structure $S \wedge T$ or the meet of the semigroups is the greatest lower bound and the largest table consistent with both original role tables where an equation holds if and only if it holds in both original structures. One way to obtain Common Structure Semigroup of role relations is by taking meta-matrices as generators where diagonal arrays record the two positional systems for $\mathcal{Q}_{FC}$ and $\mathcal{Q}_{CE}$ and zeroes elsewhere.

The construction of the shared structure is as multiplication tables or "meet semigroup," and a Common Structure Semigroup $\mathcal{Q}_{CSS}$ of the RE network role structure can have either two or four generator role relations. This is because the formal concepts include the two province characteristics as explicit roles, while the positional system product of the cumulated province hierarchy has just the transportation routes as role relations with implicit province characteristics.

Figure 6 is a graphical representation of the shared role structure $\mathcal{Q}_{CSS}$, where the Cayley graph in Fig. 6a has a force-directed layout, as with the RE role structure given in Fig. 4. The identity relation of the Common Structure Semigroup is s, which implies that the senatorial class distributes over two semigroup parts, one $\mathcal{R}$-class of the quotient semigroup, and the other part has relations between the other $\mathcal{R}$-classes except for the identity. The inclusion lattice diagram below in Fig. 6a is for the associated ordered configuration to the common semigroup with the shortest representative strings of the shared role structure. In this case, $\mathcal{Q}_{CSS}$ is the shared role structure having four generator relations recorded as meta-matrices between the two RE aggregated relational systems $\mathcal{Q}_{FC}$ and $\mathcal{Q}_{CE}$.

All members in each of the $\mathcal{R}$- and $\mathcal{L}$-classes have direct links in the partial order of role structure, and the lattice diagram allows a substantial representation of the shared system in $\mathcal{Q}_{CSS}$ taking the hierarchy among simple and representative compound strings. For instance, in terms of aggregated and shared role relations, maritime routes cover senatorial and imperial provinces but not terrestrial routes and vice versa. Although it might make sense the fact that shipping routes do not cover roads, for example, nonetheless, it may be the case that the concatenation of maritime routes "are covered" by the concatenation of terrestrial routes at this aggregated level even if we ignore the set of equations. These inclusions would be possible because of the presence of entrainments of roads and shipping routes between Roman provinces in the transport network X^+, which is reflected in the image matrix of the role structure $\mathcal{Q}_{CSS}$.

The Common Structure Semigroup $\mathcal{Q}_{CSS}$ represents a shared configuration of aggregated role structures belonging to the RE network. Since a semigroup of role relations characterizes this shared configuration, it is possible to perform a decomposition of this system in the same fashion as with the previous role structures $\mathcal{Q}_{CE}$ and $\mathcal{Q}_{FC}$. Figure 6b provides the $\mathcal{Q}_{CSS}$ semigroup of role relations with the four generators permuted and partitioned according to Green's relations. With the partition of the role table with Green's equivalences, it is more evident that the structure of $\mathcal{Q}_{CSS}$ is the union of $\mathcal{Q}_{CE}$ and $\mathcal{Q}_{FC}$ where the partition of the semigroup into $\mathcal{H}$-classes preserves the symmetries across the diagonal of the multiplication table because all string role relations in the system are undirected.

A reduction of the common semigroup structure based on Green's relations will lead us to the joint homomorphic reduction of the semigroups, even if the clustering of the strings in $\mathcal{R}$ and $\mathcal{L}$ differs. A joint homomorphic reduction implies that relations that are congruent in either semigroup are congruent in their joint

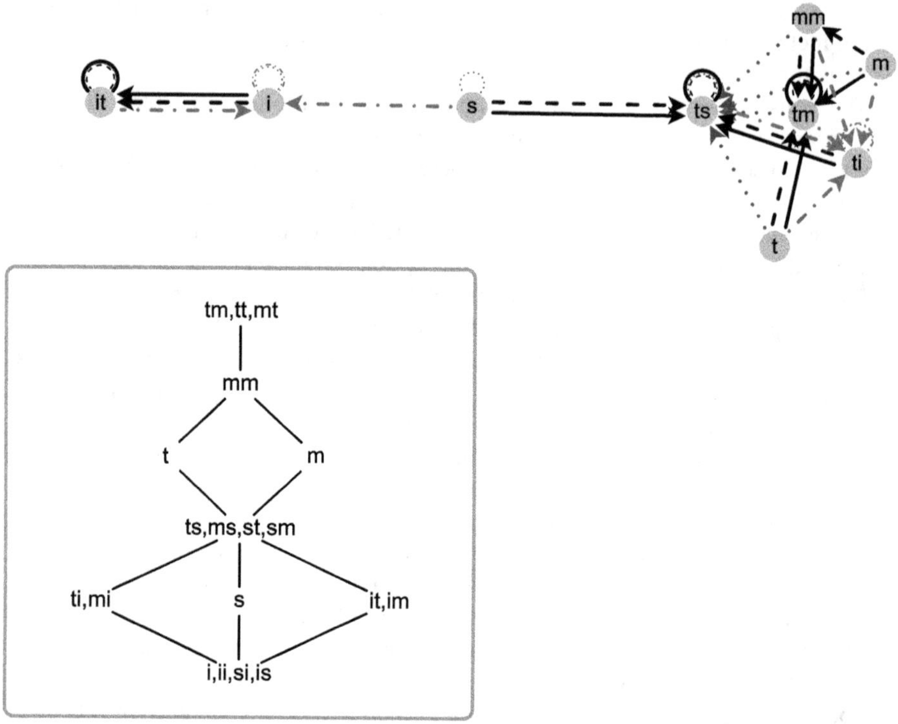

(a)

	t	tm	mm	m	s	i	ti	ts	it
t	tm	tm	tm	tm	ts	ti	ti	ts	ts
tm	tm	tm	tm	tm	ts	ti	ti	ts	ts
mm	tm	tm	tm	tm	ts	ti	ti	ts	ts
m	tm	tm	tm	mm	ts	ti	ti	ts	ts
s	ts	ts	ts	ts	s	i	ti	ts	it
i	it	it	it	it	i	i	i	it	it
it	it	it	it	it	it	i	i	it	it
ts	ts	ts	ts	ts	ts	ti	ti	ts	ts
ti	ts	ts	ts	ts	ti	ti	ti	ts	ts

(b)

Fig. 6 Common Structure Semigroup Q_{CSS} of the Roman Empire role systems with province characteristics from compositional equivalence and from formal concepts. (**a**) Cayley graph and inclusion lattice representation of the shared structure as a partially ordered common semigroup with four generators. Lattice diagram in the framed box has equations for generators and compound strings of length 2. (**b**) Egg-box diagram with the Green's relations of the common structure semigroup with representative strings that constitutes the $\mathcal{D}$-class of Q_{CSS}

reduction as well. The Common Structure Semigroup Q_{CSS} as the union of Q_{CE} and Q_{FC} in this case, and for transportation routes in X^{+} only, Q_{CSS} takes t and m as generators where the corresponding partially ordered semigroup is isomorphic to the role structure Q_{FC} in Fig. 4a from formal concepts.

6 Concluding Remarks

Algebraic representations of multilevel systems can integrate their two domains into a single structure-preserving relation within a domain. A significant aspect of the RE network is that, at an aggregated level, there is still the distinction between terrestrial and maritime routes in the role structure where just the imperial government type does have a structuring effect. The two types of imperial provinces in the multilevel network fit into separate classes of Roman provinces that are structurally equivalent, provided that the two kinds of government types also are structurally equivalent. The positional analysis of multilevel structures with related correspondences at both domains, which are equivalent entities at different stages of the multilevel configuration, facilitates a substantial interpretation of the role interlock in the network by domain experts in addressing the complexity of past human societies.

Transport systems belong to the class of spatial networks that are physically constrained, and this condition implies that the adjacency matrices for the different types of connections can bear structural zeroes that prevent linkages, in this case, in the spatial transportation system of ancient places. The same constriction of structural zeroes applies for non-overlapping affiliations in some two-mode configurations with mutually exclusive event instances. Since with multilevel networks, the levels constitute generators of the semigroup of relations with concatenations of binary ties and actor attributes representing the relational system and role structures, the implications of structural zeroes in this configuration type are topics worth exploring further.

7 Note About Software

Graphs, including Cayley graphs, and the cartographical map with the RE transportation network are portrayed with **R** (R Core Team 2021) packages **multigraph** (Ostoic 2021) and **sdam** (Ostoic et al. 2022). An initial computer program for performing algebraic analysis of social networks is **ASNET**, which was incorporated as **PACNET** module (Pattison et al. 2000) in the **StOCNET** environment (Boer et al. 2017). **ASNET** constructs partially ordered semigroups and it performs semigroup decomposition through factorization with an algorithm described in Ardu (1995). The **R** functions for semigroup decomposition are from **multiplex** package (v3.1) (Ostoic 2020) together with congruence relations, Green's relations, compositional equivalence, and for depicting lattice diagrams by invoking **Rgraphviz** (Hansen et al. 2018) that is the **R** interface of **Graphviz** (AT&T Labs Research 2019).

Acknowledgments Many thanks to Pip Pattison, Dragana Martinovic, and Melanija Mitrović for the invitation to the conference series Algebra without Borders and for their guidance. Thanks to Adéla Sobotkova from the Social Dynamics and Complexity in the Ancient Mediterranean project for the opportunity to work with historical data.

Appendix: Glossary of Notation and Definitions

In the context of social systems, a *multilevel network* X^{mlvl} for vertex sets N (domain) and M (codomain) is

$$X^{mlvl} = \langle\ N,\ M,\ R_N,\ R_M,\ R_{N\times M}\ \rangle$$

where edge sets R_N are for relations (directed or not) on N, and R_M for (undirected) relations on M that stand for n and m social entities.
A *multiplex network* X^+ adds $r > 1$ types of relations R_N.
An *affiliation network* X^B is a bipartite system with two domains N and M and constitutive relations $R_{N\times M}$ between domains.
Constitutive relations $R_{N\times M}$ describe the embeddedness of one set of entities in another set (e.g., people belonging to groups) with the two domains as levels in X^{mlvl}.
A *positional system* Y of X is a reduced network structure made of structurally equivalent entities in X, which play a similar role in Y.
A *relational system* is a relational structure produced by the concatenation of ties occurring in a multiplex network X^+.
A *role structure* Q is the product of role relations from the positional system Y.
An *abstract semigroup* is an algebraic system having a set of elements with an associated operation on it:

$$\big\langle\ S,\ *\ \big\rangle,$$

where S is the underlying or carrier set and $*$ is a binary operation on an ordered pair $*\colon\ S \times S \to S$, that for all $x, y, z \in S$ satisfies the *associative law*, $x * (y * z) = (x * y) * z$. Each semigroup S is closed under the operation, which means that the product of two or more elements in S must be also part of the semigroup.
A *monoid* T is a semigroup with identity.
The *order* of a semigroup S is the total number of unique elements in S.
A *semigroup of relations* $S(R)$ has an alphabet Σ as object set of relations on R_N associated by a composition operation

$$\big\langle\ S(R),\ \Sigma,\ \circ\ \big\rangle,$$

where any element x in $S(R)$ is expressed as a product of generating relations of Σ, $R_i \in \Sigma$ as $x = R_{N_1} \circ R_{N_2} \circ \ldots \circ R_{N_r}$, and the product of elements are described in terms of composition.
An *alphabet* $\Sigma = \left\{ R_1, R_1, \ldots, R_r \right\}$ for $\Sigma \subseteq S$ is the collection of r generator relations in the network relational structure represented by an algebraic system S.
Strings of symbols are the generators and compound relations with right multiplication in $S(R)$ represented in a table by juxtaposition, $x \circ y = xy$.
A *partially ordered semigroup* is a semigroup of relations where the composition operation is isotone

$$\langle\, S(R),\ \circ,\ \leq \rangle,$$

where $\leq$ is a natural partial order relation on strings that is reflexive, transitive, and antisymmetric.
The *Axiom of Quality* (Boorman and White 1976) or Zermelo-Fraenkel Axiom of Extensionality equates ties that link precisely the same entities in the network ensuring that S is limited to a number of unique string relations $R_1 = R_2$ implies $R_1 \leq R_2$ and $R_2 \leq R_1$.
An equivalence relation ρ on S is *left compatible* if $(x, y) \in \rho$ implies $(zx, zy) \in \rho$, and *right compatible* if $(x, y) \in \rho$ implies $(xz, yz) \in \rho$ for $x, y, z \in S$.
A *congruence* relation is an equivalence relation that is left and right compatible.
The *egg-box diagram* (Green 1951) of a semigroup S is the union of the left compatible $\mathcal{R}$ equivalence and the right compatible $\mathcal{L}$ equivalence classes that is the $\mathcal{D}$-class on S.
A *Cayley graph* is a pictographic representation of a semigroup S with respect to a given set of generator relations. The nodes of the Cayley graph are elements of S and there is a directed edge with the label x from s to $sx \in S$ for each $x \in S$ in the generator set and each $s \in S$.
Compositional equivalence (Breiger and Pattison 1986) is a type of correspondence that is built on the algebra of relational structures including local role algebras. In multiplex structure X^+ and for a relation $x \in X^+$, j and k are compositional equivalent with respect to the ith ego or $j \equiv k \bmod i$, if and only if for each role relation R^*, $R^*_{i_{xj}}$ implies $R^*_{i_{xk}}$ and $R^*_{i_{xk}}$ implies $R^*_{i_{xj}}$, whenever there exist at least one R^*_{ij}.
The *Common Structure Semigroup* (Bonacich 1980) of semigroups S and T is the common structure $S \wedge T$ in the lattice of homomorphisms of the semigroup, and the largest table consistent with both original role tables where an equation holds if and only if it holds true in both original structures.
Formal concept analysis (Ganter and Wille 1996) is a mathematical framework for the analysis of affiliation or two-mode networks as formal concepts from a formal context.
A *formal context* $\mathbb{K} = (M, N, I)$ characterizes in a cross-rectangular table of two-mode data a set of objects M and a set of attributes N with an incidence relation $I \subseteq M \times N$.

In a *formal concept* $\mathbb{C} = (A, B)$ for $\mathbb{C} \in \mathbb{K}$, objects A and attributes B are maximally contained on each other or make a maximal rectangle.
The *extent* and *intent* of , A and B, have $A' = B$ as the set of attributes common to all the objects in the intent, and $B' = A$ as the set of objects possessing the attributes to the extent.
The *Basic Theorem of Concepts Lattices* (Ganter and Wille 1996) allows us to construct the partial ordering of the concepts by the subconcept–superconcept, (A_1, B_1) and (A_2, B_2), relation of extents and intents:

$$(A_1, B_1) \leq (A_2, B_2) \quad \text{if and only if} \quad A_1 \subseteq A_2 \qquad \text{and} \qquad B_2 \subseteq B_1,$$

where the greatest lower bound of the meet and the least upper bound of the join are defined in terms of objects and attributes with an index set T

$$\bigwedge_{t \in T} (A_t, B_t) = \Big(\bigcap_{t \in T} A_t, \ (\bigcup_{t \in T} B_t)'' \Big)$$

$$\bigvee_{t \in T} (A_t, B_t) = \Big((\bigcup_{t \in T} A_t)'', \ \bigcap_{t \in T} B_t \Big).$$

A *concept diagram* is a concept lattice with the partial ordering of the formal concepts from set of inclusions among the maximal rectangles in the formal context made with Galois connections.
A *Galois connection* (Ore 1944) between power sets $P(X)$ and $P(Y)$ is derived for any subsets $A \subseteq P(X)$ and $B \subseteq P(Y)$ by A' and B':

$$\begin{aligned} A' &= y \in P(Y) \mid (x, y) \in I \qquad \text{for all } x \in A \\ B' &= x \in P(X) \mid (x, y) \in I \qquad \text{for all } y \in B \end{aligned}$$

where x and y are rows and columns in the derivation operation. $P(X)$ and $P(Y)$ result being two closed systems dually isomorphic to each other where Galois connections between the objects in the intent and the attributes in the extent produce a family of formal concepts in the formal context.

References

Ardu, S. (1995), '**ASNET**: Algebraic and Statistical Network Analysis – User Manual'.

AT&T Labs Research (2019), ***Graphviz*** *– Graph Visualization Software*. v2.40.1. https://www.graphviz.org/

Boer, P. et al. (2017), ***StOCNET****: An open software system for the advanced statistical analysis of social networks*. v1.8. Science Plus Group/ICS. http://www.gmw.rug.nl/~stocnet/

Bonacich, P. (1980), 'The 'Common Structure Semigroup,' a replacement for the Boorman and White 'Joint Reduction'', *American Journal of Sociology* **86**, 159–166.

Boorman, S. and White, H. (1976), 'Social structure from multiple networks. II. Role structures', *American Journal of Sociology* **81**(6), 1384–1446.

Boyd, J. (1991), *Social Semigroups: A unified theory of scaling and blockmodelling as applied to social networks*, George Mason University Press.

Breiger, R. and Pattison, P. (1986), 'Cumulated social roles: The duality of persons and their algebras', *Social Networks* **8**, 215–256.

Clifford, A. and Preston, G. (1967), *The Algebraic Theory of Semigroups*, American Mathematical Society.

Degenne, A. and Forsé, M. (1999), *Introducing Social Networks*, SAGE Publications.

Doreian, P., Batagelj, V. and Ferligoj, A. (2004), *Generalized Blockmodeling*, Vol. 25 of *Structural Analysis in the Social Sciences*, Cambridge University Press.

Epigraphic Database Heidelberg (2022), 'Open Data Repository'. [Online; retrieved on 10 January 2022]. https://edh.ub.uni-heidelberg.de/data/api

Ganter, B. and Wille, R. (1996), *Formal Concept Analysis – Mathematical Foundations*, Springer-Verlag.

Green, J. (1951), 'On the structure of semigroups', *Annals of Mathematics* **54**(1), 163–172.

Hansen, K. et al. (2018), ***Rgraphviz**: Provides plotting capabilities for **R** graph objects.* **R** package v2.24.0. http://bioconductor.org/packages/Rgraphviz/

Hartmanis, J. and Stearns, R. (1966), *Algebraic Structure Theory of Sequential Machines*, Prentice-Hall.

Howie, J. (1996), *Fundamentals of Semigroup Theory*, Oxford University Press.

Lévi-Strauss, C. (1949), *Les Structures Elementaire de la Parente*, Presses Universitaires de France. With an appendix by André Weil.

Mandel, M. (1983), 'Local roles and social networks', *American Sociological Review* **48**, 376–386.

Ore, O. (1944), 'Galois connexions', *Transactions of the American Mathematical Society* **55**, 493–513.

Ostoic, A., Sobotkova, A., Kase, V. and Hermankova, P. (2022), ***sdam**: Social Dynamics and Complexity in the Ancient Mediterranean.* **R** package v1.1.4. https://CRAN.R-project.org/package=sdam

Ostoic, J. A. R. (2020), 'Algebraic Analysis of Multiple Social Networks with multiplex', *Journal of Statistical Software* **92**(11), 1–41.

Ostoic, J. A. R. (2021). *Algebraic Analysis of Social Networks: Models, Methods and Applications Using R*. Hoboken. Wiley Series in Computational and Quantitative Social Sciences. Wiley.

Pattison, P. (1993), *Algebraic Models for Social Networks*, Vol. 7 of *Structural Analysis in the Social Sciences*, Cambridge University Press.

Pattison, P. and Bartlett, W. (1982), 'A factorization procedure for finite algebras', *Journal of Mathematical Psychology* **25**, 51–81.

Pattison, P. et al. (2000), 'Statistical evaluation of algebraic constraints for social networks', *Journal of Mathematical Psychology* **44**, 536–568.

R Core Team (2021), ***R**: A Language and Environment for Statistical Computing*, R Foundation for Statistical Computing, Vienna, Austria. https://www.R-project.org/

Rodrigue, J. (2020), *The Geography of Transport Systems*, 5th edn, Routledge.

Suschkewitsch, A. (1928), 'Über die endlichen Gruppen ohne das Gesetz der eindeutigen Umkehrbarkeit', *Mathematische Annalen* **99**, 30–50.

United Nations of Roma Victrix (2021), 'Roman Provinces at the Height of the Empire'. [Online; retrieved on 8 January 2021]. https://www.unrv.com/provinces/provincetable.php

Time and Sequence in Networks of Social Interactions

Lucia Falzon

1 Introduction

A social network is a set of actors or of groups of actors that are connected directly or indirectly through relational ties. Most often these ties represent dyadic relationships, such as kinship, friendship, or collaboration, and networks are constructed from data that are sourced from surveys and questionnaires. Such networks do not usually have a temporal dimension—they are static for the purpose of the study, and therefore, the network ties are either present or absent for the whole observation period. The formal representation of a network is a graph that forms the basis of structural analysis of these networks. Equivalent mathematical representations include relational algebras and adjacency matrices (Pattison 1993; Wasserman and Faust 1994; Degenne and Forsé 1999). Different representations are chosen depending on the type of analysis required, e.g., graph-theoretic measures, such as node centrality, describe their structural embedding; matrix operations allow us to calculate distance between nodes and to construct walks; compound social relations are readily determined from composition of relations.

The temporal dimension is a focus of longitudinal studies of dynamic social networks, in which networks are constructed from data collected at several points in time and network changes are recorded. The changes might be single ties forming or dissolving, or they may be more significant structural changes. The time between points at which change is recorded may be infinitesimally small or they may have a coarser granularity depending on the research objective. In longitudinal studies, data are collected at predefined time points as prescribed by the research design. Statistical models of the time-dependent processes that govern the changes are

L. Falzon (✉)
University of Melbourne, Parkville, VIC, Australia
e-mail: lucia.falzon@unimelb.edu.au

© The Author(s), under exclusive license to Springer Nature Switzerland AG 2023
M. N. Hounkonnou et al. (eds.), *Mathematics for Social Sciences and Arts*, Mathematics in Mind, https://doi.org/10.1007/978-3-031-37792-1_10

specified on the basis of a theoretical framework and as evidenced by observational data; the model parameters are estimated empirically (Robins and Pattison 2001; Snijders and Koskinen 2013; Hoffman et al. 2020). In some cases, the time at which a tie is formed is modeled (Snijders and Koskinen 2013), and in others, the time at which an actor joins a group (Hoffman et al. 2020).

The recent prevalence of new sensor technologies and online communications data has sparked new interest in networks of social interactions and relational events (Hoffman et al. 2020; Lehmann 2019; Falzon et al. 2017; Spiro et al. 2013). The dyads in these networks are constructed from observations of social interactions among a predefined set of actors. While the concepts of social relations and social interactions are related, there are key distinctions that need consideration: interactions have a short duration relative to long-term relationships; they typically have a recorded starting point and end point; and they occur sequentially (de Nooy 2015; Borgatti et al. 2013). Social processes are by their very nature dynamic and unfold over continuous time. One way to conceive of a network model of a social process is as a sequence of temporally ordered dyadic interactions. Relational event models (Butts 2008a) consider both timing and sequence to predict dyadic actions at a fine temporal scale based on the observation that the existence of a relational event at a certain point in time can change the possibility of occurrence of subsequent events.

This chapter explores the use of algebraic techniques for analyzing dynamic interactions and relational structures and discusses their utility in determining sequential patterns. In contrast to prevalent dynamic network analysis methods, such as the longitudinal studies described above, in which networks are modeled as a series of static network snapshots, this approach considers the finer granularity of dynamic interactions captured in time-stamped data on information sharing events. It enables a higher-fidelity representation of processes and flows to conduct investigations on where and when such processes were initiated and where they might terminate. In what follows, we first review algebraic representations for static network structures. Then, as an analogue of the Boolean algebra used to define operations in static networks, we introduce an algebra of temporal intervals that can be used to both analyze the underlying dynamic network structures and build meaningful measures from them. Finally, we give an overview of the different classes of temporal network measures that are currently in use and explore how they address different concepts of time.

2 Algebraic Representations of Social Networks

Algebraic representations of social networks have been widely used to facilitate network analysis in the social and behavioral sciences (Pattison 1993; Boyd 1991; Batagelj 1994). The foundational elements of such a relational algebra are a set of relational types, such as friendship or acquaintanceship, over a set of network actors, N. Each relational type is associated with a network with a fixed set of actors or nodes from N or equivalently, a directed graph, with nodes from the set N and a set of directed edges, E, each edge being an ordered pair from N. The (Boolean) adjacency matrix, A, associated with this network has entries $A_{ij} = 1$ if node i has a tie to node j, and $A_{ij} = 0$ otherwise. In the case of relations that are symmetric, e.g., kinship, the tie is non-directed, and the associated matrix is symmetric, i.e., $A_{ij} = A_{ji} \forall i, j \in N$. Other relations, such as "provides advice," have a defined direction, so that if i provides advice to j, then $A_{ij} = A_{ji}$ does not hold in general.

It is possible to define a variety of algebraic structures arising from operations over social networks (Pattison 2014). In this section, we focus principally on relational composition and Boolean addition as operations of primary interest, on the way to describing analogous algebraic constructions for dynamic networks. In what follows, we identify the relation A with its network A and its associated matrix A and use these terms interchangeably.

2.1 *Algebraic Structures on Relational Networks*

The binary operation of *relational composition* is defined for arbitrary networks A and B on a node set N as the network corresponding to the Boolean product of the matrices A and B:

$$(AB)_{i,j} = \sum_{k \in N} A_{i,k} B_{k,j},$$

where the summation denotes Boolean addition.

The matrix $AB = [AB_{i,j}]$ describes a compound relation of type $\mathcal{AB}$. The entries of AB define the presence or absence of labelled walks of type $\mathcal{AB}$ among pairs of members of the node set (Pattison 1993; Kontoleon et al. 2013).

In general, a *labelled walk* of type $\mathcal{A}_1\mathcal{A}_2 \ldots \mathcal{A}_p$ from a node i to a node j is a sequence of nodes $i = k_0, k_1, \cdots, k_p = j$ in which each pair (k_{m-1}, k_m) of adjacent nodes in the sequence is an edge in the network A_m ($m = 1, 2, ..., p$). The length of the walk is p. Labelled walks in a multiple social network are important because they describe indirect relations arising from the composition of the constituent networks. For instance, if $\mathcal{A}$ refers to friendship and $\mathcal{B}$ refers to collaboration, then labelled walks of type $\mathcal{AB}$ describe pairs (i, j) of actors for whom j is a collaborator of a friend of i. If each constituent network in a compound relation is a particular network A of type $\mathcal{A}$, then the entries of matrices AA, $(AA)A$, and so on, indicate pairs of nodes that are linked by walks in the network A of lengths 2, 3, and so on.

The union of two arbitrary networks A and B on the same node set N is defined as pairwise Boolean addition, that is,

$$(A \cup B)_{i,j} = A_{i,j} + B_{i,j} \text{ for all } i, j \in N.$$

The pair of actors (i, j) are linked by the network $A \cup B$ if and only if i is linked to j by the network A and/or the network B. We refer to the type of the network $A \cup B$ as $\mathcal{A} \cup \mathcal{B}$. If $\mathcal{A}$ and $\mathcal{B}$ refer to friendship and collaboration, for example, then two nodes are linked by a relation of type $\mathcal{A} \cup \mathcal{B}$ whenever they are linked by a friendship and/or a collaboration tie.

For any two networks A and B on the same node set N, we define a *natural order* on networks: $A \leq B$ if and only if:

$$A_{i,j} \leq B_{i,j} \text{ for all } i, j \in N.$$

It can readily be shown that the set of all binary relations on a set N forms a semigroup under the union operation defined above and is a monoid if we additionally define an identity element for this semigroup as

$$O_{i,j} = 0 \text{ for all } i, j \in N.$$

Furthermore, for all networks A, B on node set N, the union operation is commutative:

$$A \cup B = B \cup A.$$

Similarly, the set of all binary relations on this set together with relational composition forms a monoid for which the identity element is the identity network defined by

$$I_{i,j} = 1 \text{ if } i = j;\ I_{i,j} = 0, \text{ otherwise.}$$

The operations of relational composition and union and the natural order relation give rise to other algebraic structures on networks as we define below (Kontoleon et al. 2013; Pattison et al. 2015).

Definition 1 A *semiring* $(\mathbb{S}, \oplus, \otimes)$ is a set $\mathbb{S}$ equipped with two binary operations $\oplus$ and $\otimes$ with the following properties:

1. $(\mathbb{S}, \oplus)$ is a commutative monoid with identity ι.
2. $(\mathbb{S}, \otimes)$ is a monoid with identity ϵ.
3. Distributivity of $\otimes$ over $\oplus$: $\forall a, b, c \in \mathbb{S}, a \otimes (b \oplus c) = (a \otimes b) \oplus (a \otimes c)$ and $(a \oplus b) \otimes c = (a \otimes c) \oplus (b \otimes c)$.
4. Absorption: $\forall a \in \mathbb{S}, a \otimes \iota = \iota \otimes a = \iota$.

Definition 2 If $\preceq$ is an order on $\mathbb{S}$, then $(\mathbb{S}, \oplus, \otimes, \preceq)$ is an *ordered semiring* if, in addition, it satisfies:

1. If $a \preceq b$ and $c \preceq d$, then $a \oplus c \preceq b \oplus d$ for any $a, b, c, d \in \mathbb{S}$.
2. If $a \preceq b$ and $c \preceq d$, then $a \otimes c \preceq b \otimes d$ for any $a, b, c, d \in \mathbb{S}$.

The set of binary relations on the set N, with the operations $\oplus$ and $\otimes$, being union and relational composition, respectively, and with the natural order on networks as $\preceq$, is an ordered semiring. The identity element for the $\oplus$ operation is the null network, and the identity element for $\otimes$ is the identity network.

In the next section, we extend these operations to accommodate relations with a temporal dimension. A more general form of composition will be required to model and analyze flows along network walks. These operations give rise to algebraic structures that are more flexible as we define below.

Definition 3 The order relation $\preceq$ is termed a *canonical order* on $\mathbb{S}$ if it satisfies the following condition:

$$\forall a, b \in \mathbb{S}, a \preceq b \Rightarrow \exists\, c \in \mathbb{S}, \text{ such that, } a \oplus c = b.$$

Definition 4 (Baras and Theodorakopoulos 2010; Gondran and Minoux 2008) A semiring $(\mathbb{S}, \oplus, \otimes)$ (see Definition 1), in which $(\mathbb{S}, \oplus)$ is canonically ordered, is called a *dioid*. If $\otimes$ is left (right) distributive and has a left (right) identity, ϵ, then the dioid is a left (right) dioid, respectively.

In developing an extension from representations of static social relations to an algebra that allows consideration of time-sequenced flow, (Kontoleon et al. 2013) explore formal algebraic representations to establish the foundations for the computation of walk-based structures and measures. The literature on social network analysis refers to various applications of algebraic representations of social networks (Pattison 1993; Pattison et al. 2015). Our main interest here is in the algebraic approach to path-finding problems (Batagelj 1994; Baras and Theodorakopoulos 2010; Carrè 1971; Gondran and Minoux 2008).

3 Considering Time in Social Network Analysis

A considerable departure from traditional social network analysis, adopted by Kontoleon et al. (2013), was motivated by the temporal nature inherent in social interactions. Valuable information on event durations and temporal patterns and, most importantly, the sequential nature of interactions, are explicitly considered. The fundamental premise behind this reasoning is that networks of social interactions are highly dynamic, fluid structures that describe social processes continuously unfolding over networks. Therefore, modeling the network change over time by taking snapshots at regular discrete time points misses the continuous aspect of information flows.

With this change in focus comes a need to define the temporal dimension of social connections more explicitly. Network stability and concurrency of ties over the data collection period are typically assumed in classic social network analysis in which ties have a timeless quality. If we forego these assumptions, we will need to re-conceptualize conventional graph-theoretic measures (Borgatti and Everett

2006; Butts 2008b; Borgatti et al. 2013) such as centrality and cohesion. Sequential information introduces a temporal direction in networks, thus enabling tracking of information flows (diffusion) and fine-grained analysis of social processes as they play out in time (Falzon et al. 2018), which is crucial to understanding how information is transmitted and the roles that actors have in sending, receiving, or propagating information. As Moody (2002)'s study of disease transmission shows, the sequence of interactions defines paths through which diseases can be transferred. Furthermore, as Morris and Kretzschmar (1995) show, concurrent partnerships increase the rate of transmission of sexually transmitted diseases compared to sequential monogamous relationships.

3.1 An Algebra for Temporal Interval Sets

Temporal relational data records typically include the time at which interactions took place. This may be in the form of a timestamp, a single point in time, e.g., the time an email was dispatched, or it may record the time interval during which a conversation, a phone call, or a meeting occurred. So, besides knowing the chronological order of the interactions, researchers also have access to the actual time of day and the duration of interactions or sets of interactions. This extra information can be exploited to identify and measure differences of network activity over time or to explore how the role of individual actors changes over time.

Since previous contact has implications for all future interactions, labeled walks of interest in dynamic social networks must be time-ordered (Holme and Saramäki 2012). To this end, we consider network path problems (Baras and Theodorakopoulos 2010; Gondran and Minoux 2008), in which paths between two nodes are defined in terms of the weighted edges between them. In this case, the edge "weights" are actually records of times during which direct interactions occur on that edge. In the formal representation of interaction networks therefore, we define tie values (or weights) in between an actor i and another actor j by a finite set of strictly disjoint time intervals $\{[t_1, u_1], [t_2, u_2], \ldots, [t_p, u_p]\}$ denoting all the recorded timed interactions between them in the designated time period of interest, $[0, \omega]$. Each time interval $[t, u]$ is such that $t \leq u$. When $t = u$, the interval represents an instantaneous event. If there is no interaction between actors i, j over $[0, \omega]$, then the edge weight is ϕ, the empty set.

Useful relational representations in the dynamic context are ones that enable us to trace flows of material and abstract resources through the network. Following the algebraic approach of Kontoleon et al. (2013), and analogous to the Boolean operations of Sect. 2.1, we consider operations and associated algebras for sets of temporal intervals from a *base set* $\mathbb{E}$ comprising possible tie values in the dynamic case. The base set $\mathbb{E}$ is defined as follows:

Definition 5 Let $\mathbb{E}$ be the set of finite sets of disjoint, bounded, temporal intervals, that is, each element, $E = \bigcup_{i\in\{1,2,\cdots,r\}}[u_i, v_i] \in \mathbb{E}$, satisfies the following properties:

1. For all $i \in \{1, 2, \cdots, r\}$, $0 \leq u_i \leq v_i \leq \omega$, for some finite $\omega \in \mathbb{R}$, i.e., 0 is the earliest possible time point and ω is the latest one.
2. For any $[u_i, v_i], [u_j, v_j] \in E$, $[u_i, v_i] \cap [u_j, v_j] = \emptyset$.

We refer to u as the onset, v as the offset, and $(v - u)$ as the duration of the temporal interval, $[u, v]$. Zero-length intervals, i.e., for which $u = v$, represent discrete time points.

Property 2 above stipulates that each interval set in $\mathbb{E}$ must be a disjoint set. This is required to ensure a unique representation for each $E \in \mathbb{E}$, which we shall term the canonical form. If overlapping intervals result from any interval set operation, such as we define below, they are merged into a single interval.

Following Moody (2002), Kontoleon et al. (2013) introduce a composition operation that takes into account the start and end of each social interaction in a sequence to track flows through a network. Importantly, both node adjacency and the time order of interactions must be considered in the determination of potential diffusion paths. Accordingly, we define an operation that composes two interval sets and whose result is the path interval set over which transmission might be possible. For any two intervals sets, $E_1 = \bigcup_{p\in\{1..r\}}[u_p, v_p]$ and $E_2 = \bigcup_{q\in\{1..s\}}[t_q, w_q]$, in $\mathbb{E}$, we define the *composition of interval sets*, denoted by $\otimes$, as

$$E_1 \otimes E_2 = \bigcup_{\substack{p=1..r \\ q=1..s}} [u_p, v_p] \otimes [t_q, w_q],$$

where

$$[u_p, v_p] \otimes [t_q, w_q] = \begin{cases} [\max(u_p, t_q), w_q] & \text{if } \max(u_p, t_q) \leq w_q \text{ and } E_1, E_2 \neq \emptyset, \\ \emptyset & \text{otherwise.} \end{cases}$$

This definition of composition yields the longest time interval over which a potential transmission may occur via intermediary nodes. In other words, transmission is possible either when two relational events overlap or when one precedes the other, and the interval over which transmission is possible has onset equal to the later of the two interval onsets and offset equal to the offset of the second interval. For each of the constituent intervals $[u_p, v_p]$ in E_1 and $[t_q, w_q]$ in E_2, we need to account for potential transmission for all pairs of intervals making up E_1 and E_2, hence the union over both p and q.

It is easy to see that $E_1 \otimes E_2$ is an element of $\mathbb{E}$. It is the set of intervals over which potential diffusion paths may exist for the relational events represented by the intervals in E_1 that precede those in E_2.

We define a left identity for composition as the set:

$$\epsilon = \{[0, 0]\}.$$

It follows from the definition that $\epsilon \otimes E = E$ holds for all temporal interval sets $E \in \mathbb{E}$.

A natural addition operation, $\oplus$, can be defined on $\mathbb{E}$. For any two intervals $E_1, E_2 \in \mathbb{E}$,

$$E_1 \oplus E_2 = E_1 \cup E_2,$$

is the pairwise union of the intervals in E_1 and E_2,

$$E_1 \oplus E_2 = \left(\bigcup_{i \in \{1,2,\cdots,r\}} [u_i, v_i] \right) \bigcup \left(\bigcup_{j \in \{1,2,\cdots,q\}} [t_j, w_j] \right) \tag{1}$$

$$= \bigcup_{i,j} \left([u_i, v_i] \bigcup [t_j, w_j] \right). \tag{2}$$

It is clear that $\oplus$ is associative. The identity element for $\oplus$ is easily seen to be

$$\iota = \emptyset,$$

since for any $E \in \mathbb{E}$ we have

$$E \oplus \iota = \iota \oplus E = E \cup \emptyset = E. \tag{3}$$

It is also immediate that $\oplus$ is commutative and, additionally, idempotent, since $E \cup E = E$ for every $E \in \mathbb{E}$.

For any two interval sets $E_1 = \bigcup_{i \in \{1..p\}} [u_i, v_i]$ and $E_2 = \bigcup_{j \in \{1..q\}} [t_j, w_j]$, in $\mathbb{E}$, we define a natural order, $E_1 \leq E_2$, if and only if, for each $i \in \{1..p\}$, there is some $j \in \{1..q\}$, such that

$$t_j \leq u_i \leq v_i \leq w_j.$$

This order is a canonical order on $\mathbb{E}$ with respect to $\oplus$. That is, for all $E_1, E_2 \in \mathbb{E}$, if $E_1 \leq E_2$, we can find a $E_3 \in \mathbb{E}$ such that

$$E_2 = E_1 \oplus E_3,$$

and

$$E_1 \leq E_2 \text{ and } E_2 \leq E_1 \Rightarrow E_1 = E_2.$$

It is easy to show that choosing $E_3 = E_2$ satisfies the relation $E_2 = E_1 \oplus E_3$ for all $E_1, E_2 \in \mathbb{E}$, such that $E_1 \leq E_2$.

In order to guarantee that the interval sets in $E_1 \oplus E_2$ and $E_1 \otimes E_2$ are in canonical form, we require that any two overlapping intervals, $[a, b]$ and $[d, e]$, are merged into the single interval $[\min(a, d), \max(b, e)]$.

3.2 Constructing Time-Ordered Walks

The operations $\otimes$ and $\oplus$ give rise to algebraic structures that are useful for constructing time-ordered paths and walks in dynamic networks. It can easily be verified that $(\mathbb{E}, \otimes)$ is a monoid and that $(\mathbb{E}, \oplus, \otimes)$ is a left dioid. As a result, we may use the algebra of dioids (Gondran and Minoux 2008) to determine time-ordered paths in the network and compute appropriate path-based measures. Just as walks in static networks are readily determined from the Boolean products of the associated adjacency matrices, as we saw in Sect. 2.1, we can similarly construct time-ordered walks using the relational composition operations defined for dynamic networks. We begin by defining appropriate matrix operations that consider time sequence as well as node adjacency.

Consider a network with a fixed set of nodes N and a set of directed edges, $\mathcal{E}$, of ordered pairs from N. Each edge in the network is now regarded as dynamic, and the value of the tie from actor i to actor j is a finite set of n_{ij} strictly disjoint time intervals,

$$A_{ij} = \{[u_1, v_1], [u_2, v_2], ..., [u_{n_{ij}}, v_{n_{ij}}]\},$$

lying within some designated time period of interest, $[0, \omega]$. If there are no interactions between i and j, we define $n_{i,j} = 0$ and write $A_{i,j} = \emptyset$.

The *generalized adjacency matrix* (Minoux 1976; Gondran and Minoux 2008) of the dynamic network A is defined as $A = \{A_{ij} | i, j \in N\}$. Each entry in A is the set of all direct interactions between each pair of nodes. A is symmetric if the edges are bi-directional, i.e., for cases where the direction of flow goes both ways, e.g., information exchanged during a meeting between i and j. When the edges specify directional flow, e.g., i emails j, A is asymmetric.

Clearly, each entry, A_{ij}, in the generalized adjacency matrix is an element of $\mathbb{E}$. We may therefore use the interval set operations, $\oplus, \otimes$ presented above to define appropriate matrix operations for constructing paths and walks in dynamic networks. Thus the set of intervals over which time-ordered walks of length 2 from node i to node j is given by

$$(A \otimes A)_{ij} = \bigcup_{k \in N} (A_{ik} \otimes A_{kj}). \tag{4}$$

The resulting expression represents the set of intervals over which potential transmissions from actor i to actor j may occur over indirect interactions through all potential intermediary actors $k \in N$. An m-step time-ordered walk from i to j is similarly constructed from a sequence of concatenated relational intervals, each linking a pair of nodes over a finite time interval, such that it is possible for i to transmit to j through the $m - 1$ intervening nodes in the walk. Let $A \otimes A$ be denoted by A^2; then the m-step time-ordered walks in network A are given by the matrix A^m. A^m_{ij} is the interval set over which transmissions from node i to node j via $(m - 1)$ intermediary nodes are possible. As in static networks, we may compose different interaction types, e.g., A emails B who subsequently meets with C, enabling potential transmission from A to C via B over a particular time interval $[u, v]$.

The set $\mathbb{E}$, with the $\oplus$ and $\otimes$ operations, yields an accurate representation for potential flows of entities that endure in time, e.g., persistent infectious diseases (Moody 2002). It may be less accurate when we consider flows of substances that might decay over time, such as hot gossip or the infectiousness of a virus. In the next section, we detail a more general form of composition that allows the inclusion of a possible time lag or decay factor, which reflects the need articulated by Borgatti (2005) for a variety of different approaches for representing flows in social networks.

3.3 More General Compositions for Time-Ordered Walks

Interaction events are bounded by time: they have a start and end point, and sometimes they are simply instantaneous. Sequences of events determine potential trajectories through which information, resources, or viruses can be transferred from actor to actor. The formulation of time-ordered walks defined so far implicitly assumes that transmission through a network only requires temporal precedence: the interaction between i and k needs to occur before the one between k and j for something to reach j from i via k. Here, we impose a time constraint to represent a decay period, δ, after which the substance being transferred loses its potency. So two interactions may be considered a *time-ordered two-path* if they share a node and if the time elapsed between the first and second interactions is less than δ.

We formalize this by introducing an "extension operator," by which each interval that is to be composed with a subsequent one is extended by a predefined amount, δ. If we let the observation period be $[0, \omega]$, then we choose an appropriate $\delta \in [0, \omega]$ for the substance being transferred. We define an operator, $E + \delta$, acting on an element of the base set $\mathbb{E}$, that translates all of the offsets of E by δ. Let $E = \cup_{i\in\{1,2,\cdots,r\}}[\mu_i, \nu_i]$, and then for $\delta \in [0, \omega]$

$$E + \delta = \cup_{i\in\{1,2,\cdots,r\}}[\mu_i, \nu_i + \delta].$$

When the values of δ are small, only events that are close together in time are temporally dependent, while high values of delta imply that more time can elapse between observations. For example, Quintane et al. (2021) chose a δ of 8 hours, which corresponds to the length of a normal workday, to analyze email exchanges internal to an organization. In contrast, for tracking the spread of a virus, such as COVID-19, we might chose a δ of two weeks, during which an infected individual might pass on the virus to close contacts.

Kontoleon et al. (2013) call this a translation operation, noting that it is one-sided, in the sense that it only translates one end of the interval. It effectively stretches the interval. The asymmetric nature of the operator means that relational composition involving an interval modified by δ cannot be made associative. This will require the consideration of more complex algebraic structures. In particular, dioids of endomorphisms (Gondran and Minoux 2008) allow more flexible operations and enable us to solve path problems in dynamic networks.

3.4 Algebra of Endomorphisms

In this section, we define endomorphisms and, following Kontoleon et al. (2013), show how they may be used in the construction of a very general class of time-ordered walks.

Definition 6 An *endomorphism* of a monoid $(\mathbb{S}, \oplus)$ is a mapping $\phi : \mathbb{S} \to \mathbb{S}$, satisfying, $\forall s, t \in \mathbb{S}$, and ϵ_0, the identity element for $\oplus$ in $\mathbb{S}$,

$$\phi(s \oplus t) = \phi(s) \oplus \phi(t)$$

$$\phi(\epsilon_0) = \epsilon_0.$$

Let $(\mathbb{E}, \oplus)$ be a commutative canonically ordered monoid with identity element ι, and $\mathbb{H}$ the set of endomorphisms over $\mathbb{E}$. Recall that $\iota = \emptyset$ is the identity element for $\oplus$, and it obeys equation (3). Then for any $a \in \mathbb{E}$ and any $h \in \mathbb{H}$, $h(a) \in \mathbb{E}$, and for all $h \in \mathbb{H}$, we have

$$h(a \oplus b) = h(a) \oplus h(b) \quad \forall a, b \in \mathbb{E} \tag{5}$$

$$h(\iota) = \iota. \tag{6}$$

Define the operators $\oplus$ and $*$ on $\mathbb{H}$ to be

$$(h \oplus g)(a) = h(a) \oplus g(a) \qquad \forall a \in \mathbb{E} \text{ and } h, g \in \mathbb{H} \tag{7}$$

$$(h * g)(a) = g \circ h(a) = g(h(a)) \qquad \forall a \in \mathbb{E} \text{ and } h, g \in \mathbb{H}, \tag{8}$$

and the identity endomorphism on $(\mathbb{H}, \oplus)$ to be $h^{\iota}(a) = \iota$, for all $a \in \mathbb{E}$.

The operation $*$ is recognizable as functional composition, and it is clearly associative. Furthermore, $*$ is left distributive with respect to $\oplus$, and the absorbing element over $*$ is h^ι (Gondran and Minoux 2008). The structure $(\mathbb{H}, \oplus, *)$ is a semiring.

Endomorphisms enable us to construct walks in dynamic networks that respect time-ordering constraints and allow us to use more general composition rules, particularly for problems that do not fit the stricter requirements of classical path-based algebras (Minoux 1976).

Starting from the general space of endomorphisms over $\mathbb{E}$ (Kontoleon et al. 2013), we take a subset of those that allow us to formulate paths and walks in terms of weighted edges between two nodes. As we saw earlier, the edge weights are given by each entry of the generalized adjacency matrix A_{ij}.

Let $\mathbb{H}$ denote a subspace of the space of all endomorphisms over the monoid $(\mathbb{E}, \oplus)$. We define a particular endomorphism, denoted by $h_S^\delta(E) \in \mathbb{H}$ with $S \in \mathbb{E}$, acting on an element $E \in \mathbb{E}$ as

$$h_S^\delta(E) = (E + \delta) \cap S. \tag{9}$$

So $h_S^\delta(E)$ effectively composes two elements in $\mathbb{E}$, while extending the offsets of the first set of intervals by δ. Thus if

$$E = \cup_{i \in \{1,2,\cdots,r\}}[u_i, v_i] \in \mathbb{E},$$

and

$$S = \cup_{i \in \{1,2,\cdots,q\}}[x_i, y_i] \in \mathbb{E},$$

then

$$\begin{aligned} h_S^\delta(E) &= (E + \delta) \cap S \\ &= \left(\cup_{i \in \{1,2,\cdots,r\}}[u_i, v_i + \delta]\right) \cap \left(\cup_{j \in \{1,2,\cdots,q\}}[x_j, y_j]\right) \\ &= \cup_{i,j}\left([u_i, v_i + \delta] \cap [x_j, y_j]\right). \end{aligned}$$

We observe here that the composition rule given earlier, which was based on Moody (2002), is a particular case of this more general expression. In fact, if we let $\delta = \omega$,

$$\begin{aligned} h_S^\omega(E) &= \cup_{i,j}\left([u_i, v_i + \omega] \cap [x_j, y_j]\right) \\ &= \cup_{i,j}\left([u_i, \omega] \cap [x_j, y_j]\right) \\ &= \cup_{i,j}\left([\max(u_i, x_j), y_j]\right) \\ &= E \otimes S. \end{aligned}$$

Definition 7 The identity element for the composition endomorphism in Equation (9) is $h_\epsilon^0(E)$, since

$$\begin{aligned} h_\epsilon^0(E) &= \cup_{i\in\{1,2,\cdots,r\}}([u_i, v_i] \cap [0,\omega]) \\ &= E, \end{aligned}$$

and the absorbing element is defined as

$$h_\iota(E) = \iota.$$

Given Equation (9), we have, for $S_i, i = 1, \cdots n, \in \mathbb{E}$,

$$\begin{aligned} &h_{S_n}^\delta(h_{S_{n-1}}^\delta(\cdots h_{S_1}^\delta(E)\cdots))) \\ &= h_{S_1}^\delta * h_{S_2}^\delta * \cdots * h_{S_{n-1}}^\delta * h_{S_n}^\delta(E) \\ &= (E + n\delta) \cap (S_1 + (n-1)\delta) \cap (S_2 + (n-2)\delta) \cap \cdots \\ &\qquad \cap (S_2 + \delta) \cap S_n. \end{aligned} \tag{10}$$

The $\oplus$ operator over the monoid $\mathbb{E}$ induces an operation $\oplus$ over the space of endomorphisms since it is not difficult to show that, for $A, C, D \in \mathbb{E}$,

$$h_A^\delta(C) \oplus h_A^\delta(D) = h_A^\delta(C \oplus D). \tag{11}$$

Therefore, the space $(\mathbb{E}, \mathbb{H}, \oplus, *)$ forms a dioid of endomorphisms (Gondran and Minoux 2008; Minoux 1976).

Minoux (1976) likens these endomorphisms to one-step transitions that can be used to generate all the walks originating at some node to some target node. Using the operations and structures defined above, we may formulate paths and walks in terms of the entries of the generalized adjacency matrix A_{ij} that represent the interval or a set of intervals over which direct interactions occur.

We denote by E_{ij} the union of disjoint intervals over which information from i can reach j via a direct interaction or an indirect interaction via some other intermediary nodes. In other words, the set of intervals in E_{ij} is the weight of all the walks from i to j. Substituting E_{ij} and A_{jk} in Equation (9), the one-step endomorphism, $h_{A_{jk}}^\delta$, that takes us from i to k is given by

$$h_{A_{jk}}^\delta(E_{ij}) = (E_{ij} + \delta) \cap A_{jk}, \tag{12}$$

$$\forall i, j, k \in \{1, 2, \cdots, n\},$$

which we re-write as

$$h_{jk}(E_{ij}) \equiv h^{\delta}_{A_{jk}}(E_{ij}),$$

noting that E_{ij} is acted on by A_{jk}.

The matrix of one-step endomorphisms, H_A, is the generator of the subspace of endomorphisms that represent all of the possible walks in a dynamic network that has generalized adjacency matrix A.

$$H_A = \begin{bmatrix} h_{11} & h_{12} & \cdots & h_{1n} \\ h_{21} & h_{22} & \cdots & h_{2n} \\ \vdots & \vdots & \ddots & \vdots \\ h_{n1} & h_{n2} & \cdots & h_{nn} \end{bmatrix}. \tag{13}$$

Note that if $A_{ij} = \emptyset$, then $h_{ij} = h_{\iota}$, the absorbing element (Definition 7). From this, we derive the matrix H^k_A. Each element of this matrix gives us the k-step walks. Further, we define

$$H^{(k)} = \sum_{i=0}^{k} H^i, \tag{14}$$

where

$$H^0 = \begin{bmatrix} h^0_{\epsilon} & h_{\iota} & \cdots & h_{\iota} \\ h_{\iota} & h^0_{\epsilon} & h_{\iota} & \cdots \\ \vdots & \vdots & \ddots & \vdots \\ h_{\iota} & \cdots & h_{\iota} & h^0_{\epsilon} \end{bmatrix}.$$

The element $h^{(k)}_{ij}(E_{ii})$ where $E_{ii} = [0, \omega]$ gives us the union of intervals that are generated from the set of all walks of length $\leq k$ that commence in i and end in j.

The quasi-inverse of H is denoted by H^* (Gondran and Minoux 2008; Minoux 1976) and is given by

$$H^* = \lim_{k \to \infty} H^{(k)},$$

when this limit exists, i.e., $H^* = H^{(K)}$ for some finite K.

Convergence of H^* is necessary to compute network measures that are based on time-ordered walks. (Kontoleon et al. 2013) show that the walks on finite dynamic networks with the operations defined above may include cycles that have non-zero weight, which raises the possibility of infinite-length walks (Gondran and Minoux 2008). However, they prove that in the process of computing H^* each cycle can only be traversed a finite number of times, $p \geq 0$. Moreover, as the length of the cycles increases, p decreases. Therefore, H^*, and hence all the walks in the network, can be computed in a finite number of steps.

4 Conceiving of Centrality Measures in Temporal Social Networks

A basic tenet of social network theory is that locations within networks affect actor outcomes (Robins 2015; Borgatti et al. 2009). On the premise that this principle can be used to guide the development of equivalent measures, Falzon et al. (2018) explored the notion of network positions in temporal networks. They discuss whether it is feasible to aggregate social interactions over a time window to approximate static relations. As Moody et al. (2005) argued, actors in these networks are connected through streams of relational events or interactions rather than static relations. Aggregating social interactions over longer time frames makes them more similar to social relations, but it reduces the temporal information on event durations and temporal patterns, and it ignores the sequential nature of interactions. Sequences of social interactions that unfold over time enable a finer-grained investigation of chains of events that constitute social processes. However, the measures that have been developed for capturing social positions in social networks often remove timing and sequence information from interaction data. For the purposes of studying actor roles in promoting or inhibiting connectedness and controlling flow, static network concepts need rethinking (Spiro et al. 2013).

In recent years, researchers have developed various temporal network measures that distinguish between two aspects of time: the actual time of the events and the sequence of the events (see review by Holme and Saramäki 2012). In the first case, the focus is on measuring actor activity, e.g., the number of interactions on specific days of the week or hours of the day (Batagelj and Praprotnik 2016; Perra et al. 2012), or to study the rate of relation formation and dissolution (Moody et al. 2005), such as the time needed for an interaction to be reciprocated. In the second case, the focus of interest is the sequence in which social interactions unfold (Broccatelli et al. 2016; Kovanen et al. 2011) rather than the time of occurrence. In particular, the existence of paths formed by a sequence of interconnecting interactions.

Research into dynamic networks is gaining prominence across a broad range of fields. Determining influential nodes, disjoint, or overlapping communities, and other indicators of influence at the node, neighborhood, or network level while taking both time and sequence into consideration requires newly defined measures or reinterpretation of existing measures to provide insight into the systems under study (Falzon et al. 2018). Measures such as degree, brokerage, reachability, and betweenness are well studied in the static case but acquire a different significance when time and sequence are available. For example, whereas in static networks the in-degree of an actor is equated with higher prestige, in dynamic networks, it highlights the node as a receiver of information and potentially as a broker or conduit through which messages are more likely to pass *at a particular point in time*. Similarly, reachability, both in terms of nodes reached from, or nodes reaching a node, uses sequence to confirm where information can travel, and time to indicate how quickly nodes can receive information. Mantzaris et al. (2013) measure dynamic network communicability (Grindrod et al. 2011) to investigate flows of information through different parts of the brain. They showed that the measure was valuable, although not necessarily scalable. Other positional measures, for example, the various types of centrality and structural equivalence, give us an understanding of the role of specific individuals in the diffusion of information in a network. Although they have been primarily developed in the context of static relational networks, Keating (2012) shows that they can also be defined and calculated on the time-ordered reachability matrix, which represents a network of static ties derived from a dynamic network.

Taking into account the sequence of social interactions implies a definition of temporal path typically conceptualized using a notion of temporal geodesics (i.e., the shortest time-ordered path between two nodes). Moody (2002) observes that centrality measures appropriate for networks of time-sequenced interactions are based on the number and length of shortest time-ordered paths. Holme and Saramäki (2012) distinguish between two different types of temporal geodesics: the fewest number of interactions and the shortest path duration. Other authors (Grindrod and Higham 2013; Kim and Anderson 2012; Nicosia et al. 2012; Nicosia et al. 2013; Tang et al. 2010) consider a time-ordered series of network snapshots at closely spaced time points that span the whole observation interval. From these, they are able to construct temporal paths and walks that are time respecting, i.e., that do not violate the time ordering of connections. However, in this case, interactions that occur in between time points are treated as having taken place concurrently, i.e., sequence is ignored at this finer level. This leads to a discontinuity in the flow, introduced through artificial time slicing.

The algebra of dynamic networks provides enough generality so that the two temporal aspects of social interactions (time and sequence) are considered simultaneously, whereas they are typically used separately in existing temporal network investigations. Combining sequence and time together enables us to specify—based on the context—a maximum amount of time (i.e., δ as described above) to consider two social interactions as part of the same sequence (Quintane et al. 2021). For example, an email from B to A sent 30 minutes after an email from A to B can reasonably be considered as part of an AB–BA sequence and reflects reciprocation in an information exchange context. However, if the same email from B to A had been sent one week after the email from A to B (or in a more extreme case, six months after), it is not clear that it should still be considered as part of the same sequence rather than part of a new sequence.

4.1 Path-Based Measures: Temporal Implications

Temporal reachability (Falzon et al. 2018) measures the extent to which an actor can reach other nodes in a network in a certain number of steps. An actor i can reach another, j, if they are connected by at least one time-ordered path of a certain number of steps k (each step being an interaction) that originates at i and ends at j. In this case, we can say that j is reachable from i in k steps or that i can reach j via a path of length k. Thus, temporal reachability provides information about the temporal distance and temporal cohesion of an actor with the rest of the network, enabling us to trace information flows as well as their timing through the network. Temporal reachability does not depend on a specific unit of time (excepting δ), but our network analysis framework allows us to determine the duration of each k-step path connecting each pair of nodes.

Temporal betweenness (Falzon et al. 2018) measures an actor's capacity to control network flow. The measure is based on the notion of a shortest temporal path, which is the time-ordered path between two nodes that minimizes the number of sequenced interactions in the path. The only paths considered feasible are time-respecting paths, i.e., ones in which each interaction in the path sequence is also ordered in time and such that the time between two sequential interactions is less than a pre-specified threshold (i.e., δ). The temporal betweenness of an actor i is the proportion of shortest paths that go through i. We could also consider flow over the time-ordered paths that have the shortest duration as well as, or instead of, the smallest number of interconnecting interactions. These are paths that minimize the time between the first interaction in the path and the last one. Temporal betweenness calculated on the basis of paths with the shortest duration also provides information about the extent to which an actor controls the speed of information flows in a network.

Temporal Katz centrality (Grindrod and Higham 2013) is a class of dynamic centrality measures inspired by the Katz (1953) status measure for static networks. The measure is defined in terms of a combination of weighted walks, such that longer walks contribute less to the measure than shorter ones. All these measures can be modified by the determination of weighting factors, which may depend on the type of data being analyzed and the research question. Analogous to the static version of this measure actors with high Katz centrality are in a privileged position in terms of their ability to communicate with the rest of the network. An alternative version of this measure can be devised by using the duration rather than length of each walk as a weighting.

These measures describe the structural position of nodes and allow us to rank their importance in terms of their centrality, just like their counterparts in networks of static relations. They also allow us to glean some extra information such as the duration of transmissions from source to target as well as a timeline of individual activity, and how an actor's network ranking changes over time. The next measure to consider is temporal brokerage, which requires us to consider patterns of flow as well as structural positioning.

4.2 Brokerage: A Process View

Temporal order has important implications for communication and social processes, which can be conceptualized as patterns of flow and interaction sequences. One such interactive process, which details dynamic behavior, is temporal brokering (Quintane et al. 2021). Brokers' positions in social networks have been widely studied (Burt 2005, 2010; Burt et al. 2013; Everett and Borgatti 2020). In traditional social network analysis, brokerage characterizes the extent to which an individual's network of social relations spans structural holes. A structural hole exists in a network when actors A and C do not have a relationship with each other but are both directly connected to actor B (the broker). Individuals whose position in a social network spans multiple structural holes secure benefits from early access to, and control over, diverse information (Burt 2005). In contrast, temporal brokerage, as defined in Quintane et al. (2021), is not simply a network position but a social process that occurs over time, and in which opportunity is enabled as network interactions unfold sequentially among the actors A, B, and C (Gould and Fernandez 1989; Spiro et al. 2013), with B as the intermediary.

A key strength of empirical investigations on brokerage is the ability to measure and rank the brokerage positions in any network by considering the relative positions of triplets of actors. Quintane et al. (2021) developed a temporal version of this measure by studying the extent to which an individual intermediates temporal structural holes. Building on Burt's concept of measurement for brokerage as the "proportion of relationships enhanced by structural holes" (Burt 1992, p. 37), and considering Burt (2010)'s notion of constraint, a measure of how much an individual is constrained by the structure of the surrounding network, a temporal version can be defined similarly.

As we saw in Sect. 3.3, a *temporal two-path* is a pair of sequentially ordered interactions among three actors (A→B→C or C→B→A) in which the middle node is common to both interactions and the second interaction occurs within a predefined period, δ. A temporal two-path can be either open or closed, which is determined by the absence or presence (respectively) of a third interaction between A and C occurring within δ of the second interaction. A *temporal structural hole* around B exists, while the temporal two-path is open.

Definition 8 Let $O_{(2,B)}$ be the number of open temporal two-paths intermediated by B, and let $C_{(2,B)}$ the number of closed temporal two-paths intermediated by B. Then the measure of *temporal brokering* for actor B is defined as

$$\frac{O_{(2,B)} - C_{(2,B)}}{O_{(2,B)} + C_{(2,B)}}.$$

This concept of temporal brokering is a path-based measure as discussed in Sect. 4.1. The sequential nature of temporal paths allows us to distinguish between two types of different brokering processes: arbitration and collaboration. In the

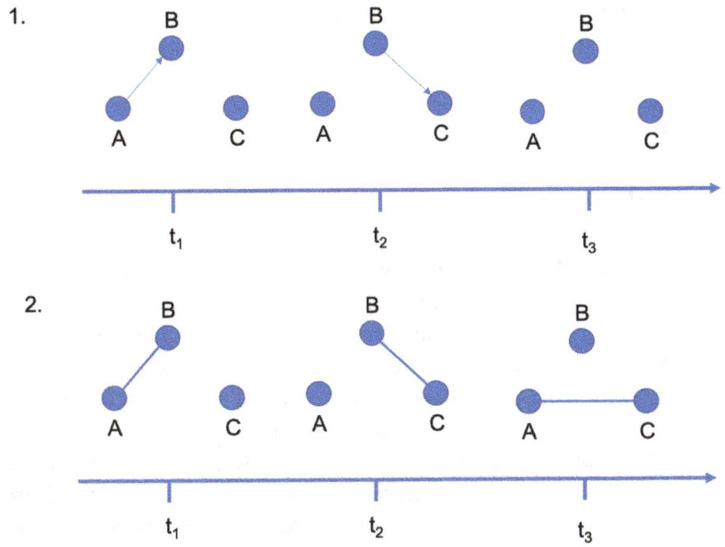

Fig. 1 The sequence of interactions describing two types of brokering (Modified from Quintane et al. 2021, Fig. 3)

former process, a broker's unconnected neighbors stay disconnected over the whole observation period, while in the latter they establish a direct connection over time.

Figure 1 shows an example of the sequence of interactions that constitute arbitration (1) and collaboration (2), further elaborated below:

1. The broker controls the flow: They bridge many open (time-ordered) 2-paths centered at B, A→B at time t_1 followed by B→C at time t_2 (t_2 is within $t_1 + \delta$).
2. The broker facilitates new connections: 2-paths close over time.
   A→B at time t_1 followed by B→C at time t_2 followed by A→C or C→A at time t_3 (t_2 is within $t_1 + \delta$ andt_3 is within $t_2 + \delta$).

We can apply this reasoning to sequences of time-stamped relational events to determine the existence and the duration of a temporal structural hole and the tendency toward an arbitration or collaboration brokering process. The temporal brokerage measure gives values on a continuum from -1 (all temporal two-paths are closed), corresponding to collaboration behavior, to 1 (all temporal two-paths are open), corresponding to arbitration (Quintane et al. 2021).

This definition captures the repetition of a specific sequence of events occurring within a given time frame—it reflects recurrent behavior and characterizes the way in which the actor behaves in a brokering process. We can also study the time it takes to form a two-path and the time it takes to close it (forming a triad), which is a useful measure for social processes. Other recurrent activity patterns may be similarly defined using the temporal network algebra described in Sect. 3, which provides the formal machinery for defining time-ordered walks and paths. In particular, it permits a useful generalization by allowing the specification of a predefined maximum lapsed time between interactions in order for them to be considered as constituents of a process.

5 Discussion

What we have presented here is one way of characterizing and analyzing networks of dynamic interactions. A different approach is adopted by Lehmann (2019), who presents six classes of communication practices and their dynamic network representations. The networks are classified by the synchronous/asynchronous nature of the communication and their fundamental network topologies, which take the form of single dyads, star structures, and cliques. The six fundamental structures are the quasi-instantaneous "topological-temporal" network structures that represent the communication patterns that occur over a very brief time slice. Lehmann (2019)'s approach considers sequences of network substructures that appear and disappear as communication events transpire in the network—the fundamental objects of analysis are the communication events. In contrast, the temporal algebra of Kontoleon et al. (2013) is concerned with the representation of the trajectory of information or other resource flowing through the network, enabling us to model the fluid nature of information flow in a continuous stream rather than the discontinuities inherent in discrete snapshots, even if infinitesimally close together in time. Two distinct forms of time are considered in these two approaches: discrete and continuous. Discrete time is used when network change is studied at discrete points in time. The outcomes from the two approaches yield very different, and complementary, insights into network dynamics. For example, as we have seen in Sect. 4, the ability to determine temporal network paths results in very fine-grained measures to study the reachability, betweenness, and brokerage capacity of individual nodes. It also allows us to detect relational processes composed of sequences of interactions over subsets of nodes (Moody et al. 2005). However, so far, our analysis has been restricted to those processes that are known from other empirical studies (like the brokerage example we presented). The fundamental structures proposed by Lehmann (2019) have been determined from actual communications data of various kinds (e.g., phone calls, online chat rooms, broadcasts, etc.) in which they varied the temporal resolution of the time slice width. In so doing, they were able to identify community structures and temporal clusters. These findings suggest ways of defining formal representations of the temporal equivalent of network components such as cohesive subgroups (Frank 1996).

A conceptually different approach to the algebraic machinery developed in Kontoleon et al. (2013) is the temporal extension to graph theory based on "stream graphs" and "link streams" provided by Latapy et al. (2018). The authors propose a formal mathematical framework in which interactions are modeled as streams of nodes and links that are active during defined periods of time. They do this by developing a generalization of graph theory in which they re-define graph-theoretic elements to simultaneously handle both the dynamics and structure of interactions. They model interactions over time as either link streams, in which nodes are present all the time but links are dynamic; or as stream graphs, in which nodes are also dynamic. The result is a formalism from the static to the dynamic in which it is possible to define graph concepts such as paths, clusters, density, degrees, cliques, etc., consistent with graph theory but in terms of streams.

An important contribution of Latapy et al. (2018)'s work is the ability to define new temporal measures that are particular to streams of nodes and links. Some useful examples include the following:

- The coverage of a stream graph is defined as the total time that all the network nodes are active as a proportion of the maximum possible time that they could be active; a coverage of 1 means that the nodes are active all the time.
- The contribution of a node to a stream graph is the proportion of time it is active.
- The number of nodes in a stream graph is the sum of contributions of all its nodes (note the temporal quality inherent in this measure).
- Similarly, the number of links in a stream graph is the sum of contributions of all its links (the proportion of time each link is active).

Similar definitions are provided for the node and link contributions of a time instant and for node and link durations. Measures that describe attributes of stream graphs and link streams, including a temporal version of density, are also derived. The relation of a subgraph to a graph in graph theory is used to define an equivalent notion of a substream of a stream. Similarly, clusters in stream graphs and link streams are equivalent to clusters in graphs (i.e., as a subset of the set of nodes in the graph). Building on from these concepts, the neighborhood of a node v is the cluster of temporal nodes that form temporal edges with v; and the degree of a temporal node is the number of nodes in this cluster. From the definition of a path in a stream graph, the authors derive expressions for subpaths; path duration and path distance; shortest paths; fastest paths and foremost paths, along with expressions for various path-based measures such as closeness, betweenness (Simard et al. 2021), reachability, path cost, etc. The outcome of these formal definitions is a rich framework of network measures such as clustering coefficient, transitivity ratio, diameter, and connectedness.

To conclude, novel approaches for the formal representation of temporal networks along with expressions for useful measures and characterizations are emerging, largely from the fields of communications networks and computer science. Time-respecting network paths are intrinsic to flow, connectivity, contagion, as well as more abstract notions of spanning trees and cuts. These concepts are essential for modeling social processes. In order to make the best possible use of these new techniques, we need a better theoretical framework that allows us to model and understand relational processes over social networks.

Appendix

The following definitions are mathematical terms that appear in the text but are not defined elsewhere.

Binary relation A *binary relation* R on a set S is a subset of $S \times S$.

Composition Let $\mathbb{R}$ be a set of binary relations on S. Any two relations $A, B \in \mathbb{R}$ may be *composed* to form a new relation, AB, on S. Each ordered pair $(i, k) \in AB$ implies that there exists a $m \in S$, such that $(i, m) \in A$ and $(m, k) \in B$.

Order A binary relation $\preceq$ on a set S is an *order* if it has the properties:

1. *Reflexivity*: $a \preceq a$ for all $a \in S$.
2. *Transitivity*: For any $a, b, c \in S$, if $a \preceq b$ and $b \preceq c$, then $a \preceq c$.
3. *Anti-symmetry*: For any $a, b \in S$, if $a \preceq b$ and $b \preceq a$, then $a = b$.

Semigroup A *semigroup* $(S, \oplus)$ is a set S endowed with a binary operation $\oplus$, satisfying:

1. *Closure*: For all $a, b \in S, a \oplus b \in S$.
2. *Associativity*: For all $a, b, c \in S$, $(a \oplus b) \oplus c = a \oplus (b \oplus c)$.

Identity The *identity* (or neutral) element, $e \in S$, for a semigroup $(S, \oplus)$ satisfies $a \oplus e = e \oplus a = a, \forall a \in S$.

Absorption An element (zero element), $\varepsilon \in (S, \oplus)$, is said to be *absorbing* if it satisfies: $a \oplus \varepsilon = \varepsilon \oplus a = \varepsilon, \forall a \in S$.

Monoid A semigroup $(S, \oplus)$, with an identity element, e, is called a *monoid*.

Left identity The *left identity* element, $e \in S$, for an operation, $\otimes$ on S, satisfies $e \times a = a, \forall a \in S$.

Right identity The *right identity* element, $e \in S$, for an operation, $\otimes$ on S, satisfies $a \times e = a, \forall a \in S$.

Left distributive For operations, $\oplus$, $\otimes$ on S, we say that $\otimes$ is left distributive over $\oplus$ if, for $a, b, c \in S$: $a \otimes (b \oplus c) = (a \otimes b) \oplus (a \otimes c), \forall a \in S$.

Right distributive For operations, $\oplus$, $\otimes$ on S, we say that $\otimes$ is right distributive over $\oplus$ if, for $a, b, c \in S$: $(a \oplus b) \otimes c = (a \otimes c) \oplus (a \otimes c), \forall a \in S$.

References

Baras, J. and Theodorakopoulos, G. (2010), Path problems in networks, *in* 'Synthesis Lectures on Communication Networks', Vol. 3, Morgan and Claypool, pp. 1–77.

Batagelj, V. (1994), 'Semirings for social networks analysis', *The Journal of Mathematical Sociology* **19**(1), 53–68.

Batagelj, V. and Praprotnik, S. (2016), 'An algebraic approach to temporal network analysis based on temporal quantities', *Social Network Analysis and Mining* **6**(1), 1–22.

Borgatti, S. P. (2005), 'Centrality and network flow', *Social Networks* **27**, 55–71.

Borgatti, S. P. and Everett, M. G. (2006), 'A graph-theoretic perspective on centrality', *Social Networks* **28**(4), 466–484. https://www.sciencedirect.com/science/article/pii/S0378873305000833

Borgatti, S. P., Everett, M. G. and Johnson, J. C. (2013), *Analyzing Social Networks*, Sage, Thousand Oaks, CA.

Borgatti, S. P., Mehra, A., Brass, D. J. and Labianca, G. (2009), 'Network analysis in the social sciences', *Science* **323**(5916), 892–895.

Boyd, J. P. (1991), *Social semigroups: A unified theory of scaling and blockmodelling as applied to social networks*, George Mason University Press, Fairfax, VA.

Broccatelli, C., Everett, M. and Koskinen, J. (2016), 'Temporal dynamics in covert networks', *Methodological Innovations* **9**, 1–14.

Burt, R. S. (1992), *Structural holes: The social structure of competition*, Harvard University Press.

Burt, R. S. (2005), *Brokerage and closure: An introduction to social capital.*, Oxford University.

Burt, R. S. (2010), *Neighbor networks: Competitive advantage local and personal.*, Oxford University Press.

Burt, R. S., Kilduff, M. and Tasselli, S. (2013), 'Social network analysis: Foundations and frontiers on advantage', *Annual Review of Psychology* **64**(1), 527–547. PMID: 23282056.

Butts, C. (2008*a*), 'A relational event framework for social action', *Sociological Methodology* **38**(1), 155–200.

Butts, C. (2008*b*), 'Social network analysis: A methodological introduction', *Asian Journal of Social Psychology* **11**(1), 13–41. https://onlinelibrary.wiley.com/doi/abs/10.1111/j.1467-839X.2007.00241.x

Carrè, B. (1971), 'An algèbra for network routing problems', *J.Inst. Maths Applics* **7**, 273–294.

de Nooy, W. (2015), 'Structure from interaction events', *Big Data & Society* **2**(2), 1–4.

Degenne, A. and Forsé, M. (1999), *Introducing social networks*, SAGE., London.

Everett, M. G. and Borgatti, S. P. (2020), 'Unpacking Burt's constraint measure.', *Social Networks* **62**, 50–57.

Falzon, L., McCurrie, C. and Dunn, J. (2017), Representation and analysis of Twitter activity: A dynamic network perspective, *in* 'Proceedings of the 2017 IEEE/ACM International Conference on Advances in Social Networks Analysis and Mining 2017', ACM, pp. 1183–1190.

Falzon, L., Quintane, E., Dunn, J. and Robins, G. (2018), 'Embedding time in positions: Temporal measures of centrality for social network analysis', *Social Networks* **54**, 168–178.

Frank, K. A. (1996), 'Mapping interactions within and between cohesive subgroups', *Social Networks* **18**(2), 93–119. https://www.sciencedirect.com/science/article/pii/037887339500257X

Gondran, M. and Minoux, M. (2008), *Graphs, Dioids and Semirings: New Models and Algorithms*, Operations Research/Computer Science Interfaces, Springer, New York, NY, USA.

Gould, R. V. and Fernandez, R. M. (1989), 'Structures of mediation: A formal approach to brokerage in transaction networks', *Sociological Methodology* **19**, 89–126.

Grindrod, P. and Higham, D. J. (2013), 'A matrix iteration for dynamic network summaries', *SIAM Review* **55**(1), 118–128. http://dx.doi.org/10.1137/110855715

Grindrod, P., Parsons, M. C., Higham, D. J. and Estrada, E. (2011), 'Communicability across evolving networks.', *Physical Review. E, Statistical, Nonlinear, and Soft Matter Physics* **83**(4 Pt 2), 046120.

Hoffman, M., Block, P., Elmer, T. and Stadtfeld, C. (2020), 'A model for the dynamics of face-to-face interactions in social groups', *Network Science* **8**(S1), S4–S25.

Holme, P. and Saramäki, J. (2012), 'Temporal networks', *Physics Reports* **519**(3), 97–125.

Katz, L. (1953), 'A new status index derived from sociometric analysis', *Psychometrika* **18**(1), 39–43. http://ideas.repec.org/a/spr/psycho/v18y1953i1p39-43.html

Keating, L. (2012), A test of time: Time-aggregated networks and time-ordered networks in behavioural ecology, M.sc. in mathematical modelling and scientific computing, Linacre College, University of Oxford.

Kim, H. and Anderson, R. (2012), 'Temporal node centrality in complex networks', *Phys. Rev. E* **85**, 026107:1–8. http://link.aps.org/doi/10.1103/PhysRevE.85.026107

Kontoleon, N., Falzon, L. and Pattison, P. (2013), 'Algebraic structures for dynamic networks', *Journal of Mathematical Psychology* **57**(6), 310–319. Social Networks. http://www.sciencedirect.com/science/article/pii/S0022249613001089

Kovanen, L., Karsai, M., Kaski, K., Kertész, J. and Saramäki, J. (2011), 'Temporal motifs in time-dependent networks', *Journal of Statistical Mechanics: Theory and Experiment* (11), 11005.

Latapy, M., Viard, T. and Magnien, C. (2018), 'Stream graphs and link streams for the modeling of interactions over time', *Social Network Analysis and Mining* **8**(1), 61. https://doi.org/10.1007/s13278-018-0537-7

Lehmann, S. (2019), *Fundamental structures in network temporal communication*, Temporal Network Theory, Springer-Nature, New York, NY, USA.

Mantzaris, A. V., Bassett, D. S., Wymbs, N. F., Estrada, E., Porter, M. A., Mucha, P. J., Grafton, S. T. and Higham, D. J. (2013), 'Dynamic network centrality summarizes learning in the human brain', *Journal of Complex Networks* **1**(1), 83–92.

Minoux, M. (1976), 'Structures algébriques généralisées des problèmes de cheminement dans les graphes', *R.A.I.R.O. Recherche Opérationelle* **10**(6), 33–62.

Moody, J. (2002), 'The importance of relationship timing for diffusion', *Social Forces* **81**(1), 25–56.

Moody, J., McFarland, D. A. and Bender-deMol, S. (2005), 'Dynamic network visualization: Methods for meaning with longitudinal network movies', *American Journal of Sociology* **110**(4), 1206–1241.

Morris, M. and Kretzschmar, M. (1995), 'Concurrent partnerships and transmission dynamics in networks', *Social Networks* **17**(3), 299–318. Social networks and infectious disease: HIV/AIDS. https://www.sciencedirect.com/science/article/pii/037887339500268S

Nicosia, V., Tang, J., Mascolo, C., Musolesi, M., Russo, G. and Latora, V. (2013), *Graph Metrics for Temporal Networks*, pp. 15–40.

Nicosia, V., Tang, J., Musolesi, M., Russo, G., Mascolo, C. and Latora, V. (2012), 'Components in time-varying graphs', *Chaos* **22**(2). http://scitation.aip.org/content/aip/journal/chaos/22/2/10.1063/1.3697996

Pattison, P. (1993), *Algebraic models for social networks*, Cambridge University Press, New York.

Pattison, P. (2014), *Relation algebras and social networks*, SAGE Publications Ltd, pp. 447–458.

Pattison, P., Falzon, L. and Kontoleon, N. (2015), Algebra of social networks, *in* J. D. Wright, ed., 'International Encyclopedia of the Social & Behavioral Sciences Behavioral Sciences', 2nd edition edn, Vol. 22, Elsevier, pp. 478–484.

Perra, N., Gonçalves, B., Pastor-Satorras, R. and Vespignani, A. (2012), 'Activity driven modeling of time varying networks', *Scientific Reports* **2**(1), 469. https://doi.org/10.1038/srep00469

Quintane, E., Wood, M., Dunn, J. and Falzon, L. (2021), 'Temporal brokering: A measure of brokerage as a behavioral process', *Organizational Research Methods* . https://doi.org/10.1177/10944281211002911

Robins, G. (2015), *Doing Social Network Research: Network-based Research Design for Social Scientists*, SAGE Publications Ltd.

Robins, G. and Pattison, P. (2001), 'Random graph models for temporal processes in social networks', *The Journal of Mathematical Sociology* **25**(1), 5–41. https://doi.org/10.1080/0022250X.2001.9990243

Simard, F., Magnien, C. and Latapy, M. (2021), 'Computing betweenness centrality in link streams', *CoRR* **abs/2102.06543**. https://arxiv.org/abs/2102.06543

Snijders, T. and Koskinen, J. (2013), *Longitudinal Models*, Structural Analysis in the Social Sciences, Cambridge University Press, pp. 130–140.

Spiro, E. S., Acton, R. M. and Butts, C. T. (2013), 'Extended structures of mediation: Re-examining brokerage in dynamic networks', *Social Networks* **35**(1), 130–143.

Tang, J., Musolesi, M., Mascolo, C., Latora, V. and Nicosia, V. (2010), Analysing information flows and key mediators through temporal centrality metrics, *in* 'Proceedings of the 3rd Workshop on Social Network Systems', SNS '10, ACM, New York, NY, USA, pp. 3:1–6. http://doi.acm.org/10.1145/1852658.1852661

Wasserman, S. and Faust, K. (1994), *Social network analysis: Methods and applications*, Vol. 8, Cambridge University Press.

Algebraic Structures and Social Processes

John Levi Martin

1 Introduction

There has long been an interest among social scientists in the use of algebraic structures to analyze social data. This is quite reasonable, as certain social processes may be understood as intrinsically algebraic (such as kinship as studied by Weil [1969] and H. White [1963]) and perhaps even cognition (Piaget 1970). But in addition to such *explanatory* uses of algebraic structures (where the structure is believed to shed light on a particular process), there are also algebraic approaches that arise in the quest for data reduction and interpretation that do not correspond to particular social processes, such as Galois analysis (Duquenne 1995, 1996; Freeman and White 1993; Ganter and Wille 1999). I shall call such models *descriptive* because they give a parsimonious overview of, and perhaps insight into, the meaning of an observed multiway distribution of data (i.e., the pattern of interconnections and implications) without explaining their generation. In recent years, the social sciences have seen more of an interest in such descriptive approaches and less attention to explanatory ones. Yet it may be that the explanatory are required if we are to build strong theory.

Here I wish to consider a set of different, inter-related, explanatory algebraic structures for two-mode (asymmetric) binary data. In all cases, we start with observed data, make assumptions about some underlying process, and then recreate (to an extent that varies across method) the structure of the generating process. I show that, depending on our understanding of the social and cognitive processes involved in the generation of the data, very different types of structures may be appropriate, or the same structure may have very different interpretations. To do

J. L. Martin (✉)
Department of Sociology, University of Chicago, Chicago, IL, USA
e-mail: jlmartin@uchicago.edu

© The Author(s), under exclusive license to Springer Nature Switzerland AG 2023

M. N. Hounkonnou et al. (eds.), *Mathematics for Social Sciences and Arts*, Mathematics in Mind, https://doi.org/10.1007/978-3-031-37792-1_11

this, I summarize results from previously published works which will be referred to for proofs of propositions made. In all cases, I emphasize an intuitive and practical orientation to the methods. Definitions of conventional algebraic terms are given in an appendix; these terms are italicized upon first use in both the text and in the appendix for easy location.

2 Data

A typical data table in sociological analysis is a "two-mode" matrix that links persons (as one mode) to some other entity which may be considered an attribute of persons (as the other mode). As a running example, we can consider a typical survey of subjective phenomena such as attitudes and beliefs. The algebraic approaches all begin with dichotomous or binary data, and so we will assume that these phenomena are measured either as "present" (e.g., some belief is held, some object is valued, some fact is known) as opposed to "absent." Thus if we have N people who answer M different items, we may represent our data as a matrix $\mathbf{X}$ with N rows and M columns, where the value in the ith row and the jth column is 1 if person i responds affirmatively to item j, and 0 otherwise. Here I will use a hypothetical data set consisting of 6 persons and 7 subjective variables; these data are presented in Table 1. We wish to investigate the sorts of algebraic structures implicit in such data that could be revealing of the social and/or cognitive processes involved.

It is worth emphasizing that many algebraic structures make assumptions about the nature of the data involved. In particular, the most common approaches (such as Galois lattice analysis) assume that if the *intersections* of observations (in this case, rows of Table 1) have not been observed, they are still possible. (We will call this a "closure" assumption—the data involved are "closed" under intersection.) While we commonly ignore the significance of implied but unobserved patterns, at a number of points we will suggest that they may be of use in deciding between alternate structural models.

Table 1 Example data

		Beliefs/information						
		A	*B*	*C*	*D*	*E*	*F*	*G*
Persons	1	1	1	1	1	1	1	1
	2	0	1	0	1	0	1	1
	3	0	0	1	0	1	0	0
	4	0	0	0	1	1	1	1
	5	0	0	0	0	0	1	0
	6	0	0	0	0	0	0	1

3 Social Structures of Influence

The set of approaches to be outlined in this chapter may be summarized as in Fig. 1; we show schematically how the underlying $N \times M$ data may be the basis for algebraic structures that can be treated as Boolean matrices. Our original matrix is portrayed in the upper center; we begin by considering the transformation schematized by the arrow leading to the matrix in the upper right.

One way of positing an underlying structure that might generate the data would be to propose that there is an underlying relation of transmission through social networks. Of course, with a simple data matrix, it would be impossible to recreate an underlying network of transmission unless two conditions are met. The first is that the existing data includes several beliefs at a sufficient number of different states of progress of diffusion through the structure, and the second is that the structure has to be of a particularly simple form. However, there is one sort of social structure that has been found by researchers to be a likely recurrent form where influence is significant (see Martin 2009, Chapter 5). This is a hierarchy which (1) involves

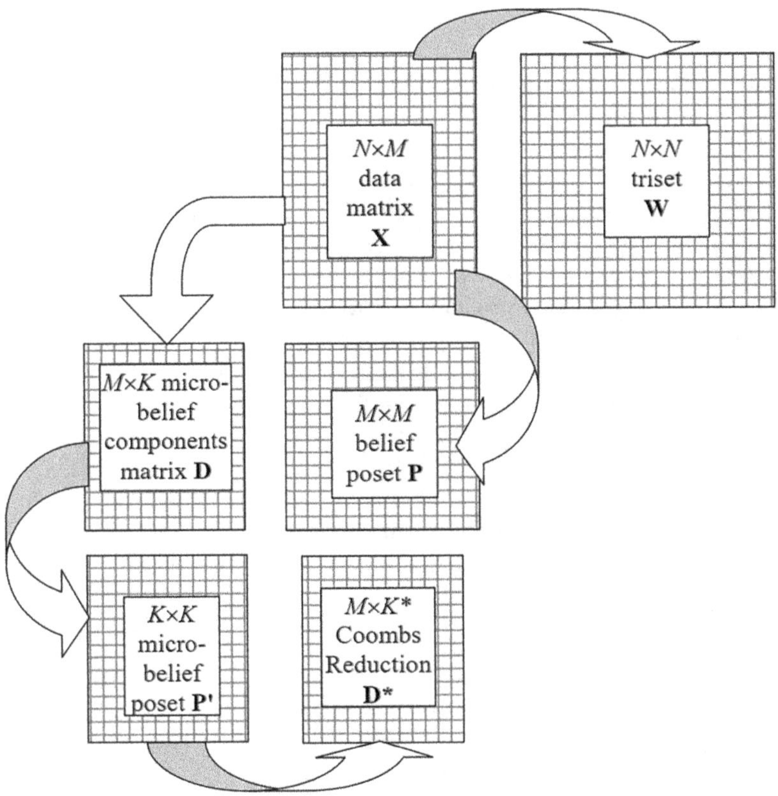

Fig. 1 Representation of the family of approaches

Fig. 2 A triset

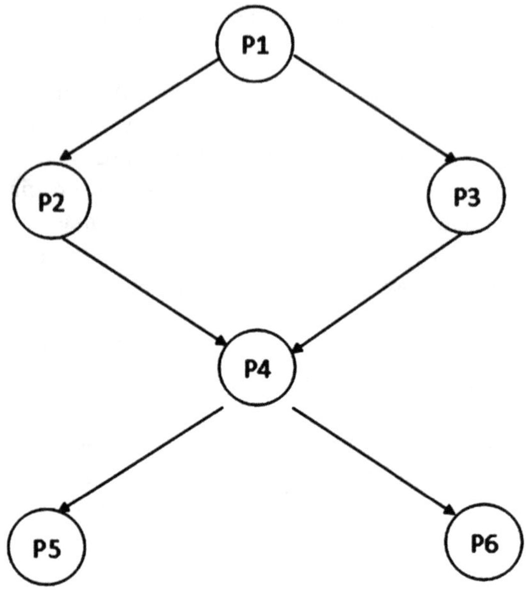

a vertical ranking of persons, but (2) may also include horizontal differentiation, and (3) does not necessarily involve relations implied by *transitivity*; indeed, these relations may be suppressed.

Consider the social network graphed in Fig. 2. An arrow indicates a path of influence or information transmission from the superior to the inferior node, each node representing a person. Relations implied by transitivity are not assumed to be present. This graph can be denoted in matrix form as **Y** in which $y_{ij} = 1$ if and only if P*i* ➔ P*j* in Fig. 2. This structure has two of the properties of a *partial order*, namely, reflexivity and antisymmetry, but not transitivity. Such structures would seem reasonable for cases in which information or influence needs to be distributed rapidly and efficiently, and it is interesting that they have reappeared in empirical studies of influence despite their mathematical awkwardness. To formalize, consider a binary relation ➔ on some set $I = \{\alpha_1, \alpha_2, \alpha_3 \ldots\}$; where it does not cause confusion this may be abbreviated to $I = \{1,2,3 \ldots\}$. Together, I and the relation ➔ may be denoted Y, also expressible in matrix form as **Y** in which $y_{ij} = 1$ if α_i ➔ α_j in Y and 0 otherwise. Further, let this binary relation ➔ have the following properties:

1. It is reflexive (α_i ➔ α_i for all i)
2. It is antisymmetric (if α_i ➔ α_j and α_j ➔ α_i, $\alpha_i = \alpha_j$);
3. It is acyclic (if α_i ➔ α_j there is no possible path from α_j to α_i). Formally, this may be said as follows: if α_i ➔ α_j in Y, there is no subset T $\{t_1, t_2, t_3 \ldots t_\mathrm{M}\}$ of I, in which the following condition is satisfied: $y_{jt_1} y_{t_1t_2} \cdots y_{t_{M-1}t_M} y_{t_M i} = 1$.

Such a binary relation, written in matrix form, implies that there is some permutation of the rows of **Y** such that $y_{ij} = 0$ if $i < j$; that is, **Y** can be triangulated (or it is "upper triangular"; cf. Hage and Harary 1983, p. 99). Hence we may call *Y* a "triset." (We will see this structure reappearing as a "biorder" in our final section.)

Given such a triset, we assume that (1) any belief can be transmitted from one person (P*i*) to another (P*j*) without loss to the transmitter if $y_{ij} = 1$; (2) a person who holds a belief never loses the ability to transmit this information (in the case of information, people do not forget; in the case of control, exercising control does not weaken one); (3) no sources outside the network can transmit a belief to a person; (4) only maximal persons (those who are influenced by no one) can generate beliefs. Any belief's distribution across persons is a column in a data matrix such as that in Table 1, which we will call "social belief states." Finally, let a "non-redundant" triset have no direct relations that, when eliminated, do not change the set of possible observed states, and also let a "mediated" triset be a triset in which ➔ is wholly non-transitive—there are no redundant direct paths where there are intermediaries.

Martin (2002) shows that in this case, the set of possible "social belief states" forms a lattice that is closed under *union* and sketches the relation between sets of related social structures that all produce the same lattice of observations. More significantly, given a set of observed social belief states (e.g., the columns in Table 1), it is possible to reproduce a triset that could have generated the observations (and this triset is necessarily a non-redundant one whose relation to redundant trisets can be specified). Given a lattice closed under union with states representing columns in a table such as Table 1, we focus on the "join irreducible elements"—those that cannot be represented as the *join* (*least upper bound*) of two or more elements of the lattice, neither of which is the element in question. Every *join irreducible element* (or *JIRE*) *covers* only one other element in the lattice and can be associated with the unique observation that it contains and the element that it covers lacks. This labeling is present in Fig. 3, which presents the social belief state lattice implied by the triset of Fig. 2, where the JIREs are placed in double outlined boxes, and the unique labeling shown.

Three things become apparent here (for proof, Martin 2002): first, that the triset of Fig. 2 can be reproduced from the relations of covering between the JIREs; second, that there is a doubling of the structure where there is an "in-tree" in the triset (one person receiving influence from two or more persons); and third, that the observed data of Table 1 are all contained herein. (The JIRE labeled "1" corresponds to belief A, that labeled 2 to "B," and so on for those labeled 3, 4 (left), 4 (right), 5 (left), and 6 (left) seriatim.) Thus this is one possible structure that could have produced the observed data. In contrast, other structures that have sometimes been used to derive social structure from such data such as the Galois lattice investigated below could *not* have produced the observations, if the process of influence is that outlined here (and we are not aware of any *other* processes being proposed to link observations to underlying structure).

We noted that the "in-tree" structure generates a doubling in the lattice of observed states (here because the in-tree is of order 2). Martin (2006) demonstrates that where such in-trees are absent in the triset there is a self-dual structure to

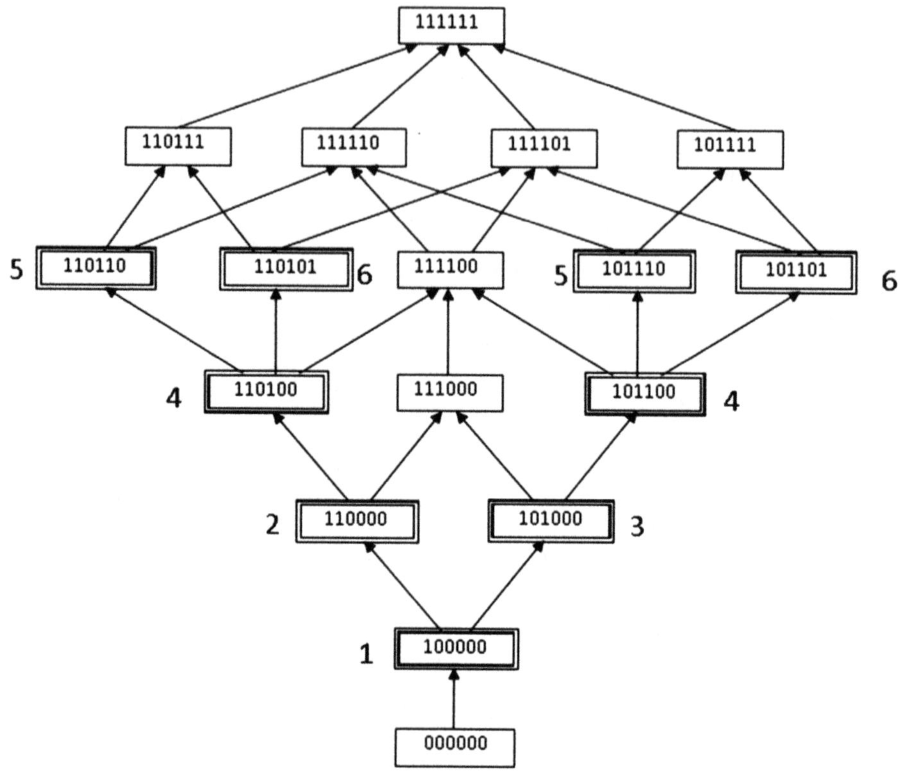

Fig. 3 Social belief state lattice

the resulting lattice coming from the fact that one person holding a belief is both necessary and sufficient for the possibility of holding of the belief in any of the persons that she influences. The resulting lattice is *distributive*, and there is a one-to-one mapping between the JIREs and the *MIREs* (the *meet-irreducible elements*) such that the relation between persons in the triset can be read either from the JIREs or the MIREs.

4 Partial Orderings of Items

The above approach is appropriate for cases when there are social structures of influence but no intrinsic ordering among the subjective beliefs or responses. This absence of inter-item structure might make sense for some sorts of domains, but not for others. Certainly, it would seem implausible where the items are test questions that tap skills, such that some items are harder to answer than others; a similar type of structure might also arise for certain sets of attitude or belief items. It is commonly found that "hardness" also exists in the measure of opinions that have quantitative aspects. For example, those persons who do not believe that a woman should have a right to abortion when the health of the mother is in danger (a relatively "easy" item on an abortion scale) are unlikely to indicate that they believe that a woman should have a right to abortion when she is poor and "doesn't want any more children." Such inter-item relations would mean that a person could not necessarily accept *social* influence from another without having the proper *cultural* or *mental* prerequisites.

In some cases, it may be possible to use algebraic approaches to study structures in which there is *both* interpersonal transmission *and* an internal structure to the items, but this is difficult and requires exponentially more data than the sort found in Table 1 (see Martin 2006). For this reason, we now go on to examine cases in which there is no social bottleneck to the acquisition of beliefs (all persons are exposed to new beliefs), but only the inner bottleneck of relations between the items. Thus (see Fig. 1) instead of taking our $N \times M$ matrix and using it to develop an $N \times N$ matrix of (influence) relations between persons, we use it to develop an $M \times M$ matrix of (precedence) relations between items.

Here we assume that the relation between items is that one item can "precede" another, where the relation of precedence is reflective, antisymmetric, and transitive. Hence, by construction, the set of all items is a partial order. Thus, in contrast to the most commonly used approaches for the investigation of sets of items in the social sciences, which assume an underlying order, but probabilistic or fuzzy relations between softer and harder items, here we assume deterministic relations (if item i precedes item j, no one can hold item j who does not also hold item i), but not a complete order.

In this case, it can be shown (see Wiley and Martin 1999, Theorem 2) that the partial ordering among the items generates a distributive lattice of possible states (rows in a matrix like that seen in Table 1). For the case of Boolean vectors such as we have here, this means that the lattice is closed under both intersection (which is equivalent to meet) and union (equivalent to join). Further, from a set of observed states (e.g., our Table 1), it is possible to recreate the strongest partial order of the items compatible with the observations via attention to the MIREs of the resulting lattice, as any MIRE can be labeled by the unique item added by the state that covers it, the relation between these labels then indicating the partial order of items (Ibid., Theorem 3). There is, however, a simpler formula. If our observed data are $\mathbf{X}$, and we use the matrix $\mathbf{P}$ as a representation of the partial order of the items in which $p_{ij} = 1$ iff item i precedes item j, 0 otherwise, we can construct $\mathbf{P}$ as

$$\mathbf{P} = \overline{\left\lceil \overline{(\mathbf{X}^t)}\mathbf{X} \right\rceil}$$

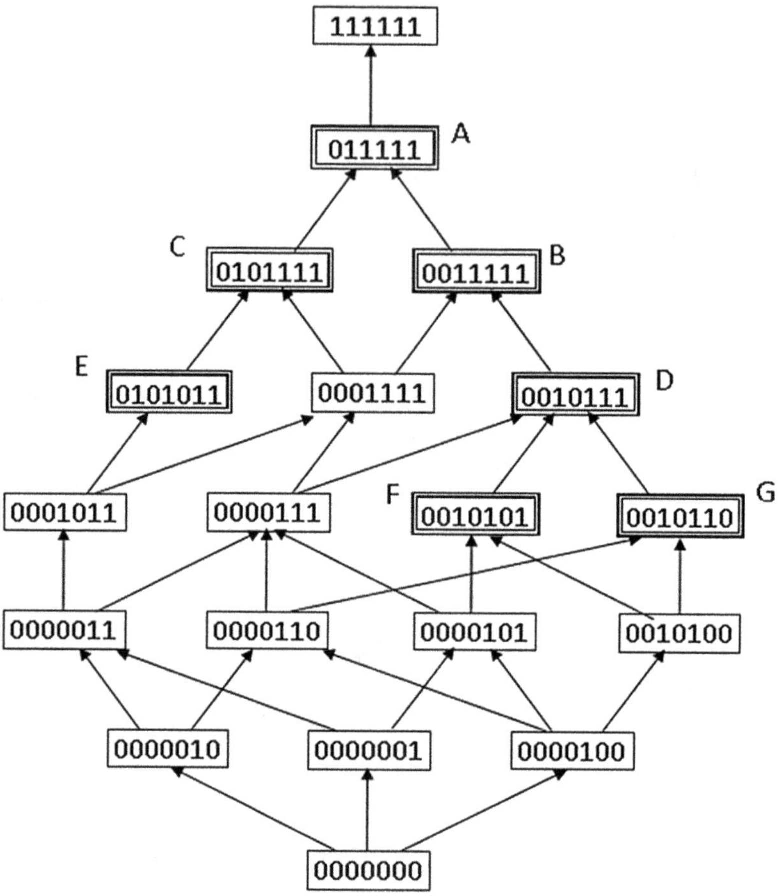

Fig. 4 Distributive belief state lattice

where the t superscript indicates transpose and the bar complementation (exchanging 0 s and 1 s).[1]

For the example data, the set of all the rows observed in Table 1, augmented with any observed unions or intersections, produces a lattice of states as graphed in Fig. 4. All the MIREs are placed as double outlined boxes, with the item-labeling discussed above added. The relations of precedence between the items (**P**) can be retrieved either from the formula above or by examination of the resulting lattice; the set of precedence relations are sketched in Fig. 5. This may be understood as

[1] The logic of this construction is straightforward—item i precedes item j, if there is no row k in **X** such that $x_{ik} = 0$ and $x_{jk} = 1$. The other formulae presented below are also all susceptible of such an intuitive natural language restatement.

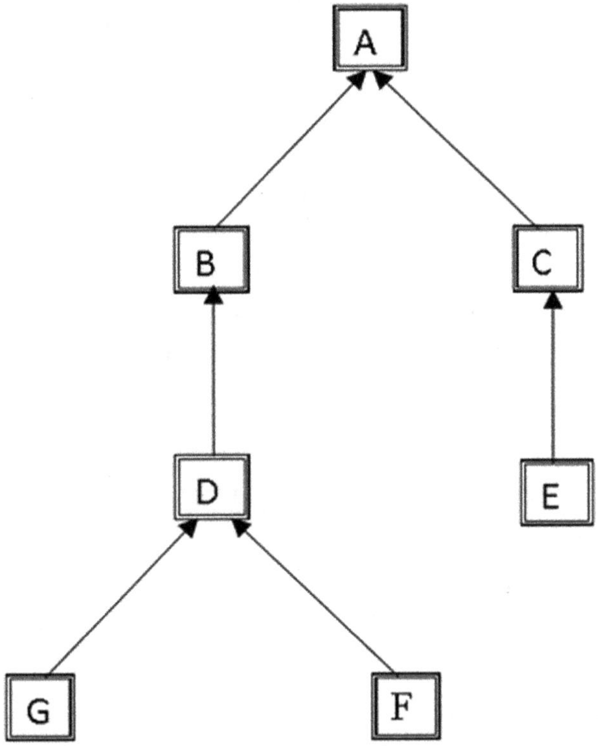

Fig. 5 Item poset retrieved from Fig. 4

a parsimonious representation of asymmetric relations between items that has only some of the ordering of a linear scale.

5 Microbelief Representations

Let us compare the above structure to the popular Galois lattice representation of the data in Table 1 (Fig. 6). The points are given the dual labeling in which inclusions relations between columns can be read from the letters (a path from bottom-up indicates a relation of inclusion; dually, a path from top down represents a relation of $\leq$), and inclusion relations between rows are indicated by numbers (a path from top down indicates a relation of inclusion; dually, a path from bottom up represents a relation of $\leq$). We can see that the partially ordered set (Fig. 5) is preserved in the Galois lattice. However, the Galois lattice is more parsimonious than the belief state lattice of Fig. 4 (having only 9 possible states, while the belief state lattice has a full 19).

If our goal is simply a parsimonious representation of dual order relations in the data, we might well prefer the Galois representation. But if we are interested in

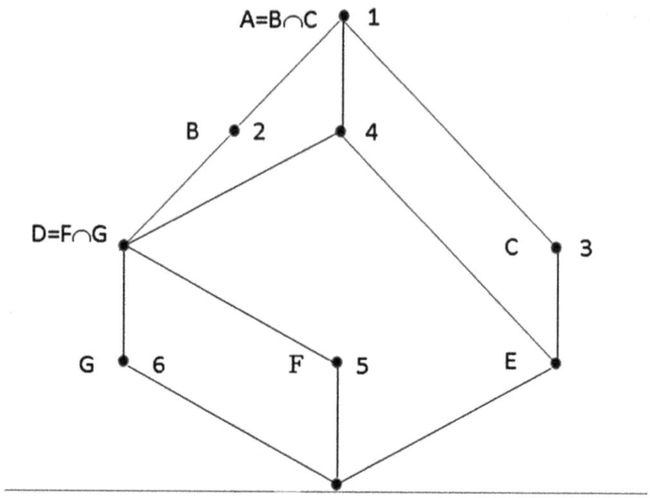

Fig. 6 Galois lattice for Table 1

the social-psychological processes involved—and we had collected a fair amount of data—we might wonder why the ten patterns logically compatible with the partial ordering but omitted in Fig. 6 were not observed.

One possible reason might be that the actual response process does not involve each item being considered as a whole unit, but rather, that there are shared components (which we shall here call "microbeliefs") that are common to more than one item (now called "macrobeliefs"). If a macrobelief (say, α_1) is composed of three microbeliefs (say, β_1, β_2, and β_3), a respondent will answer α_1 in a positive direction if and only if she possesses all three of the required microbeliefs. While the former are latent, and only the latter manifest, we can use an algebraic decomposition to recover a set of microbeliefs compatible with the observed macrobelief structure (Haertel and Wiley 1993).

Let us take our observed data from Table 1, ensure that it is closed under intersection (but not, as in our previous analysis of Fig. 4, also closed under union). We label the states here in the same way as we labeled the belief state lattice of Fig. 4 and hence term this a macrobelief state lattice (Fig. 7). Once again, the MIREs are presented in double outlined boxes. It can be seen that this structure is homologous to that Galois lattice of Fig. 6; it is the interpretation that will be different. Rather than the structure coming from the interpenetrating ordering of rows and columns, it emerges from the overlap in the pattern of microbeliefs.

Let us propose that in fact these seven manifest macrobeliefs are conglomerates of only five different microbeliefs. The fact that the number of microbeliefs is the same as the number of MIREs—and hence the cardinality of the row-basis of the Boolean complement of the matrix **X** (Kim 1982, pp. 5–7)—is not accidental. We present this hypothesized "microbelief inclusion matrix" in Table 2. We will denote this matrix **D** with elements d_{ij}, where $d_{ij} = 1$ iff macrobelief α_j incorporates

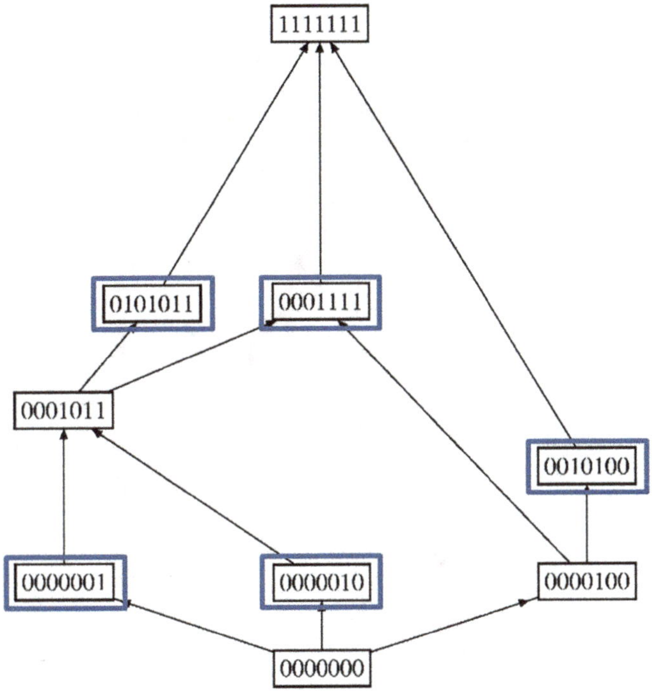

Fig. 7 Macrobelief state lattice

Table 2 Microbelief inclusion matrix (D)

	A	B	C	D	E	F	G
a	1	1	1	1	1	1	0
b	1	1	1	1	1	0	1
c	1	1	0	1	0	1	1
d	1	0	1	0	1	0	0
e	1	1	1	0	0	0	0

microbelief β_i, 0 otherwise. If we provisionally assume that, in the population, all possible combinations of microbeliefs are held (thus the poset of microbeliefs is an *antichain* consisting of all $2^5 = 32$ combinations), a set of possibilities which we denote **Z**, we can construct **X** via the formula $\mathbf{X} = \overline{\left[\overline{\mathbf{Z}}\mathbf{D}\right]}$, and any such **X** is necessarily a lattice closed under intersection.

Following Haertel and Wiley, Martin and Wiley (2000; Theorem 4) demonstrate that given **X**, the most parsimonious **D** can be produced by taking the subset of rows of **X** that are MIREs and taking the complement of this submatrix. Thus (see Fig. 1) we go from the $N \times M$ observed data to the $M \times K$ microbelief inclusion matrix **D**, where K is the rank of $\overline{\mathbf{X}}$. That the **D** matrix in Table 2 does in fact imply the lattice of Fig. 7 can be verified by the reader. We have, however, assumed so far that there is no structure among the microbeliefs. We can understand that some sorts of

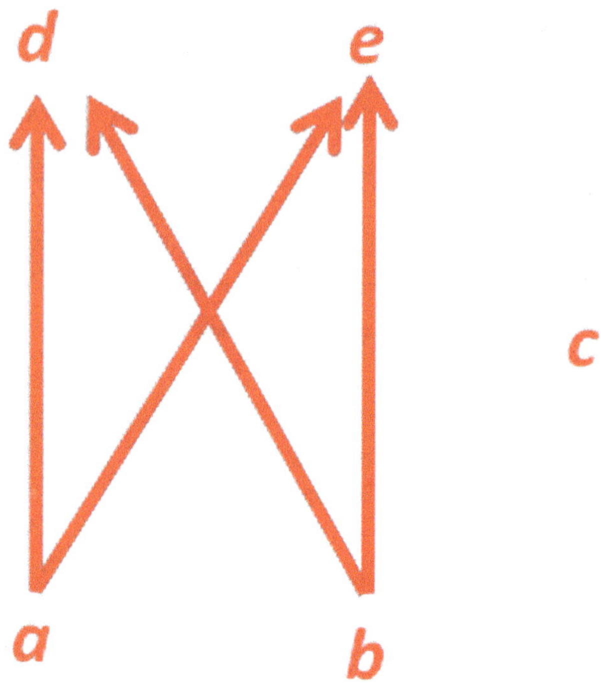

Fig. 8 Microbelief poset

structure would prove problematic for the observation of the macrobelief structure. For example, suppose that there was a precedence relation such that $d \rightarrow b$; no one could hold microbelief b who did not already hold microbelief d. Consider the MIRE 0101011 (the double-outlined box to the upper left of Fig. 7). Inspection of Table 2 demonstrates that the set of macrobeliefs here implies that respondents hold the microbelief set $\{a, b, c, e\}$. But if $d \rightarrow b$, then anyone who held these would also hold d, and the set $\{a, b, c, d, e\}$ would map into the universal upper bound instead (i.e., state 0101011 cannot be observed). This precedence relation, then, is incompatible with the observed structure and thus negates the possibility of the generating structure proposed. Martin and Wiley (Theorem 5) show that the set of all possible precedence relations between microbeliefs that are compatible with the original observations can be derived via the formula

$$\mathbf{P}' = \overline{\left\lceil \overline{\mathbf{D}}\mathbf{D}^t \right\rceil}.$$

Thus we now go from the $M \times K$ microbelief inclusion matrix $\mathbf{D}$ to the $K \times K$ microbelief poset $\mathbf{P}'$ (see Fig. 1). The $M \times M$ *macro*belief poset $\mathbf{P}$ can be produced directly from $\mathbf{D}$ as well. For our example, $\mathbf{P}'$ can be shown in graph form as in Fig. 8. However, this is not the only possible interpretation of an underlying latent structure.

6 Coombs Factorizations

We have been assuming that the latent components are dichotomous, but once we establish a poset among them, we can probably understand intuitively that it might also be possible to conceive of the latent space as being composed of ordinal dimensions. For the case at hand, one possible factorization would involve three dimensions: one that stretches {0, a, d}, another that stretches {0, b, e}, and a third that stretches {0, c}. Of course, that is not the only possible three-dimensional solution; one which arranged the dimensions {0, a, e}, {0, b, d}, {0, c} would produce the same observed data, and there are four and five dimensional solutions as well.

This idea of an ordinal factor analysis turns out to be identical to the discrete, noncompensatory model of response originally proposed by Clyde Coombs (1964). Coombs had sketched a process in which any item appeared as a "corner" in a multidimensional space (his example is reproduced in Fig. 9). Coombs' numbered regions each can be shown to correspond to a row in a belief state lattice; here, these regions are turned into nodes connected by lines indicating "covering" relations; the

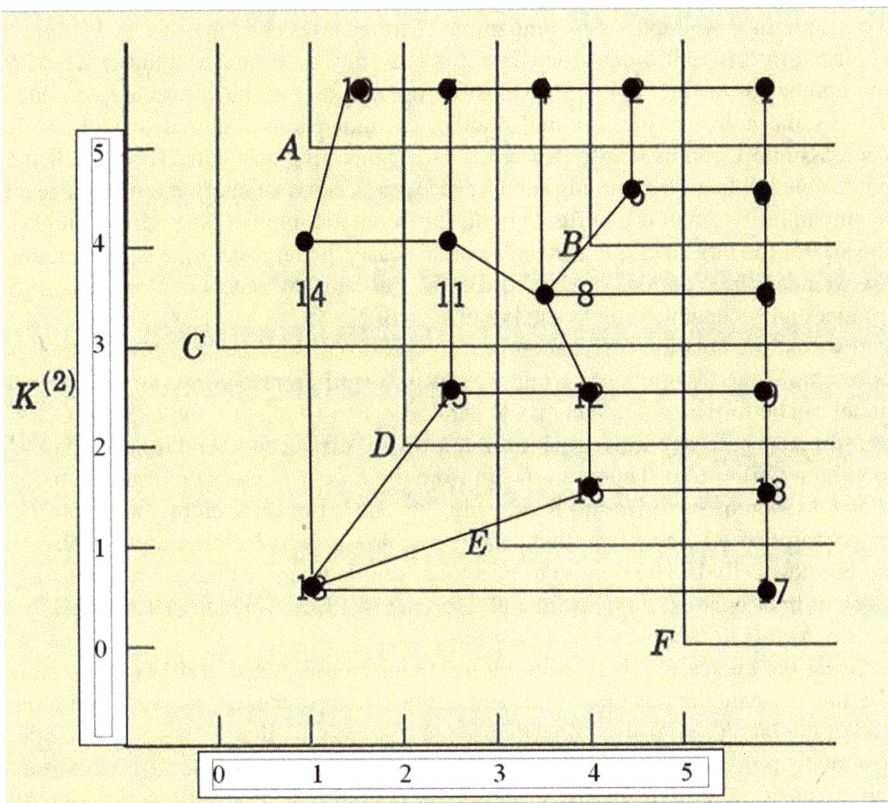

Fig. 9 Coombs's example

universal lower bound is to the lower left and the universal upper bound to the upper right, hence one state is greater than another if it is to the right of or above another. It will be noted that the elements that are along the upper boundary and the right boundary are all MIREs; again, this is not an accident: any interior element is the intersection (meet) of MIREs.

Coombs knew how to go from a model in a latent ordinal space to a set of observed possibilities, but not how to go the other way, nor how to estimate the minimum number of dimensions needed to recreate any observed structure. Based on the previous logic, Martin (2014; Theorem 2) showed that the minimal dimensionality can be derived from the matrix $\mathbf{P}'$ previously defined. Any two elements in a poset $\mathbf{P}'$ are said to be incomparable if neither precedes the other, and the set of all the elements of $\mathbf{P}'$ along with their incomparability relations is known as an incomparability graph. If we construct

$$\mathbf{S} = \overline{\left[\mathbf{P}' + \mathbf{P}'^{t}\right]}$$

where addition is Boolean, we have a matrix representation of this incomparability graph. That for the running example (Table 1) is portrayed in Fig. 10. Any members of a clique in this graph, as incomparable (neither precedes the other in $\mathbf{P}'$), must be placed in different dimensions. For this reason, the chromatic number K^* of $\mathbf{S}$ (the number of distinct colors that must be used so that no tie connects two nodes of the same color) is the minimal number of dimensions required to recreate $\mathbf{X}$ if we assume Coombs's spatial process as a generating structure. (I will call the dimensional reduction assuming such a process a "Coombs factorization.") Since an incomparability graph is a perfect graph, its chromatic number is its clique number (the size of the largest clique). Because in this case, the largest clique is 3, we know that we need three dimensions to recreate $\mathbf{X}$. Further, any acceptable coloring of $\mathbf{S}$ produces an acceptable dimensional reduction of $\mathbf{X}$.

If we use the notation $C(\mathbf{S})$ to denote a coloration (in this case, a vector of integers representing the assignment of our K MIREs to different colors), a dimensional model can be formalized as a matrix $\mathbf{R}$ defined $r_{ij} = 1$ iff $p'_{ij} = 1$ and $C(\mathbf{S})_i = C(\mathbf{S})_j$; less formally precise but perhaps more intuitively the second condition might also be written $C(i) = C(j)$. Thus we now go from the $K \times K$ microbelief poset $\mathbf{P}'$ to the $M \times K^*$ Coombs factorization $\mathbf{R}$ (see Fig. 1). The set of all such factorizations (in matrix form of $\mathbf{R}$), and hence that of all possible Coombs factorizations of any $\mathbf{X}$, can be shown (Ibid., Theorem 4) to be a meet-semilattice in which the relation of $\leq$ is one of finer/coarser; more technically, for any two Boolean matrices $\mathbf{A}$ and $\mathbf{B}$, we can say $\mathbf{A} \leq \mathbf{B}$ iff there is no $i, j \mid a_{ij} > b_{ij}$.

Given the choice of which coloration scheme to use, Martin (2014) shows how one may recreate the position of all items in the latent space. Of course, as with the case of the latent social structures that we first examined, there is a set of potential generating processes that will produce the same data; in this case, it is important that the position of items in this space satisfy certain conditions of discriminability discussed in Martin (2014). These positions can be understood as a set of "Discrete

Fig. 10 MIRE Incomparability Matrix

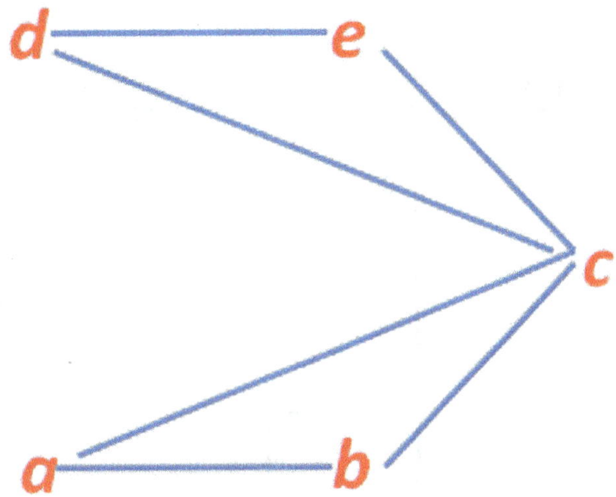

Table 3 Discrete item curves (D)

	A	B	C	D	E	F	G
K^1	2	1	2	1	2	1	0
K^2	2	2	1	1	1	0	1
K^3	1	1	1	1	0	1	1

Item Curves," organized as a matrix D in which the macrobeliefs are again columns, and the rows correspond to dimensions in the space. Each item's position is given according to an ordinal scale; it corresponds to the point of the "corner" dividing the space into two regions, one in which the item is answered in a negative direction, and the other in which the item is answered in a positive direction. Thus Table 3 contains the discrete item curves for the example with the coloration corresponding to $\{0, a, d\}$, $\{0, b, e\}$, $\{0, c\}$. The microbelief analysis (which followed Haertel and Wiley 1993) may be understood as a special case of the uniquely maximal dimensional reduction $\{0, a\}$,$\{0, b\}$, $\{0, c\}$,$\{0, d\}$, $\{0, e\}$.

7 Coombs Factorizations and the Biorder Approach

Such a discrete, noncompensatory factor analysis of binary matrices has been developed in another manner in what is known as the "biorder" approach by Doignon et al. (1984), Doignon and Falmagne (1984), Chubb (1986), and Koppen (1987), who saw this as a solution to the problem posed by Coombs. Here the goal was to use the minimum number of distinct dimensions to embed each item. For example, consider persons asked three items, with observed responses $\mathbf{x}_1 = [1,0,0]$, $\mathbf{x}_2 = [0,1,0]$, and $\mathbf{x}_3 = [0,0,1]$ which we of course must augment with the universal upper and universal lower bounds ([1,1,1], [0,0,0]) to compose our $\mathbf{X}$. The biorder

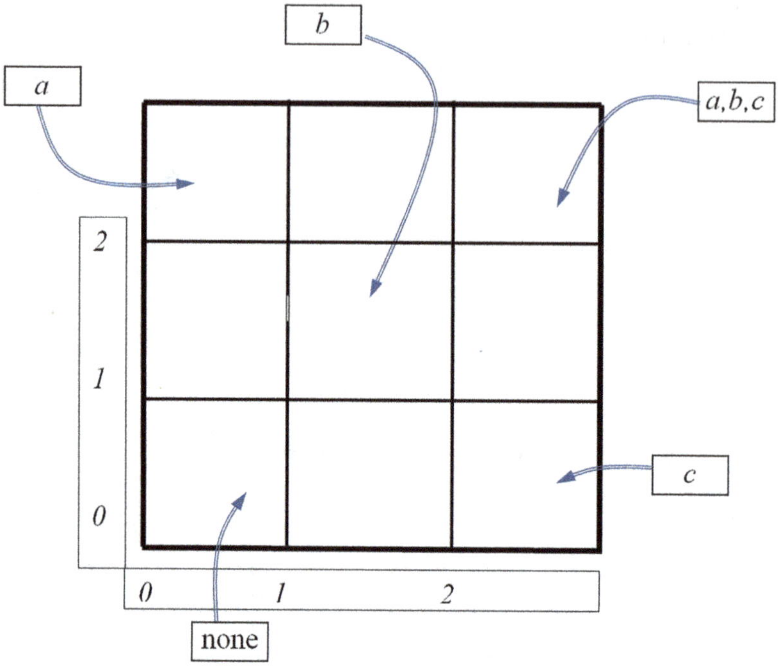

Fig. 11 Biorder representation of simple data

solution would require two traits, each with three levels, producing nine possible regions (see Fig. 11 for one possible factorization). Item *a* requires that a subject exceed the second threshold on the vertical dimension (which is why neither the second nor the third subject answers this positively); item *c* requires that a subject exceed the second threshold on the horizontal dimension (which is why neither the first nor the second subject answers this positively), while item *b* requires that a subject exceed the first threshold on both the vertical and the horizontal dimension (which is why neither the first nor the third subject answers this positively).

What will be noticed is that there are nine possible regions, only five of which are actually occupied. If we had, say, only asked three people our items, it would not be surprising that we did not find the response patterns [1,1,0] or [0,1,1], even though these would be implied as possible given a *spatial* interpretation *ala* Coombs of the underlying response process (portrayed in Fig. 12; the shaded regions are unobserved). However, if we had had many respondents, we might think that the biorder approach, although terse in terms of dimensions, was not orienting us to the actual response process.

In contrast, the Coombs factorization as laid out here would retrieve three dichotomous dimensions (Fig. 13). In this solution, there are no "blank" regions, and while there are more dimensions, there are fewer total occupied cells. Indeed, Martin (2016) demonstrates that while the biorder reduction is always minimal in

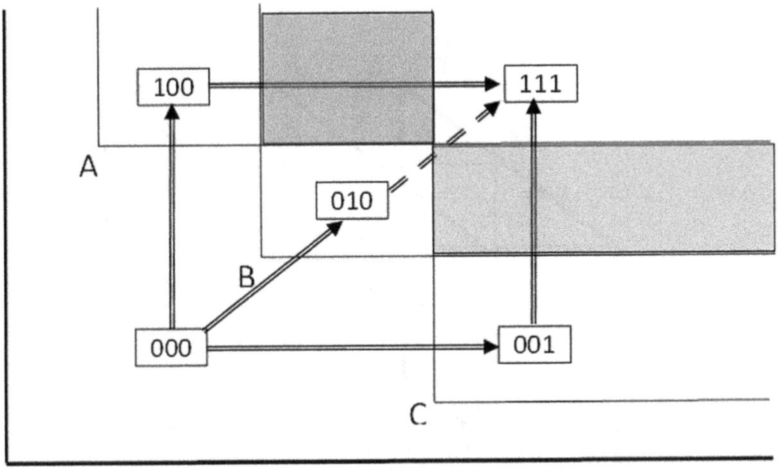

Fig. 12 Coombs representation of biorder solution

terms of dimensions, the Coombs factorization is always minimal in terms of the total number of thresholds. Further, the two can be shown (Ibid,. Theorem 3) to be equivalent for the case in which **X** is a score-graded lattice; that is, if element $\mathbf{x}_1$ covers $\mathbf{x}_2$ in **X**, then $\sum_i x_{1i} = \sum_i x_{2i}+1$. Such a score-graded lattice (indeed, a lower-semimodular lattice) will arise, for one, if the thresholds of the discrete item curves (D) are such that no two items occupy the same position on any dimension (Theorem 2). This is a somewhat nice property, because as the number of respondents grows, the chances that two items cannot be distinguished on any dimension would tend to decrease; since the solution of the biorder problem is *NP*-hard and that of the Coombs factorization is a simple polynomial problem of finding the clique number of an incomparability graph, this means that for large data sets we may imagine the Coombs solution to be appropriate and tractable.

If we return to the macrobelief state lattice that arises when we augment our observed patterns (Table 1) with any intersections (Fig. 7), we may notice that it is a planar graph (no lines cross in the two-dimensional representation). This might suggest to us that there must be a two-dimensional biorder representation, and this is indeed the case. Table 4 presents two biorders whose intersection recreates the data in Table 1. It can be seen that for each there is a permutation of the rows and columns such that the matrix is upper triangular and that for any two rows, one precedes the other. The proper permutations for each become apparent in Fig. 14, which recasts this in the Coombs spatial representation.

In Fig. 14, each item is represented as a discrete item curve, in this case, a corner whose lines go to the top and to the right; all individuals with trait values inside the corner answer the item in a positive direction. Each observed state of Table 1 is located in the proper region. (It is the initial label, and not the longer representation of the vector, that indicates which region some pattern is in. Thus

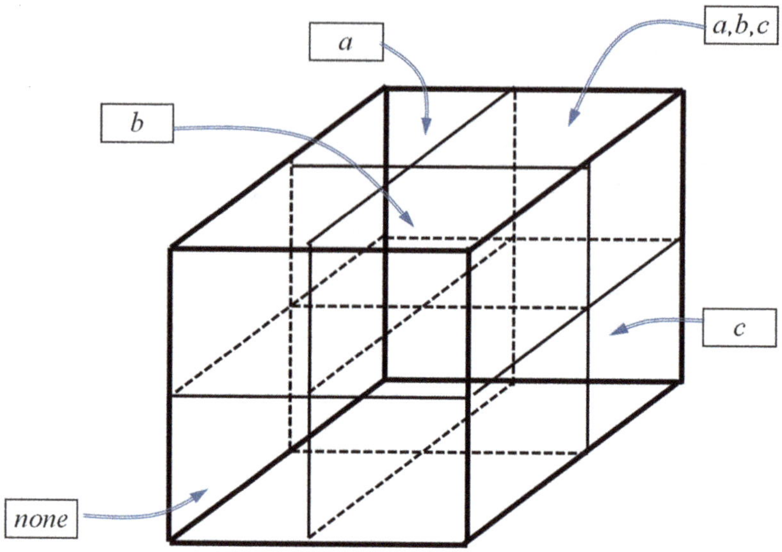

Fig. 13 Coombs factorization of example data

Table 4 Biorders recreating Table 1

Table (a) First dimension

		A	B	C	D	E	F	G
ABCDEFG	*p*1	1	1	1	1	1	1	1
BDFG	*p*2	0	1	0	1	0	1	1
CE	*p*3	0	1	1	1	1	1	1
DEFG	*p*4	0	1	0	1	1	1	1
F	*p*5	0	0	0	0	0	1	1
G	*p*6	0	0	0	0	0	0	1

Table (b) Second dimension

		A	B	C	D	E	F	G
ABCDEFG	*p*1	1	1	1	1	1	1	1
BDFG	*p*2	0	1	1	1	1	1	1
CE	*p*3	0	0	1	0	1	0	0
DEFG	*p*4	0	0	1	1	1	1	1
F	*p*5	0	0	1	0	1	1	0
G	*p*6	0	0	1	0	1	1	1

pattern *p*2, corresponding to the second row in Table 1 in which beliefs B, D, F, and G are held, corresponds to the small square whose lower left corner has the label B.) The implied but unobserved states that are the intersection (meet) of observed states are labeled with *m*'s as opposed to *p*'s. Then, on both the first (horizontal) and second (vertical) dimensions, the thresholds implied by the discrete item curves are continued via dotted lines to axes, which are then labeled both by the row and the column(s) in Table 1 to which they correspond. These, it will be noted, contain

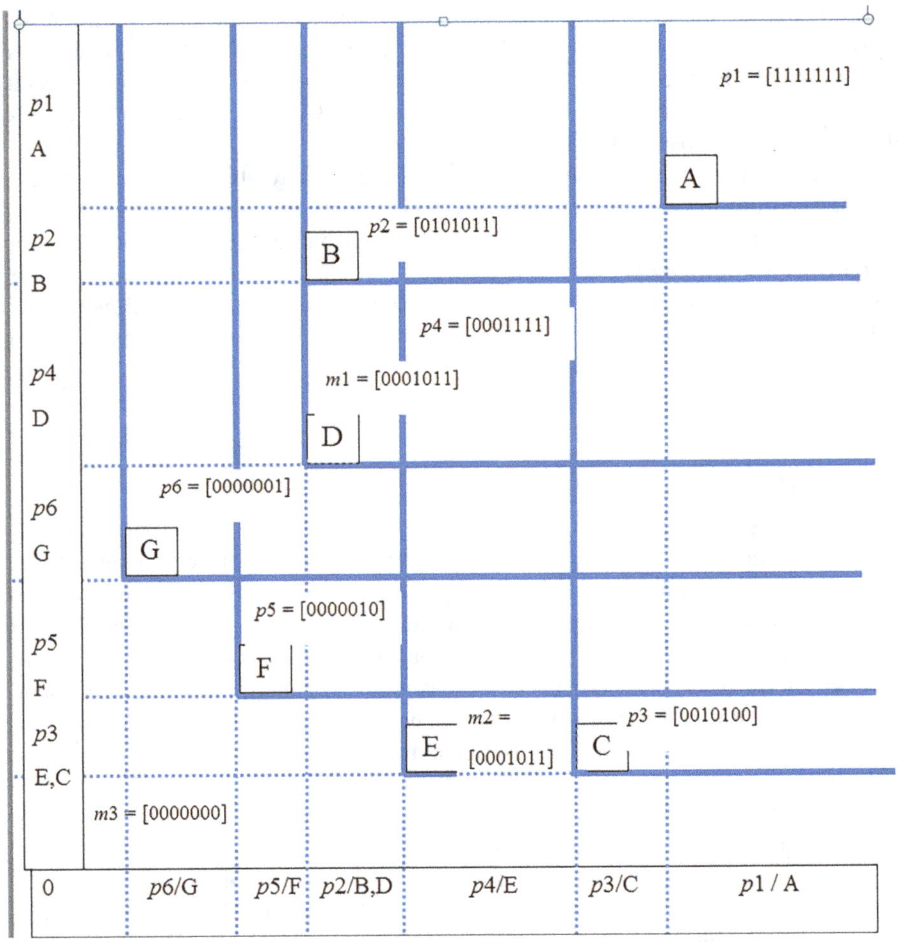

Fig. 14 Coombs representation of biorder solution

the information on the proper permutation of the rows and columns in Table 4(a, b), respectively, to demonstrate that they are in fact biorders.

It can be seen that Fig. 14 will in fact re-create the macrobelief state lattice of Fig. 7. If we connect any two observed states (e.g., *p*5 and *p*4) where one is either above or two the right of the other (or both), and then eliminate any transitively implied ties, we will produce the same covering relations indicated in the Hasse diagram of Fig. 7; such lines have been omitted only for clarity. While Fig. 7 went from the center-bottom of the page to the center-top, the structure here goes from the lower-left to the upper-right.

It will also be noted that the two dimensions each have seven distinct values, suggesting a total of 49 possible states. Of course, the non-compensatory nature of the response process means that many of these would map onto the same observed states. However, the Coombs interpretation of the biorder solution implies that we would expect 17 distinct states (when only six were observed in Table 1) and thus augments the observed states with 11 unobserved (one of which is the universal lower bound, ∅). In contrast, the three-dimensional Coombs reduction only adds three states to the observed (one of which is ∅).

More important than judging parsimony is the interpretability of the representation. The question of whether there are three or only two dimensions underlying the response process given some set of data is one that may be substantively very weighty. If we are interested in using data to formulate theoretical models of the response process, and the distribution of the data suggests a higher dimensional solution is appropriate, we would not want to prefer a lower dimensional solution merely for the sake of parsimony.

8 Conclusion

We have examined a variety of methods that attempt to infer structures underlying typical Boolean data matrices common to the social and behavioral sciences. Thus rather than look for elegant ways of compacting data, we highlighted procedures that are oriented to answering substantive questions about data-generating processes. In some cases, the data-generating process is assumed to be interpersonal, one of social influence. In other cases, the data-generating process is assumed to be intrapersonal, about the inherent psycho- and socio-logical relations of implications between sets of beliefs.

But in all cases, the aim is to formulate structural models for data that are not merely elegant, but that cast light on substantive processes. They are therefore not applicable to all Boolean matrices, and even when they initially appear to yield tractable models, these must be examined closely for their internal consistency and external plausibility before concluding that the structural model is appropriate. But where such models do in fact appear appropriate, they may not only cast considerable light on particular questions of interest but contribute to that vision of a structural social science so tantalizingly held out to us by Lévi-Strauss and others.

Acknowledgments Several figures are re-used from Martin (2006, 2014, 2016) by the kind permission of the publishers. I am grateful to the editors for comments that improved this contribution.

Appendix: Glossary

A *partially ordered set* (or "poset," for short) is a set of elements $\{\delta_1, \delta_2, \delta_3, \ldots\}$ and a binary relation denoted $\leq$, which satisfies the following three conditions:

(i) *Transitivity*: $\delta_1 \leq \delta_2$, $\delta_2 \leq \delta_3$ implies $\delta_1 \leq \delta_3$;
(ii) *Reflexivity*: $\delta_1 \leq \delta_1$;
(iii) *Antisymmetry*: $\delta_1 \leq \delta_2$ and $\delta_2 \leq \delta_1$ implies $\delta_1 = \delta_2$.

A poset in which the maximal number of possible relations of $\leq$ are present is known as a *chain* (a linear order), and a poset in which no relations of $\leq$ are present is known as an *anti-chain*.

Consider a poset A, consisting of elements δ_1, δ_2, δ_3, etc. together with a binary relation $\leq$ as defined in the text. The *lower bound* of a pair of elements, δ_1 and δ_2, in A is an element δ_3, such that $\delta_3 \leq \delta_1$ and $\delta_3 \leq \delta_2$. Similarly, the *upper bound* of a pair of elements, δ_1 and δ_2, in A is an element δ_3, such that $\delta_1 \leq \delta_3$ and $\delta_2 \leq \delta_3$. The *greatest lower bound* or *meet* of any two elements δ_1 and δ_2 in A, denoted $\delta_1 \wedge \delta_2$, is a unique element δ_3 in A such that $\delta_3 \leq \delta_1$ and $\delta_3 \leq \delta_2$ and there is no δ_4 in A such that $\delta_3 \leq \delta_4 \leq \delta_1$ and $\delta_3 \leq \delta_4 \leq \delta_2$. Where our elements are Boolean vectors (say, $\mathbf{x}$ and $\mathbf{y}$), meet is equivalent to *intersection*: thus $\mathbf{x} \wedge \mathbf{y} = \mathbf{x} \cap \mathbf{y} = \{\mathbf{z} \mid z_i = x_i y_i\}$. Similarly, the *least upper bound* or *join* of any two elements, δ_1 and δ_2 in A, denoted $\delta_1 \vee \delta_2$, is a unique element δ_3 in A such that $\delta_1 \leq \delta_3$ and $\delta_2 \leq \delta_3$, and there is no δ_4 in A such that $\delta_1 \leq \delta_4 \leq \delta_3$ and $\delta_2 \leq \delta_4 \leq \delta_3$. Where our elements are Boolean vectors $\mathbf{x}$ and $\mathbf{y}$, join is equivalent to *union*: thus $\mathbf{x} \wedge \mathbf{y} = \mathbf{x} \cup \mathbf{y} = \{\mathbf{z} \mid z_i = x_i + y_i$, where $+$ is Boolean$\}$. A *lattice* is then a poset that is closed under the binary operations of meet and join; that is, for any two elements δ_1 and δ_2 in A, $\delta_1 \vee \delta_2 \in A$; $\delta_1 \wedge \delta_2 \in A$. A lattice is *distributive* if for any three elements δ_1, δ_2, and δ_3 in A, $\delta_1 \wedge (\delta_2 \vee \delta_3) = (\delta_1 \wedge \delta_2) \vee (\delta_1 \wedge \delta_3)$ and $\delta_1 \vee (\delta_2 \wedge \delta_3) = (\delta_1 \vee \delta_2) \wedge (\delta_1 \vee \delta_3)$.

An element δ_1 of some lattice L is *meet-irreducible* if $\delta_1 = \delta_2 \wedge \delta_3$ implies that either $\delta_3 = \delta_1$ or $\delta_2 = \delta_1$, for all δ_2, δ_3 in L. The universal upper-bound I (or the vector [1,1 . . . 1] in the cases here) may be considered a trivially meet-irreducible element [*MIRE* for short] because $I = \delta_1 \wedge \delta_2$ implies $\delta_1 = I$ and $\delta_2 = I$. An element δ_1 of some lattice L is *join-irreducible* if $\delta_1 = \delta_2 \vee \delta_3$ implies that either $\delta_3 = \delta_1$ or $\delta_2 = \delta_1$, for all δ_2, δ_3 in L. The universal lower-bound Ø may be considered a trivially join-irreducible element [*JIRE* for short] because $Ø = \delta_1 \vee \delta_2$ implies $\delta_1 = Ø$ and $\delta_2 = Ø$. An element δ_1 in some poset S is said to *cover* some element δ_2 if $\delta_2 \leq \delta_1$ and there is no δ_3 in $\underline{S}$ such that $\delta_2 \leq \delta_3 \leq \delta_1$.

References

Chubb, C. (1986). Collapsing binary data for algebraic multidimensional representation. *Journal of Mathematical Psychology, 30*, 161–187.

Coombs, C. H. (1964). *A Theory of Data*. New York: John Wiley and Sons.

Doignon, J.-P., Ducamp, A., & Falmagne, J.-C. (1984). On realizable biorders and the biorder dimension of a relation. *Journal of Mathematical Psychology, 28*, 73–109.

Doignon, J.-P., & Falmagne, J.-C. (1984). Matching relations and the dimensional structure of social choices. *Mathematical Social Sciences, 7*, 211–229.

Duquenne, V. (1995). Models of possessions and lattice analysis. *Social Science Information, 34*, 253–267.

Duquenne, V. (1996). On lattice approximations: syntactic aspects. *Social Networks, 18*, 189–200.

Freeman, L.C., & White, D.R. (1993). Using Galois lattices to represent network data. *Sociological Methodology 1993*.

Ganter, B., & Wille, R. (1999). *Formal Concept Analysis: Mathematical Foundations*. Berlin, Springer Verlag.

Haertel, E. H., & Wiley, D. E. (1993). Representations of ability structures: implications for testing. In Frederiksen, N., Mislevy, R. & Bejar, I. (Eds.) *Test Theory for a New Generation of Tests*. Hillsdale, NJ: Erlbaum.

Hage, P., & Harary, F. (1983). *Structural Models in Anthropology*. Cambridge: Cambridge University Press.

Kim, K. H. (1982). *Boolean Matrix Theory & Applications*. New York: Marcel Dekker, Inc.

Koppen, M. G. M. (1987). On finding the bidimension of a relation. *Journal of Mathematical Psychology, 31*, 155–178.

Martin, J. L., & Wiley, J. (2000). Algebraic representations of beliefs and attitudes ii: microbelief models for dichotomous belief data. *Sociological Methodology, 30*, 123–164.

Martin, J. L. (2002). Some algebraic structures for diffusion in social networks. *Journal of Mathematical Sociology* 26, 123–146.

Martin, J. L. (2006). Jointness and duality in algebraic approaches to dichotomous data. *Sociological Methods and Research, 35*, 159–192.

Martin, J. L. (2009). *Social Structures*. Princeton University Press.

Martin, J. L. (2014). Spatial processes and Galois/concept lattices. *Quality and Quantity, 48*, 961–981.

Martin, J. L. (2016). The dimensionality of discrete factor analyses. *Quality and Quantity 50*, 2451–2467.

Piaget, J. (1970). *Structuralism*. New York: Basic Books.

Weil, A. (1969). On the algebraic study of certain types of marriage laws (Murngin system). Pp. 221–229 in Lévi-Strauss, C., *The Elementary Structures of Kinship*. Boston: Beacon Press.

White, H. C. (1963). *An Anatomy of Kinship*. Englewood Cliffs, NJ: Prentice Hall.

Wiley, J., & Martin, J. L. (1999). Algebraic representations of beliefs and attitudes: partial order models for item responses. *Sociological Methodology, 29*, 113–146.

Printed in the USA
CPSIA information can be obtained
at www.ICGtesting.com
LVHW021131180224
772141LV00003B/191

9 783031 377914